DEVELOPMENTS IN SOIL MECHANICS—1

DEVELOPMENTS IN SOIL MECHANICS—1

Edited by

C. R. SCOTT
B.A., M.I.C.E., M.I. Struct. E

*Senior Lecturer, Department of Civil Engineering,
The City University, London, UK*

APPLIED SCIENCE PUBLISHERS LTD
LONDON

APPLIED SCIENCE PUBLISHERS LTD
RIPPLE ROAD, BARKING, ESSEX, ENGLAND

ISBN: 0 85334 771 9

WITH 15 TABLES AND 173 ILLUSTRATIONS

Printed in Great Britain by Galliard (Printers) Ltd Great Yarmouth

PREFACE

In the last few years, there have been considerable developments in the field of soil mechanics, and in its application to ground engineering. Although most of these developments have been published in journals and conference proceedings, the enormous volume of such literature means that it is often difficult for the practising engineer to find and select the information he requires. As a result, many of these techniques are not being as fully used as they might be, and it is hoped that this volume, by bringing together reports of this work in a number of fields, will help to make the necessary information more readily available.

The greater availability of powerful computers has meant that sophisticated methods of numerical analysis, such as the finite difference, finite element, and boundary element techniques, can now be generally used for the economical solution of analytical problems whose complex boundary conditions would previously have made them totally intractable. As a result, a much clearer insight has been obtained into the behaviour of many structures. In parallel with—and partly because of—this development, there has been increasing interest in field observations of structures and in the determination from field observations and *in situ* measurements of reliable soil models and soil parameters to use in analysis. As several of the authors have pointed out, sophisticated analysis can only be justified by reliable data. In addition to all this, there have been many technical developments in design and construction, particularly in the fields of piling, diaphragm walling, and ground anchorage systems.

v

In selecting the topics for discussion here, we have attempted to choose those which appear most likely to make a significant impact on engineering practice in the next few years. The authors of the various chapters are all actively engaged in research and development in the fields they discuss, which cover a wide range of soil mechanics and its applications.

CONTENTS

LIST OF CONTRIBUTORS

P. K. BANERJEE
Department of Civil and Structural Engineering, University College, Cardiff CF1 1XL, UK.

J. B. BURLAND
The Building Research Station, Garston, Watford WD2 7JR, Herts, UK.

R. W. COOKE
The Building Research Station, Garston, Watford WD2 7JR, Herts, UK.

T. H. HANNA
Department of Civil and Structural Engineering, The University, Sheffield S1 3JD, Yorks, UK.

J. A. HOOPER
Ove Arup and Partners, 13, Fitzroy Street, London W1P 6BQ, UK.

B. K. MENZIES
Department of Civil Engineering, University of Surrey, Guildford GU2 5XH, Surrey, UK.

R. T. MURRAY
Transport and Road Research Laboratory, Crowthorne RG11 6AU, Berks, UK.

D. J. NAYLOR
Department of Civil Engineering, University College of Swansea, Singleton Park, Swansea SA2 8PP, UK.

N. E. SIMONS
Department of Civil Engineering, University of Surrey, Guildford GU2 5XH, Surrey, UK.

Chapter 1

FINITE ELEMENT METHODS IN SOIL MECHANICS

D. J. NAYLOR

Department of Civil Engineering, University College of Swansea, UK

SUMMARY

This chapter presents the essential principles of the finite element formulations most widely used in soil mechanics applications. These may be used to solve many classes of problems, such as those involving loads and displacements, steady seepage, consolidation and soil dynamics. The essential features of the most important non-linear techniques are described and discussed, as well as some simple techniques which greatly enhance the usefulness of analyses based on effective stress. The necessary stress–strain laws are discussed in Chapter 2.

1.1 INTRODUCTION

The finite element method is an immensely powerful and versatile tool for numerical analysis, which allows complicated boundary shapes and variations in material properties to be accommodated without difficulty. It can be used to solve many problems, including those involving loads and displacements, steady seepage, consolidation and soil dynamics. It is also possible to handle both linear and non-linear material properties, which may be formulated in terms of either total or effective stress.

In the following sections, the basic finite element equations are derived from first principles for a very simple example. An outline is then given of the more general and formal derivation of the same equations. All the formulations presented here can be derived with equal validity in terms of one, two or three dimensions.

1.2 GEOMETRIC PRELIMINARIES

The region to be analysed (Fig. 1) is first divided into a number of sub-regions (the finite elements) each of which is defined by the prescribed co-ordinates of a number of node points. The nodes are usually on the element boundary, although some Lagrangian elements have nodes within the element.[18] In principle, the element may be of any shape and, within certain limits, may have any number of nodes. A particularly useful element for two-dimensional soils problems is the eight-noded quadrilateral element with parabolic sides shown in Fig. 2.

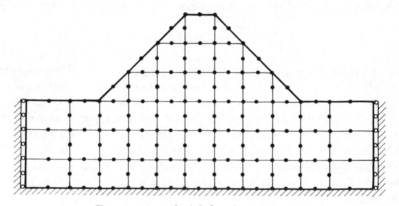

FIG. 1.   A typical finite element mesh.

We next require some relation between quantities (such as strain, distributed load, etc.) which vary smoothly across the element, and their values at the element nodes. *Shape functions* provide this relation.

1.2.1 Shape Functions
Let V represent a quantity whose local value at a point (x, y) within a two-dimensional element is to be related to nodal values V_i. The shape function, N_i, for the node i is defined by

$$V(x, y) = \Sigma N_i(x, y) V_i \tag{1}$$

The summation is taken over the nodes associated with the element.

The same shape function need not be used for all the quantities V, although this is usual. If the same shape functions used to define the

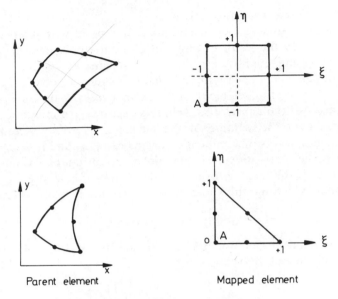

Parent element   Mapped element

FIG. 2.   Curvilinear co-ordinates.

shape of the element are also used to define the variation of the basic unknown then the element is said to be *isoparametric*.

1.2.2 Local Co-ordinates

It would be cumbersome and restrictive to define the shape functions in terms of the global co-ordinates x, y. Instead, they are defined in terms of local co-ordinates. Rectangular elements may be conveniently defined by co-ordinates ξ, η which vary between $+1$ and -1 (Fig. 2). A transformation is therefore used to 'map' the element from its real (possibly curved) shape in the x, y co-ordinate system to a square of side 2 units in the local ξ, η reference system.

The shape function for node A of the eight-noded quadrilateral shown in Fig. 2 is of the form

$$N_A = -\tfrac{1}{4}(1 - \xi)(1 - \eta)(1 + \xi + \eta)$$

It will be noted that this function has a value $+1$ at A and zero at all other nodes, and varies parabolically in the directions of the ξ and η axes. The shape functions for the six-noded triangle shown in Fig. 2 may be found in reference 16.

1.2.3 Co-ordinate Transformations

If the value of some quantity is required at a point within an element, we first specify the point in terms of ξ, η and apply eqn. (1) with V representing the quantity. If the x, y co-ordinates of the point are required, eqn. (1) is used with V equal to x and y in turn.

When formulating element stiffness matrices, and sometimes at other stages of the analysis as well, it is necessary to integrate certain quantities over the volume of the element. These integrations are formulated in terms of the local co-ordinates (ξ, η) and it is necessary to relate the derivatives of the global co-ordinates (x, y) to the derivatives of the local co-ordinates. Thus, for a plane stress or strain situation in which the thickness in the z direction is taken as unity, we require

$$d(\text{vol}) = dA = f(d\xi, d\eta) \tag{2}$$

Let the infinitesimal area dA represented by $d\xi \cdot d\eta$ in the mapped element be defined by the vectors

$$\mathbf{a} = \left[\frac{\partial x}{\partial \xi} \cdot d\xi, \frac{\partial y}{\partial \xi} \cdot d\xi\right]^{T}$$

$$\mathbf{b} = \left[\frac{\partial x}{\partial \eta} \cdot d\eta, \frac{\partial y}{\partial \eta} \cdot d\eta\right]^{T}$$

in real space (Fig. 3). Then the cross product of $\mathbf{a}$ and $\mathbf{b}$ (i.e. $\mathbf{a} \times \mathbf{b}$) is a vector normal to the x, y plane and of magnitude equal to dA. Whence

$$dA = |J| \, d\xi \, d\eta \tag{3}$$

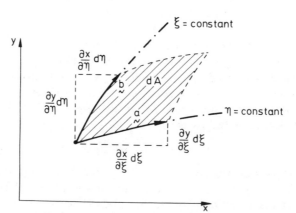

FIG. 3.   Element area transformation.

in which $|J|$ is the determinant of the Jacobian matrix

$$J = \begin{bmatrix} \dfrac{\partial x}{\partial \xi} & \dfrac{\partial x}{\partial \eta} \\ \dfrac{\partial y}{\partial \xi} & \dfrac{\partial y}{\partial \eta} \end{bmatrix}$$

In deriving the Jacobian components it will be necessary to use eqn. (1) to relate differentials. Thus

$$\frac{\partial x}{\partial \xi} = \frac{\partial}{\partial \xi} \sum N_i x_i = \sum \frac{\partial N_i}{\partial \xi} x_i$$

In general any dependent variable V differentiated with respect to an independent variable s is related to its nodal values by

$$\frac{\partial V}{\partial s} = \sum \frac{\partial N_i}{\partial s} \cdot V_i \tag{4}$$

the sum being taken over the nodes associated with the element as previously.

1.3 FORMULATIONS FOR LOAD-DISPLACEMENT PROBLEMS

The *displacement* formulation is usual for this class of problem. This will be set up from first principles for the simple example. The *direct stiffness* method will then be used to set it up for the general case. Finally a brief discussion of alternative formulations and methods will be given.

1.3.1 A Simple Example

Consider the example shown in Fig. 4. To simplify matters the stress distribution in the ground before the $100 \, \text{kN/m}^2$ load is applied will not be included in the analysis. This is permissible when—as here—a linear elastic representation of the soil is used, since superposition can be applied. The forces, R, acting on the ends of the elements are therefore caused only by the applied loads.

We first write down by inspection the force–displacement relationships for each element. This is done by fixing all the degrees of freedom in the element (two in this example) except one and then computing the forces which would be developed at all the degrees of freedom if the unfixed one were given a unit positive displacement.

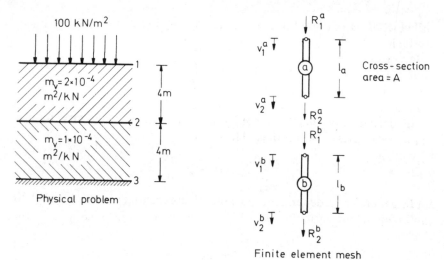

Finite element mesh

FIG. 4. Extensive uniformly distributed load applied to ground surface (drained).

These forces fill one column of a stiffness matrix. In this way we obtain for either element

$$\frac{A}{l^e m_v^e}\begin{bmatrix} 1 & -1 \\ -1 & 1 \end{bmatrix}\begin{Bmatrix} v_1^e \\ v_2^e \end{Bmatrix}=\begin{Bmatrix} R_1^e \\ R_2^e \end{Bmatrix} \tag{5}$$

where l is the length of the element and m_v is the coefficient of volume compressibility (Fig. 4); e identifies the element, i.e. a or b (A is the same in both).

In writing down this equation a stress–strain law has been invoked. The next stage is to connect the elements by recognising that nodes linking elements have a unique displacement. Node 2 links the two elements, therefore $v_2^a = v_1^b$. We denote v_1^a, $v_2^a = v_1^b$, and v_2^b by v_1, v_2 and v_3 respectively. Next, equilibrium between nodal forces is satisfied by equating the vector sum of the element nodal forces to the external forces (if any) at each node, i.e. $R_1 = R_1^a$, $R_2 = R_2^a + R_1^b$, and $R_3 = R_2^b$. Summing the element nodal forces allows the element equations to be *assembled* into a single matrix equation. With the values given on Fig. 4 this is

$$\begin{bmatrix} 1\,250 & -1\,250 & 0 \\ -1\,250 & 3\,750 & -2\,500 \\ 0 & -2\,500 & 2\,500 \end{bmatrix}\begin{Bmatrix} v_1 \\ v_2 \\ v_3 \end{Bmatrix}=\begin{Bmatrix} R_1 \\ R_2 \\ R_3 \end{Bmatrix} \tag{6}$$

The boundary conditions are $R_1 = 100$ KN, $R_2 = 0$, and $v_3 = 0$. Introducing these and solving the equations gives $v_1 = 0 \cdot 12$ m, $v_2 = 0 \cdot 04$ m and $R_3 = -100$ KN. Note that R_3 satisfies the requirement of equilibrium with the applied loads. Note also that eqn. (6) is symmetric and banded (with a half band width of 2).

Equation (6) is an example of the general matrix equation

$$\mathbf{K}\boldsymbol{\delta} = \mathbf{R} \qquad (7)$$

in which $\mathbf{K}$ is the global (overall or assembled) stiffness matrix of order $q \times q$ where q denotes the total number of degrees of freedom in the mesh, and $\boldsymbol{\delta}$ and $\mathbf{R}$ are respectively the displacement and applied load vectors each of length q. For two-dimensional continua, q will be twice the number of nodes in the mesh, and $\boldsymbol{\delta}$ will contain alternately x and y components of displacement, i.e. $\boldsymbol{\delta} = [u_1, v_1, u_2, v_2, \ldots]^{\mathrm{T}}$.

1.3.2 The Direct Stiffness Method

The setting up of the element stiffness equations in the simple example was straightforward because a relationship between nodal force and displacement could be written down directly. In the case of two- and three-dimensional continua this cannot be done and it is necessary to invoke the *principle of virtual work*. This equates the work done by a set of arbitrary (but small) displacements of the element nodes to the energy absorbed within the element. Denoting the set of element nodal forces (applied loads plus any due to initial stresses or strains—see eqn. (12)) by $\mathbf{F}^e$, and virtual displacements and the corresponding strains by a superposed bar ($\bar{\boldsymbol{\delta}}, \bar{\boldsymbol{\varepsilon}}$), the virtual work relation is

$$(\mathbf{F}^e)^{\mathrm{T}}\bar{\boldsymbol{\delta}} = \int \boldsymbol{\sigma}^{\mathrm{T}}\bar{\boldsymbol{\varepsilon}} \, \mathrm{d(vol)} \qquad (8)$$

in which the integration is taken over the element volume.

Stresses may be related to strain by the matrix equation

$$(\boldsymbol{\sigma} - \boldsymbol{\sigma}_0) = \mathbf{D}(\boldsymbol{\varepsilon} - \boldsymbol{\varepsilon}_0) \qquad (9)$$

The subscript zero indicates initial values. $\mathbf{D}$ is a modulus matrix which may be variable (see Chapter 2).

Strain is related to the local displacement components by the standard small strain relations. For plane stress or strain applications these are

$$\boldsymbol{\varepsilon} = \left\{ \begin{array}{c} \varepsilon_x \\ \varepsilon_y \\ \gamma \end{array} \right\} = \left\{ \begin{array}{c} \dfrac{\partial u}{\partial x} \\ \dfrac{\partial u}{\partial y} \\ \dfrac{\partial u}{\partial y} + \dfrac{\partial v}{\partial x} \end{array} \right\} \tag{10}$$

Using eqn. (4) to relate the displacement derivatives to the element nodal displacement vector, $\boldsymbol{\delta}^e$, gives

$$\boldsymbol{\varepsilon} = \mathbf{B}\boldsymbol{\delta}^e \tag{11}$$

in which $\mathbf{B}$ is a matrix of order $2n \times 3$, n being the number of nodes in the element, and is made up of a row of n submatrices.

$$\mathbf{B}_i = \left[\begin{array}{cc} \dfrac{\partial N_i}{\partial x} & 0 \\ 0 & \dfrac{\partial N_i}{\partial y} \\ \dfrac{\partial N_i}{\partial y} & \dfrac{\partial N_i}{\partial x} \end{array} \right]$$

Elimination of $\boldsymbol{\sigma}$ in eqn. (8) by eqn. (9), and then $\boldsymbol{\varepsilon}$ by eqn. (11) gives

$$(\mathbf{F}^e)^T \bar{\boldsymbol{\delta}} = \int (\mathbf{D}\mathbf{B}\boldsymbol{\delta}^e + \boldsymbol{\sigma}_0 - \mathbf{D}\boldsymbol{\varepsilon}_0)\mathbf{B}\bar{\boldsymbol{\delta}} \, \mathrm{d}(\mathrm{vol})$$

Since the $\bar{\boldsymbol{\delta}}$ are arbitrary, every $\bar{\boldsymbol{\delta}}$ coefficient on the left-hand side can be equated to the corresponding coefficient on the right-hand side, whence

$$\mathbf{F}^e = \mathbf{K}^e \boldsymbol{\delta}^e + \mathbf{F}^e_\sigma - \mathbf{F}^e_\varepsilon \tag{12}$$

in which

$$\mathbf{K} = \int \mathbf{B}^T \mathbf{D} \mathbf{B} \, \mathrm{d}(\mathrm{vol})$$

$$\mathbf{F}_\sigma = \int \mathbf{B}^T \boldsymbol{\sigma}_0 \, \mathrm{d}(\mathrm{vol})$$

$$\mathbf{F}_\varepsilon = \int \mathbf{B}^T \mathbf{D} \boldsymbol{\varepsilon}_0 \, \mathrm{d}(\mathrm{vol})$$

Alternatively, eqn. (12) may be expressed

$$\mathbf{K}^e \boldsymbol{\delta}^e = \mathbf{R}^e \tag{13}$$

in which $\mathbf{R}^e = \mathbf{F}^e - \mathbf{F}^e_\sigma + \mathbf{F}^e_\varepsilon$ represents the *applied* loads. $\mathbf{F}^e_\sigma$ and $\mathbf{F}^e_\varepsilon$ represent the element nodal forces which are equivalent (as determined by virtual work) to the initial stresses and strains respectively. In the simple example these two vectors were null.

In non-linear applications it will be necessary to include the nodal forces equivalent to the initial stresses since the stiffness properties will depend on the actual stress level at any stage in the analysis. Even in linear applications, where the principle of superposition obviates the need for this, it will often be expedient to include the initial stresses. It is then possible, for example, to relate the elastic stresses to a failure criterion. (This can be useful even though the stresses are not relaxed when the criterion is violated).

Assembling eqn. (13) for all elements leads to eqn. (7).

1.3.3 Equivalent Nodal Forces

So far it has been assumed that the loads are applied as discrete forces at the nodes. It is common in soils applications for the loads to be distributed either over the surface or through the body of the material. Gravity and seepage forces are examples of the latter. The principle of virtual work can be used to derive nodal forces equivalent to the distributed forces.[18] The nodal force component in, say, the x direction at a node i on a side of an element loaded with a distributed pressure having an x component p_x is

$$R^p_{xi} = \int N_i p_x \, \mathrm{d}A \tag{14}$$

The integration is taken over the area of the loaded side. These values are assembled for all the loaded sides (and in all component directions), care being taken if internal element boundaries are loaded to integrate along the side of only one of the adjoining elements (otherwise they will cancel out). The resulting load vector is denoted $\mathbf{R}_p$. Details of this procedure and how it may be programmed are given in reference 6.

The corresponding expression for a distributed body force having component b_x in the x direction is

$$R^b_{xi} = \int N_i b_x \, \mathrm{d(vol)} \tag{15}$$

the integration this time being taken over the volume of the loaded element.

Denoting load applied directly to nodes by $\mathbf{R}_n$ we have the total applied load vector

$$\mathbf{R} = \mathbf{R}_n + \mathbf{R}_p + \mathbf{R}_b \qquad (16)$$

1.3.4 Disequilibrium

The principle of virtual work says nothing about equilibrium. The only use we have made of equilibrium has been in the assembly of the element nodal forces to obtain the total force vector. In general there is local disequilibrium. This is illustrated in Fig. 5. The resultant force vector $\mathbf{X}$ of the stresses integrated over AB will not in general equal $-\mathbf{F}$. Furthermore there will not generally be equilibrium of stresses across element boundaries.

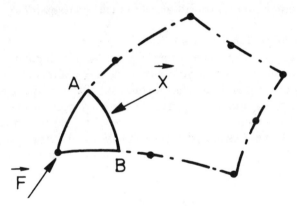

FIG. 5.   Stress near an element node.

1.3.5 Compatibility

It is usual to ensure compatibility of displacements along the common elementary boundaries. Parabolic isoparametric elements, for example, automatically achieve this since the deformed shape of the common element edge is the same parabola in both elements. Generally the requirement is that the shape functions defining displacement variation should match along the common boundary. Thus the eight-noded quadrilateral of Fig. 2 is fully compatible with the six-noded triangle. Neither, however, would be compatible with a linear element.

If elements are compatible it can be shown that for a linear elastic displacement formulation the total strain energy will be less than the strain energy obtained from an exact solution. This *lower bound* solution results in displacements which are generally on the low side.

1.3.6 Alternative Methods and Formulations

So far we have considered only the displacement formulation and set it up by means of the direct stiffness method. There are two other formulations and at least two other methods. (Not all the methods, however, can be applied to each formulation.) We shall simply mention the alternative formulations since they do not appear to offer any improvement over the displacement formulation. The alternative methods will be considered in more detail since they both throw light on the derivation given above and are needed in the next section to set up the matrix equations for field problems.

The principal alternative methods are *variational* methods and *weighted residual* methods. The former incorporate a variational principle to define a *functional*. Mathematically, a functional is a 'function of functions'. In the finite element context it is a scalar function of the discretised unknowns, δ. It is usually in two parts, one of which is integrated over the volume of the mesh (the domain) and the other which is integrated over the boundary. The key to the variational approach is that the functional, χ, has a stationary value when the unknowns have their optimum value, i.e. when

$$\frac{\partial \chi}{\partial \delta} = 0 \tag{17}$$

Application of this condition leads to eqn. (7).

The principle of minimum potential energy is used to form the functional from which the displacement formulation can be obtained. The principle of complementary energy is used to set up the *equilibrium* formulation. The essential feature of this formulation is that the shape functions define the variation of force across the elements. Forces or stress functions become the unknown nodal quantities. The 'stiffness' matrix **K** is replaced by a flexibility matrix. The method is described in reference 5. There are also hybrid or 'mixed' formulations which lie between the two.

The functional for the displacement formulation represents the sum of the potential energy of the nodal forces plus the internal strain energy. Using our previous notation it has the form

$$\chi = \tfrac{1}{2}\delta^{T}\mathbf{K}\delta - \mathbf{R}^{T}\delta \tag{18}$$

which on differentiating with respect to δ yields eqn. (7).

The method of weighted residuals obtains eqn. (7) directly from the governing differential equation in discretised form. The particular form of the method attributed to Galerkin leads to precisely the same equations as would be obtained using the minimum potential energy variational method or the direct stiffness method. Its chief use, however, is in field problems. It has the advantage that it does not require a variational principle, and also that it is easy to understand.

The reader is referred to finite element texts for further information. That by Huebner[8] has a particularly comprehensive list of references.

1.4 DIFFUSION AND DYNAMIC FORMULATIONS

A very wide range of physical phenomena are described by only a few governing equations. Steady seepage, uncoupled consolidation, and dynamic applications form a part of this range. Some applications in other fields which have identical formulations are described at the end of this section.

With some slight qualification all these have governing equations which are special cases of the differential equation

$$\nabla^2 \psi = A + B \frac{\partial \psi}{\partial t} + C \frac{\partial^2 \psi}{\partial t^2} \tag{19}$$

In this equation ψ stands for excess pore pressure or pore pressure head in seepage and consolidation applications. In dynamics it is a measure of strain or displacement. A, B and C are constants in linear applications. For steady seepage, eqn. (19) becomes the Laplace equation with $A = B = C = 0$. For uncoupled consolidation, it becomes the diffusion equation with $A = C = 0$ and $B \neq 0$. For undamped dynamics applications, $A = B = 0$ and $C \neq 0$. If there is viscous damping, $B \neq 0$. If there is non-viscous (frictional) damping (as with driven piles[14]) $A \neq 0$.

The 'slight qualification' relates to the left-hand side of eqn. (19). $\nabla^2 \psi$ presumes isotropic properties. In seepage problems, the soil is often assumed to have different permeabilities in the co-ordinate directions. Thus in a stratified two-dimensional deposit in which the x or y direction is parallel to the stratification and the permeabilities k_x, k_y are constant, $k\nabla^2 \psi$ is replaced by $k_x \partial^2 \psi / \partial x^2 + k_y \partial^2 \psi / \partial y^2$.

There are close analogies between diffusion and structural formulations. Head (or more generally potential) corresponds to force. The
condition of continuity is equivalent to equilibrium. These analogies
are brought out in the derivations which follow.

The terms on the right-hand side of eqn. (19) have physical inter-
pretations. They can all be thought of as contributing to forces or flow
occurring within the domain. A is a time-independent contribution
(flow generated or absorbed, or an externally applied body force
field). $B \partial \psi / \partial t$ is a time-dependent contribution (storage of fluid, or
viscous damping), and $C \partial^2 \psi / \partial t^2$, which only enters into dynamic
formulations, represents inertial loading.

1.4.1 A Steady Seepage Example (Fig. 6)

The head in the underlying aquifer is at the ground surface. It is
required to find the head at the junction of the layers, and the vertical
seepage flow.

The downward seepage, Q, in a column of cross-section A is given
by Darcy's law as

$$\frac{Q}{A} = k \frac{\mathrm{d}h}{\mathrm{d}x}$$

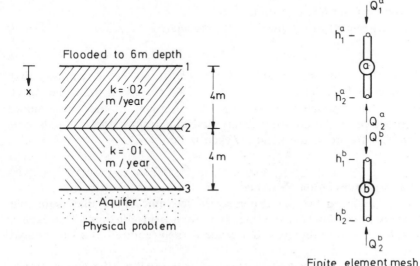

FIG. 6.   Steady seepage through two layers.

in which k is the vertical permeability, x is positive downwards and h is the piezometric head above the ground surface.

Observing that the head varies uniformly across each element, and also that $Q_1^e = -Q_2^e$ in which 'e' stands for 'element' (either a or b), we can write for either element

$$\frac{Ak^e}{l^e}\begin{bmatrix} 1 & -1 \\ -1 & 1 \end{bmatrix}\begin{Bmatrix} h_1^e \\ h_2^e \end{Bmatrix} = \begin{Bmatrix} Q_1^e \\ Q_2^e \end{Bmatrix} \tag{20}$$

Substituting the values shown in Fig. 6 and assembling eqn. (20) for the two elements (exactly as we did in the first example) we obtain

$$10^{-3}\begin{bmatrix} 5 & -5 & 0 \\ -5 & 7.5 & -2.5 \\ 0 & -2.5 & 2.5 \end{bmatrix}\begin{Bmatrix} h_1 \\ h_2 \\ h_3 \end{Bmatrix} = \begin{Bmatrix} Q_1 \\ Q_2 \\ Q_3 \end{Bmatrix} \tag{21}$$

in which Q_1, Q_2, Q_3 are the external flows, i.e. $Q_1 = Q_1^a$, $Q_2 = Q_2^a + Q_1^b$, $Q_3 = Q_2^b$, and h_1, h_2, h_3 the nodal heads, i.e. $h_1 = h_1^a$, $h_2 = h_1^b = h_2^a$, $h_3 = h_2^b$.

The boundary conditions are $h_1 = 6$ m, $h_3 = 0$ m and $Q_2 = 0$. We can therefore solve eqn. (21) to obtain $h_2 = 4$ m and $Q_1 = -Q_3 = 10$ litres/m^2/year.

The parallels with the structural problem are clear. Head corresponds to displacement, flow to force. The condition of continuity of flow corresponds to equilibrium.

We now generalise eqn. (21) into the matrix equation

$$\mathbf{H}\,\mathbf{h} = \mathbf{Q} \tag{22}$$

This corresponds to the stiffness eqn. (7). It will be shown later that the form of $\mathbf{H}$ is very similar to that of $\mathbf{K}$. Note that the order, q, of eqn. (22) is the same as the number of nodes in the mesh since, irrespective of the number of physical dimensions, there is only one degree of freedom (the head) at each node.

1.4.2 A Consolidation Example

The load of the first example is applied quickly so that negligible drainage occurs during loading. It is then held constant. The excess head h relative to the ground surface is then required as a function of time.

Since the variation of h across the mesh will not be linear, rather poor results can be expected from a mesh having only two linear elements. The main purpose, however, is to show the formulation, and two elements suffice for this.

We separate the flow through each element into two parts: a steady flow component Q and a storage component q (Fig. 7). To simplify the notation we omit the superposed e. Q_1 and Q_2 thus relate to a typical element until the context indicates otherwise. The element equations derived previously for steady seepage (eqn. (20)) apply to the steady flow part.

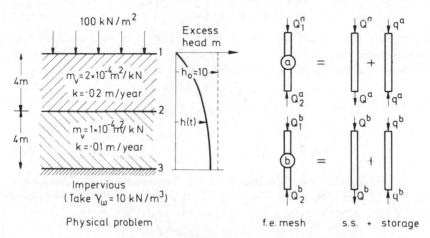

FIG. 7. Two-layer one-dimensional consolidation problem.

Thus

$$\frac{kA}{l}\begin{bmatrix} 1 & -1 \\ -1 & 1 \end{bmatrix}\begin{Bmatrix} h_1 \\ h_2 \end{Bmatrix} = \begin{Bmatrix} Q \\ -Q \end{Bmatrix}$$

Adding the storage component and noting that $Q_1 = Q + q$, $Q_2 = -Q + q$, we have

$$\frac{kA}{l}\begin{bmatrix} 1 & -1 \\ -1 & 1 \end{bmatrix}\begin{Bmatrix} h_1 \\ h_2 \end{Bmatrix} + \begin{Bmatrix} q \\ q \end{Bmatrix} = \begin{Bmatrix} Q_1 \\ Q_2 \end{Bmatrix} \tag{23}$$

The fluid stored in a time interval Δt is equated to an increase in the voids in the mesh. This in turn is related to the change in the average vertical effective stress, $\Delta\bar{\sigma}'$, in time Δt, i.e.

$$2q\Delta t = -m_v lA\Delta\bar{\sigma}'$$

In this case the vertical total stress is constant* so by the principle of effective stress $\Delta\bar{\sigma}' = -\gamma_w\Delta\bar{h}$. Writing $c_v = k/\gamma_w m_v$ (the coefficient of consolidation) and rearranging

$$q = \frac{kAl}{2c_v} \cdot \frac{\Delta\bar{h}}{\Delta t} \qquad (24)$$

It is assumed that $\bar{h} = \frac{1}{2}(h_1 + h_2)$. To facilitate substitution in eqn. (23) we write eqn. (24) twice as the matrix equation.

$$\begin{Bmatrix} q \\ q \end{Bmatrix} = \frac{kAl}{4c_v} \begin{bmatrix} 1 & 1 \\ 1 & 1 \end{bmatrix} \begin{Bmatrix} \dot{h}_1 \\ \dot{h}_2 \end{Bmatrix} = \mathbf{C}^e\dot{\mathbf{h}}^e$$

in which $\dot{h} = \Delta h/\Delta t$ and the superposed e is now introduced to identify the elements.

Denoting the left-hand coefficient matrix of eqn. (23) by

$$\mathbf{H}^e = \frac{kA}{l} \begin{bmatrix} 1 & -1 \\ -1 & 1 \end{bmatrix}$$

we can write eqn. (23) as

$$\mathbf{H}^e\mathbf{h}^e + \mathbf{C}^e\dot{\mathbf{h}}^e = \mathbf{Q}^e \qquad (25)$$

A central difference scheme in time is now introduced. Denoting h at time $(t + \frac{1}{2}\Delta t)$ by h^+ and h at $(t - \frac{1}{2}\Delta t)$ by h^-, and letting unprimed quantities relate to time t, we can express eqn. (25) as

$$\frac{1}{2}\mathbf{H}^e(\mathbf{h}^{e-} + \mathbf{h}^{e+}) + \mathbf{C}^e\frac{(\mathbf{h}^{e+} - \mathbf{h}^{e-})}{\Delta t} = \mathbf{Q}^e$$

Multiplying by $2\Delta t$ and re-arranging,

$$(\Delta t\mathbf{H}^e + 2\mathbf{C}^e)\mathbf{h}^{e+} = 2\Delta t\mathbf{Q}^e - (\Delta t\mathbf{H}^e - 2\mathbf{C}^e)\mathbf{h}^{e-} \qquad (26)$$

Assembling for all the elements

$$(\Delta t\mathbf{H} + 2\mathbf{C})\mathbf{h}^+ = 2\Delta t\mathbf{Q} - (\Delta t\mathbf{H} - 2\mathbf{C})\mathbf{h}^- \qquad (27)$$

This is a recurrence relationship. To start, the h at time $t = 0$ must be known. Knowing all the Q (at all times) except on the head controlled

---

*This assumption is usually only valid in one dimension. The 'Rendulic' method assumes it applies in three dimensions[10]. It has the advantage that the governing equations are uncoupled. If the assumption is not made the problem becomes coupled and is defined by simultaneous differential equations (see Chapter 4).

boundaries, eqn. (27) is solved for the head h at $(t + \Delta t)$ (that is, h^+). If desired, the mid-difference Q values (at time t) can be avoided by writing $Q = \frac{1}{2}(Q^+ + Q^-)$. All values in eqn. (27) are then related to times $t - \frac{1}{2}\Delta t$ or $t + \frac{1}{2}\Delta t$.

Substituting the values shown on Fig. 7 and assembling the equations for the two elements gives

$$\mathbf{H} = 10^{-3} \begin{bmatrix} 5 & -5 & 0 \\ -5 & 7 \cdot 5 & -2 \cdot 5 \\ 0 & -2 \cdot 5 & 2 \cdot 5 \end{bmatrix}$$

and

$$\mathbf{C} = 10^{-3} \begin{bmatrix} 2 & 2 & 0 \\ 2 & 3 & 1 \\ 0 & 1 & 1 \end{bmatrix}$$

The initial conditions are $h_2 = h_3 = 10$ m. The boundary conditions are $h_1 = 0$, $Q_2 = Q_3 = 0$ at all times. Figure 8 shows the results. It also shows the exact solution, which because c_v is the same in both

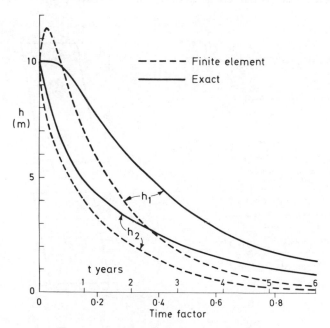

FIG. 8.   Dissipation of excess head, h.

elements is the well known Terzaghi solution. As expected this very coarse mesh does not give good answers. None the less it can be seen to be trying!

The above method is known as the Crank–Nicolson method. It is characterised by a parameter $\alpha = 1/2$ where α determines the value of a typical variable V as $\alpha V^+ + (1 - \alpha)V^-$. In this case V stands for h, $\dot{h}$ and Q in turn. $\alpha = 0$ and $\alpha = 1$ define explicit and implicit methods respectively. The methods are unconditionally stable only if $\alpha \geqslant 0.5$. The Crank–Nicolson method is therefore at the limit of the stable methods. It is useful for general application. Smith,[15] however, suggests that it is worth using a higher order finite difference approximation in time (the above are linear) on the basis that a significant improvement in accuracy can be achieved at small cost. Zienkiewicz and Lewis[20] propose a least squares method which also appears to offer improved accuracy, but at some computing cost.

1.4.3 A Dynamic Example (Fig. 9)

The soil is saturated. A known horizontal motion is applied to the 'rock' which is presumed to underly the two soil layers. We wish to

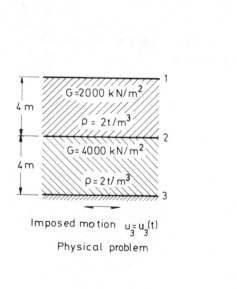

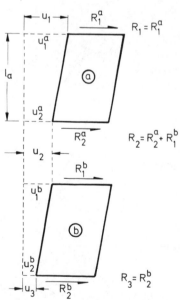

f.e. mesh (cross-section $A = 1$)

FIG. 9.   Propagation of shear waves.

know the horizontal movement (the response) at levels 1 and 2 as a function of time. Damping is neglected.

The problem is made static by applying D'Alembert's principle. The condition for equilibrium of a point of mass m displaced u from its equilibrium position is

$$\lambda u = R - m\ddot{u} \tag{28}$$

in which λ is the spring stiffness and R is an externally applied force. Thus it is clear that the formulation of the dynamic problem basically involves the addition of an inertia term (and in general a damping term as well—see Section 1.4.4) to the applied forces.

In evaluating the inertia term for the element matrix equation we assume that the soil mass is concentrated at the soil nodes. The force displacement relations for a typical element then become

$$\frac{AG^e}{l^e}\begin{bmatrix} 1 & -1 \\ -1 & 1 \end{bmatrix}\begin{Bmatrix} u_1^e \\ u_2^e \end{Bmatrix} = \begin{Bmatrix} R_1^e \\ R_2^e \end{Bmatrix} - \frac{\rho Al}{2}\begin{bmatrix} 1 & 0 \\ 0 & 1 \end{bmatrix}\begin{Bmatrix} \ddot{u}_1^e \\ \ddot{u}_2^e \end{Bmatrix}$$

This is abbreviated to

$$\mathbf{K}^e\mathbf{u}^e = \mathbf{R}^e - \mathbf{M}^e\ddot{\mathbf{u}} \tag{29}$$

$\mathbf{K}^e$ is the element stiffness matrix. It is not the same as that derived in the first simple example (Section 1.3.1) because horizontal shear stiffness is involved rather than vertical stiffness. $\mathbf{M}^e$ is the element mass matrix. It is called a 'lumped' mass matrix because the mass is lumped together at the nodes.

As in the consolidation example a central difference time stepping scheme is used. This time, however, second differences are required. Let the subscripts $-$, 0, $+$ represent values at times $t - \Delta t$, t, and $t + \Delta t$ respectively.

Then we can write

$$\ddot{u} \simeq \frac{u^- - 2u^0 + u^+}{\Delta t^2}$$

Substituting into eqn. (29), and assuming the $\mathbf{u}$ on the left-hand side relates to time t

$$\mathbf{K}^e\mathbf{u}^{e0} = \mathbf{R}^{e0} - \frac{M^e}{\Delta t^2}(\mathbf{u}^{e-} - 2\mathbf{u}^{e0} + \mathbf{u}^{e+})$$

Rearranging and assembling for all elements we have

$$\mathbf{M}\mathbf{u}^+ = \Delta t^2\mathbf{R}^0 - (\Delta t^2\mathbf{K} - 2\mathbf{M})\mathbf{u}^0 - \mathbf{M}\mathbf{u}^- \tag{30}$$

Knowing the values of $\mathbf{u}$ at time $t - \Delta t$ and time t, and also $\mathbf{R}$ at time t, eqn. (30) can be solved to give u at time $t + \Delta t$. To get started we specify the initial displacement $\mathbf{u}_i^0$ and velocity $\dot{\mathbf{u}}_i$. This allows us to obtain $\mathbf{u}_i^-$ from the finite difference relation

$$\dot{\mathbf{u}}_i = \frac{\mathbf{u}_i^+ - \mathbf{u}_i^-}{2\Delta t}$$

$\mathbf{u}_i^- = \mathbf{u}_i^+ - 2\Delta t \dot{\mathbf{u}}_i$ is then substituted in eqn. (30) with $\mathbf{u}^0 = \mathbf{u}_i^0$. To obtain $\mathbf{u}_i^+$ the term $\mathbf{Mu}_i^+$ has to be added to both sides. This is done for the first step only.

Substituting the values shown on Fig. 9 we obtain

$$\mathbf{K} = \begin{bmatrix} 500 & -500 & 0 \\ -500 & 1500 & -1000 \\ 0 & -1000 & 1000 \end{bmatrix}$$

$$\mathbf{M} = \begin{bmatrix} 4 & 0 & 0 \\ 0 & 8 & 0 \\ 0 & 0 & 4 \end{bmatrix}$$

The boundary conditions are $u_1 = u_2 = 0$ at time zero, u_3 is specified as a function of time (the full line in Fig. 10) and $R_1 = R_2 = 0$ since no external loads are applied. Selecting a time step $\Delta t = 0.02$ sec (this must not exceed a critical value, see below) the response of nodes 1 and 2 is then obtained by successive solution of equation (30). Since $\mathbf{M}$ is a diagonal matrix, $\mathbf{u}$ is obtained explicitly. The expressions for u_1 and u_2 are

$$u_1^+ = 1\cdot950u_1^0 + 0\cdot050u_2^0 - u_1^-$$
$$u_2^+ = 0\cdot025u_1^0 + 1\cdot925u_2^0 + 0\cdot05u_3^0 - u_2^-$$

(31)

The results are plotted in Fig. 10. In this example the bed rock displacement has been specified, but it is more common in practice for accelerations to be specified.

For this method to be unconditionally stable the time step must not exceed the critical value T_n/π where T_n is the smallest natural frequency of the discretised system. In this example it is $0\cdot126$ sec. Since the critical time step reduces as the finite element mesh is refined a very large number of time steps may be required in a real

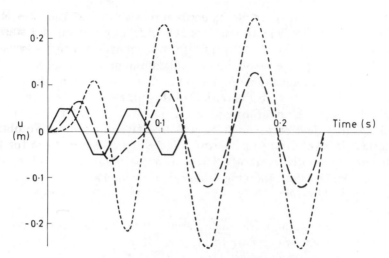

FIG. 10. Response of soil layers to prescribed base motion ---- node 1
(surface); -- node 2; — node 3 (base-prescribed).

situation. None the less, since the equations do not have to be solved simultaneously due to the mass matrix being diagonal, this is usually quite practical. It is common to have several thousand time steps. It is quite feasible to solve non-trivial problems on a mini-computer, an example being the processing of several thousand time steps of a 20 degrees of freedom problem on a PDP11 in about half an hour.

An alternative to the lumped mass matrix is the so called *consistent* mass matrix.[3] In this example it may be derived by assuming the mass to be concentrated at the centre, although this simple picture does not apply generally. The consistent element mass matrix is

$$\mathbf{M}^e = \frac{\rho A l}{4} \begin{bmatrix} 1 & 1 \\ 1 & 1 \end{bmatrix} \tag{32}$$

The disadvantage of this formulation is that the equations to be solved are no longer diagonal, with a consequent loss in efficiency in the time stepping process. To overcome this disadvantage the consistent mass matrix may subsequently be diagonalised. A description of ways of doing this is given in the next section when the general form for the mass matrix is given.

We have used this simple example to derive one of four formulations for the finite element analysis of dynamics problems. As well as being straightforward it appears to be eminently suited to soils problems such as the response of a fill dam to an earthquake loading. It has the advantage shared by other time stepping methods that non-linearity (that is, the variation in the stiffness matrix with stress or strain) can be incorporated at small cost. It suffers the disadvantage that time steps must be less than a critical value to ensure stability. It is referred to as an *explicit direct step* method.

There are *implicit* methods. These have the advantage that they can be made stable, thus allowing longer time steps. They take longer per time step, however. A recent one is the Wilson–Newmark 'θ' method.[3,17]

A powerful method for linear applications is the *modal decomposition* or *mode superposition* method.[3] It involves finding a number of fundamental frequencies. One of its attractions is that not all of these are required. The number will depend on the geometry and the accuracy required.

Damping may be included in the above formulations. This is straightforward in time stepping schemes. An additional term is added to the basic matrix eqn. (29). Its form is given in the next section. With modal decomposition it is necessary to diagonalise the damping matrix before it can be included.

1.4.4 General Formulations for Steady Seepage, Consolidation, and Dynamic Problems

Although we were able to formulate the finite element equations for these problems directly for very simple examples, it is not possible to do this for the general case. Nor is it possible to use a method equivalent to the direct stiffness method in the displacement formulation. Recourse must be made to either a variational method or the Galerkin method of weighted residuals. Space does not allow these methods to be described here. The reader is referred to the literature.[3,5,8,13,18,21] Instead, the general forms for the matrices in eqns. (22, 25 and 29) are given below.

Steady Seepage
The matrix equation is

$$\mathbf{Hh} = \mathbf{Q} \qquad (22)$$

$\mathbf{H}$ is assembled from element stiffness matrices which have the general form

$$\mathbf{H}^e = \int \mathbf{B}_s^T \mathbf{D}_s \mathbf{B}_s \, d(\text{vol}) \tag{33}$$

in which, for two-dimensional applications, $\mathbf{B}_s$ is a row of n column matrices where n is the number of nodes in an element:

$$\mathbf{B}_{si} = \left[\frac{\partial N_i}{\partial x}, \frac{\partial N_i}{\partial y} \right]^T \qquad (i = 1, n)$$

and

$$\mathbf{D}_s = \begin{bmatrix} k_x & 0 \\ 0 & k_y \end{bmatrix}$$

k_x, k_y being the permeabilities in the x and y directions. If x and y do not coincide with the directions of anisotropy, the off-diagonal terms are non-zero. The dimensions of $\mathbf{H}$ are (length)^2/time. $\mathbf{h}$ is the vector of piezometric head (e.g. in metres), there being one value per node. $\mathbf{Q}$ is the vector of inflows at each node. Note that the Q value associated with each node is a scalar quantity. It has dimensions (length)^3/time. Surface integrals analogous to those used for surface pressures may be needed to convert distributed surface flows to Q values.

Consolidation
The assembled matrix equation corresponding to eqn. (25) is

$$\mathbf{H}\mathbf{h} + \mathbf{C}_s \dot{\mathbf{h}} = \mathbf{Q} \tag{34}$$

$\mathbf{H}$ and $\mathbf{Q}$ are precisely as in the steady seepage case. $\mathbf{C}_s \dot{\mathbf{h}}$ represents the rate of storage of fluid. $\mathbf{C}_s$ is assembled from element matrices whose (i, j)th element $(i, j = 1, n)$ has the form

$$\mathbf{C}_{s(ij)}^e = \int C \gamma_w N_i N_j \, d(\text{vol}) \tag{35}$$

in which, for incompressible pore fluid and soil *particles*, C is the bulk compressibility of the soil *skeleton*.

For isotropic elastic soils C is related to various elastic constants as follows.

One dimension (no lateral strain)

$$C = m_v = \frac{(1 + \nu)(1 - 2\nu)}{(1 - \nu)E} = \frac{(1 - \nu)}{3(1 - \nu)K_b}$$

Two dimensions (plane strain)

$$C = \frac{2(1 + \nu)(1 - 2\nu)}{E} = \frac{2(1 + \nu)}{3K_b}$$

Three dimensions

$$C = \frac{3(1 - 2\nu)}{E} = \frac{1}{K_b}$$

(36)

where K_b is the elastic bulk modulus of the soil skeleton.

Dynamic Applications

Generalising eqn. (29) to include all displacement components in $\boldsymbol{\delta}$, assembling, and adding a damping term results in

$$\mathbf{K}\boldsymbol{\delta} = \mathbf{R} - \mathbf{C}_d \dot{\boldsymbol{\delta}} - \mathbf{M}\ddot{\boldsymbol{\delta}}$$

(37)

$\mathbf{K}$ is the structural stiffness matrix assembled from element stiffness matrices the form of which was derived in Section 3.2., i.e.

$$\mathbf{K}^e = \int \mathbf{B}^T \mathbf{D} \mathbf{B} \, d(\text{vol})$$

(38)

$\mathbf{R}$, the vector of applied nodal forces (which includes the nodal equivalents of distributed body forces and surface pressures), is also the same as in the static case.

The damping matrix $\mathbf{C}_d$, which was not included in the example for simplicity, is assembled from element damping matrices, each composed of n^2 submatrices each of which in two dimensions is a 2×2 matrix having the form

$$\mathbf{C}^e_{d(ij)} = \int \mu N_i N_j \begin{bmatrix} 1 & 0 \\ 0 & 1 \end{bmatrix} d(\text{vol}) \qquad (i, j = 1, n)$$

(39)

μ is a damping constant. If x and y do not coincide with the directions of anisotropy, the off diagonal terms are non-zero.

The consistent element mass matrix is similarly defined. Its submatrices have the form

$$M^e_{(ij)} = \int \rho N_i N_j \begin{bmatrix} 1 & 0 \\ 0 & 1 \end{bmatrix} d(\text{vol}) \qquad (i, j = 1, n)$$

(40)

in which ρ is the bulk density. It can be verified that this will not result in a diagonal matrix. However, as mentioned above, good results can be achieved by making the matrix diagonal. This may be done by making the off diagonal terms zero, and making the new diagonal terms proportional to those of the consistent mass matrix but increased so that their sum is equal to the total mass, i.e. the sum of all the terms in the consistent matrix.[7] Alternatively a diagonal matrix may be obtained by carrying out the numerical integration of eqn. (40) by sampling at the nodes.[18] This results in $N_i N_j = 1$ when $i = j$ otherwise $N_i N_j = 0$. This method should not be used for parabolic or higher order elements since the corner node masses become negative and cause instability.

In the simple one-dimensional example, M_{ij}^e in eqn. (40) becomes a scalar and has the value $\frac{1}{4}\rho Al$, whence eqn. (32) may be derived.

1.4.5 Concluding Remarks

The basic similarity in the formulations for the static load-deformation problem and the three classes of problem considered in this section is now apparent. The 'stiffness' matrices **K** or **H** both involve matrix products of shape function derivatives and material properties contained in the 'modulus' matrices **D** or $\mathbf{D}_s$.

Static load-deformation and steady seepage formulations have in common that they are both boundary value problems (their governing differential equations are elliptic). The techniques for their solution are almost identical. Indeed, if provision is made for setting up the matrices $\mathbf{B}_s$ and $\mathbf{D}_s$ as an alternative to **B** and **D**, a finite element program designed for structural applications can also be used for steady seepage.

Dynamic and consolidation formulations have in common that they are bounded in space but are initial value problems in time. The identical form, apart from a physical constant, of the consolidation 'storage' matrix $\mathbf{C}_s$, the dynamic damping matrix $\mathbf{C}_d$, and the consistent mass matrix **M** should be noted.

There are many applications in the physical sciences which have the same mathematical formulations as the three considered in this section. It is worth while for the soils engineer to be aware of them as he can then 'cash in' on expertise from these other fields. Table 1 lists some of them.

TABLE 1
FIELDS OF APPLICATION OF THE GOVERNING EQUATIONS

Equation	Field	ψ (SI units)
Laplace (E)* $\nabla^2\psi = 0$	Seepage Heat conduction Electrostatics Hydrodynamics (irrotational flow)	Head, h, (m) Temperature (°C) Electrical potential (V) Stream function (sec^{-1})
Poisson (E)* $\nabla^2\psi = A$	Viscous flow in uniform channel Torsion of prismatic bar Uniformly loaded membrane (e.g. soap bubble)	Velocity† (m/sec) Stress function† (N) Out-of-plane deflection (m)
Diffusion (P)* $\nabla^2\psi = A + B(\partial\psi/\partial t)$	Consolidation (uncoupled) As for Poisson and Laplace but conditions transient. A = 0 implies no generation or absorption of flow	As above
Damped wave (H)* $\nabla^2\psi = A + B(\partial\psi/\partial t) + C(\partial^2\psi/\partial t^2)$	Motion in elastic media (2- or 3-D) Pile (usually A ≠ 0, B = 0) Column of fluid Vibration of taut string Electric transmission line	Volumetric strain/vorticity‡ Displacement (m) Deflection (m) Electric potential (V) or current (A)

*E = elliptic, P = parabolic, H = hyperbolic, in one-dimensional form, i.e. $\nabla^2\psi = (\partial^2\psi/\partial x^2)$. Note that $\nabla^2\psi$ can be shorthand for $\alpha_x(\partial^2\psi/\partial x^2) + \alpha_y(\partial^2\psi/\partial y^2)$, α_x, $\alpha_y \neq 1$ if a geometric transformation is used, i.e. $(\partial^2\psi/\partial\bar{x}^2) = \alpha_x(\partial^2\psi/\partial x^2)$, etc.

†Across section normal to axis.

‡Involve simultaneous governing differential equations, one for ψ = volumetric strain, the rest (two or three) for ψ = vorticity components. When discretised, ψ represents the displacement components.

1.5 NON-LINEAR TECHNIQUES

Attention is restricted to load-deformation problems. Central in the finite element formulation is the non-linear stress–strain law. This can be expressed in the form

$$\Delta\sigma = D\Delta\varepsilon \tag{41}$$

in which D is the variable modulus matrix whose components most commonly depend on stress (usually effective) but may also depend on strain, and/or the stress history. Δ may indicate a large difference or—and this particularly applies to elastic-plastic laws—it may approximate the differential. Equation (41) forms the subject of Chapter 2.

Non-linear techniques can be broadly divided into those in which all the load can be applied at once, and those in which it is applied in a series of small increments. The former category involves iterations or a series of re-solutions akin to iterations. The equivalent load, secant modulus, and visco-plastic methods come into this category. The load can, however, be applied incrementally with iterations within each increment. The tangential modulus method is the obvious example of the second category. Increments must be small, approximating the differential, and eqn. (41) is required in differential form (or if not it must be differentiable).

Engineers are trained to think in terms of geometric concepts rather than in the abstract. For this reason diagrams representing the behaviour of single degree of freedom systems are used to help explain the techniques. A word of warning, however: these diagrams should be used only as a framework on which to hang ideas. They can be misleading if used to predict multi-degree of freedom behaviour.

1.5.1 Equivalent Load Methods

In these methods a set of artificial nodal forces are derived which, when applied to a linear elastic counterpart of the actual body, produce the correct nodal displacements.

Element stiffness matrices are calculated and assembled once only to produce the initial elastic overall stiffness matrix (K_0). A number of re-solutions is then required. Each of these costs a fraction—in the order of one fifth—of the cost of a solution involving re-calculation of element stiffnesses and assembly. Consequently many more re-solutions can be carried out for a given computing cost.

There are two basic methods. These are usually called the 'initial stress' and 'initial strain' methods.[18] To avoid confusion with the actual initial stresses and strains they will be called here the *stress transfer* and *strain transfer* methods.

Both methods start with an elastic analysis. Let this produce strains $\Delta\varepsilon_1$. Point A_1 on Fig. 11 is defined. The stress change from O to A_1, $\Delta\sigma_1$, may then be calculated from the known form of the curve OA_1. The point E_1 represents the point that would be reached if the body had the elastic properties represented by K_0. The next stage is to calculate the artificial load to be added to the actual load to correct the discrepancy due to E_1 not lying on OA_1, and this is where the stress transfer and strain transfer methods differ.

The stress transfer method bases the corrective force on the stress represented by A_1E_1. Let τ_1 denote this stress. Then

$$\tau_1 = D_0\Delta\varepsilon_1 - \Delta\sigma_1$$

in which D_0 is the elastic modulus used in the formulation of K_0. (It need not relate to the physical properties of the material, although it will often be an initial modulus, and in the case of elastic-plastic materials is invariably the constant modulus for the elastic region.) The corrective force ψ_1—called a *residual*—is assembled from element contributions

$$\psi_1^e = \int B^T\tau_1 \, d(vol) \tag{42}$$

in which B is the element matrix of shape function derivatives (see Section 1.3.2) and integration is over the element volume. ψ_1 becomes the right-hand side for the first re-solution. Re-solution yields strains $\Delta\varepsilon_2$ from which τ_2 and ψ_2 are obtained in the same way as were τ_1 and ψ_1. The process continues until the force residuals are small enough not to cause significant error. (The criterion for this is problem-dependent and is found by trial.)

The strain transfer method computes the force residuals on the basis of the strain difference represented by E_1B_1 in Fig. 11(b). This is denoted by γ_1. The force residuals, denoted now by ψ_1^*, are assembled from element contributions

$$\psi_1^{*e} = \int B^T D_0\gamma_1 \, d(vol) \tag{43}$$

ψ_1^* becomes the right-hand side for the first re-solution. Re-solution yields strains $\Delta\varepsilon_2^*$, from which γ_2 and ψ_2^* are obtained in the same

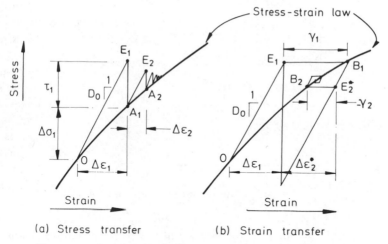

FIG. 11.   The stress transfer and strain transfer methods.

way as were γ_1 and ψ_1^*. The process continues until the residuals are small.

The stress transfer method is preferable if strains are largely controlled. An extreme case is that in which the displacements of all nodes are specified. The point A_1 in Fig. 11 then represents the correct solution. Conversely the strain transfer method is preferable if the stresses do not change much during the re-solutions. This would occur if the boundary conditions were mainly stress.

The methods can be speeded up or slowed down by multiplying the residual vector (ψ or ψ^*) by a scalar 'accelerator' α before re-solution. A value of α equal to about 1.8 can typically halve the time required for solution in a stress transfer method application. Instability may arise with α greater than 1. In the rare cases where instability arises without the accelerator (e.g. in the stress transfer analysis of a 'locking' material —i.e. having a concave upwards stress–strain curve) a value of α less than 1 may be used to stabilise the analysis.

There is a case for applying a judicious blend of the stress and strain transfer methods.

1.5.2 Secant Modulus Method
This is essentially an iterative method. The full load is applied in each iteration. The method is illustrated by Fig. 12 which depicts a modu-

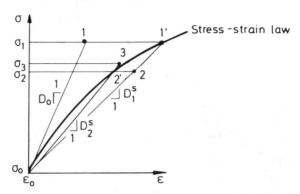

FIG. 12.   The secant modulus method.

lus matrix, **D**, dependent on stress only. The first analysis using an initial stiffness based on $\mathbf{D}_0$ results in stresses $\boldsymbol{\sigma}_1$. A secant modulus $\mathbf{D}_1^S$ is then obtained as the average modulus over the loading range $\boldsymbol{\sigma}_0$ to $\boldsymbol{\sigma}_1$. It is used to compute a new overall stiffness and a second analysis is carried out to yield $\boldsymbol{\sigma}_2$. The process is continued until the change in stress and strain from one iteration to the next becomes negligible.

1.5.3 Tangential Modulus Method
The load is applied in increments which ideally are small. The tangential stiffness is evaluated from the tangential modulus matrix ($\mathbf{D}^T$) at the start of each increment (Fig. 13). In its crudest form $\mathbf{D}^T$ is evaluated for the known stress and strain conditions at the *start* of the increment. A refinement is to estimate the stress and strain changes likely to occur during the increment from the previous increment and then calculate the average tangential modulus (either the mean of the two ends or the modulus evaluated at the centre of the increment). A further refinement is to iterate within each increment to improve the average tangential stiffness.

The tangential method works well for monotonic loading along smooth stress–strain curves which always have a symmetric and positive definite $\mathbf{D}^T$ matrix. It can sometimes be used when these conditions are violated. For instance a non-symmetric matrix, such as occurs with non-associative plasticity (see Chapter 2, Section 2.4), can be used, provided a non-symmetric solution routine is available (the solution cost will be increased by a factor of about 3). A $\mathbf{D}^T$

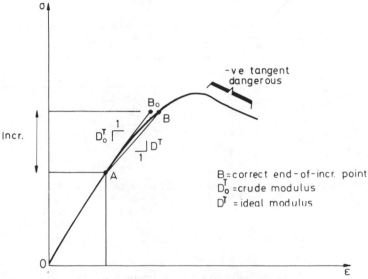

FIG. 13. The tangential modulus method.

matrix with negative diagonals such as is associated with strain softening can also sometimes be successfully handled but there is a danger here that numerical conditioning errors will occur.

The method can also be used when the curve has a slope discontinuity (bi-linear elasticity, elastic-plastic laws). Some error must be tolerated at the discontinuity but this can be made quite small.

1.5.4 Visco-plastic Method

This method may be applied to materials which are both viscous and plastic, or—and to date it has mainly been used in this second role—to materials which are plastic only. In this case an artificial viscosity is introduced. It is related only to the length of the 'time' steps used in the solution scheme. This time is only real if the method is used in its first role as applying to truly viscous materials. Otherwise it is an artifice used to control a 'marching' scheme which, as in the equivalent load methods, involves a series of re-solutions using a constant stiffness matrix.

The method is explained by reference to the single degree of freedom model illustrated in Fig. 14. Initially the system is in equilibrium under load P_0 which is insufficient to cause the slider to slide.

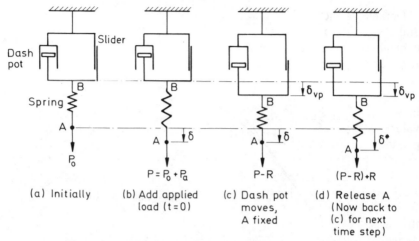

FIG. 14.   One degree of freedom visco-plastic model.

The load P is then applied causing an immediate extension δ of the spring, but no movement at B since the dash-pot requires time to move. It is supposed that the slider can sustain a load Y before it slides. If $P < Y$ no sliding occurs and the calculation is complete. Otherwise sliding will occur. It will be at a rate governed by the excess of P over Y and the viscous characteristics of the dash-pot. A time stepping process now begins. A time step, Δt, is chosen by empirical or other means. The velocity of B is computed as a function of $P - Y$ and the known dash-pot properties, and multiplied by Δt to give the displacement δ_{vp} at B. The imposition of this displacement with the node A fixed marks the first of two stages in the time stepping process. It causes a reaction $R = K\delta_{vp}$, where K is the spring stiffness. The second stage is to release node A. This results in δ increasing by δ_{vp} to δ^*. This completes the time step. Before proceeding to the next one, Y is modified to take into account any strain hardening. The process continues until $\delta^* - \delta \to 0$, which in this simple model will only happen if there is strain hardening.

The parallels between the one degree of freedom model and the finite element prototype are as follows. P and δ represent the nodal forces and displacements. The force in the spring represents the internal stress field. δ_{vp} represents visco-plastic strains and $\delta - \delta_{vp}$ elastic strains (δ in the model serves a dual role in that it stands for both strains and displacements).

Details of the computational procedure are given in references 4 and 19. The expression for the visco-plastic strains is, however, central to the calculation and will be given. The visco-plastic strain *rates* are given by

$$\dot{\varepsilon}_{vp} = \mu \frac{F}{F_0} \mathbf{a}_q \qquad (44)$$

in which μ is the 'viscosity', F is a yield function, F_0 a reference value to make F/F_0 dimensionless and $\mathbf{a}_q$ the gradient vector to a plastic potential (see Chapter 2, Section 2.4 for definitions and expressions for F and $\mathbf{a}_q$). Equation (44) is a flow rule.

1.5.5 Assessment of Techniques

It is difficult to come to conclusions about the relative efficiencies of the different methods. Computing times are job- and tolerance-dependent. A change in the maximum permitted force residual, for example, can have a major effect on the number of re-solutions necessary.

The equivalent load methods can be applied to virtually all forms of stress–strain law. When unaccelerated they are usually convergent, though may be extremely slow when there is severe non-linearity, i.e. when a large change in stiffness has to be brought about over a large region. A case in point is the use of the critical state model (Chapter 2, Section 2.5) to simulate soil being loaded beyond its pre-consolidation pressure. Under such conditions equivalent load methods are not competitive.

The secant modulus method, like the tangential method, requires a non-symmetric solver if laws with non-symmetric modulus matrices are to be handled. It has the advantage over the tangential method that it can deal with strain-softening stress–strain laws (provided, of course, the secant modulus remains positive). It is not, however, as efficient as the tangential method.

The tangential method, when it can be used, is a potential winner, particularly for strongly non-linear problems. It is probably worth incorporating predictor techniques so that an average tangential stiffness rather than the start-of-increment tangential stiffness is obtained. This costs little and allows larger increments to be used. To do this for elastic-plastic laws, however, involves rather complex coding.

A disadvantage shared by the tangential and secant modulus methods is that they cannot readily incorporate changed stiffness on loading such as occurs in elastic-plasticity. The stress transfer and visco-plastic methods do this automatically, provided K_0 is the elastic unloading stiffness.

The visco-plastic method provides a more efficient alternative to the stress transfer method for elastic-plastic analysis. It has the advantage that the stress–strain law is needed in the form of the inverse of eqn. (41), which is simpler.

1.6   THE ROLE OF THE EFFECTIVE STRESS METHOD

The effective stress method separates the pore fluid from the soil skeleton. This gives the method versatility. If the pore pressure changes are known (e.g. from piezometer measurements), or can be obtained independently, the method can be used without reference to the state of drainage. This is particularly relevant to the monitoring of structures, e.g. the stability of a fill dam during construction.

The method can be used for undrained analysis. The excess pore pressure is obtained explicitly. Furthermore it is obtained irrespective of the stress–strain law, unlike total stress methods which can only compute the excess pore pressure directly when the soil is saturated and the skeleton elastic and isotropic. The excess pore pressure is then approximately equal to the change in mean total stress. Total stress methods can be used to obtain the excess pore pressures indirectly by means of pore pressure parameters.

1.6.1 Analyses with Known Pore Pressure Change
The finite element mesh models the soil skeleton. The pore pressure change is treated as an external loading. It is a distributed body force and has a local intensity (force/unit volume)

$$\mathbf{p} = -\nabla(\Delta u) \tag{45}$$

in which $\mathbf{p}$ is a vector in actual space and Δu is the known pore pressure increase at the typical point. It is not, however, necessary to evaluate the gradient of Δu (an error prone process) since it can quite easily be shown by the principle of virtual work that the element nodal force set $\mathbf{P}^e$ corresponding to $\mathbf{p}$ has an x component for node i

$$P_{ix}^e = -\int \frac{\partial N_i}{\partial x} \Delta u \, \mathrm{d(vol)} \tag{46}$$

in which N_i is the shape function for node i and the integration is over the element, with similar expressions for the other co-ordinate directions. The $\mathbf{P}^e$ are assembled to give a load vector $\mathbf{P}$, and a conventional finite element analysis performed.

Apart from the monitoring role mentioned above, the method may be used to assess a change in water table level, or a change from one seepage pattern to another (these would be obtained independently by flow net), or to assess the effect of dissipation of excess pore pressures.

This last application is an interesting one as the alternative is to carry out a much more complicated consolidation analysis. The initial excess pore pressure distribution—the removal of which provides the loading—may be obtained by an undrained analysis (see Section 1.6.2). The mesh or its properties may then be altered to incorporate, for instance, some structural change such as the construction of a foundation raft,[12] and the dissipation analysis carried out. The method, however, is only rigorous for a linear elastic soil skeleton. If a soil model other than mildly non-linear is used error will result due to the wrong effective stress path being followed. The extent of this error needs to be investigated.

1.6.2 Excess Pore Pressure Calculation

The 'soil skeleton' finite elements have 'pore fluid' finite elements attached to them. They both experience the same strains. (This implies no relative movement of pore fluid and soil skeleton, which is taken here as a definition of 'undrained'.) Effective stress stiffness properties are assigned to the soil skeleton component, and these define the modulus matrix $\mathbf{D}'$. An equivalent pore fluid compressibility K_e is assigned to the pore fluid component.* Invoking the principle of effective stress and the above mentioned equality of strains it is easily shown[11] that the total stress modulus matrix

$$\mathbf{D} = \mathbf{D}' + K_e \mathbf{I} \qquad (47)$$

in which for plane strain applications

$$\mathbf{I} = \begin{bmatrix} 1 & 1 & 0 \\ 1 & 1 & 0 \\ 0 & 0 & 0 \end{bmatrix}$$

---

*K_e is approximately related to the actual pore fluid modulus, K_f, and soil particle (as opposed to skeleton) bulk modulus, K_s, by $1/K_e = n/K_f + (1-n)/K_s$ in which n = porosity. This is an approximation to a more general expression given by Bishop.[1]

D is used to obtain the element stiffness matrices in the usual way, and the analysis proceeds to obtain the displacement field and strains, $\Delta\varepsilon$. The total stress change is then given by

$$\Delta\boldsymbol{\sigma} = \mathbf{D}\Delta\varepsilon$$

which is where the analysis would end if it were in terms of total stress. The breakdown of **D** by eqn. (47), however, allows direct determination of the effective stress change

$$\Delta\boldsymbol{\sigma}' = \mathbf{D}'\Delta\varepsilon \tag{48}$$

and the change in pore pressure (the excess pore pressure)

$$\Delta u = K_e\Delta\varepsilon_v \tag{49}$$

in which $\Delta\varepsilon_v$ is the volumetric strain change.

For saturated soils K_e is large, some orders of magnitude greater than the bulk modulus of the soil skeleton (except for very dense, well-graded, granular soils, which may have a bulk modulus approaching that of water, which is about $2\,\mathrm{GN/m^2}$). In the same way as occurs in the elastic analysis of materials with a Poisson's ratio close to 0.5, analyses with a large K_e are liable to result in stresses and pore pressures of low accuracy. Fortunately this does not occur with isoparametric elements using reduced integration provided the stresses and pore pressures are calculated at the Gauss integrating points.[11] This underlines the importance of eight-noded parabolic elements using 2×2 Gauss integration (i.e. at the points with local co-ordinates $\xi, \eta = \pm 1/\sqrt{3}$; see Section 1.2) which are excellent elements quite apart from this consideration. It is therefore not necessary to resort to the different formulations needed to solve completely incompressible materials.[2] Saturated soils can be analysed perfectly satisfactorily by the cheaper displacement method.

It is not critical what value is assigned to K_e when it represents a saturated soil. In the same way that a Poisson's ratio of 0.49 will give much the same answer as a ratio of 0.499 so also will a ten-fold increase in K_e have little effect provided it is much greater than K, the soil skeleton bulk modulus. Unfortunately this useful finding does not apply to partly saturated soils where the moduli may be of the same order.

No restrictions have been put on the form of **D'**. It can represent any non-linear stress–strain law, thus justifying the generality claimed above for the prediction of excess pore pressures.

1.6.3 Comment on the Effective and Total Stress Alternatives

Most non-linear stress–strain laws incorporate the Mohr–Coulomb failure parameters. In terms of effective stress these, i.e. c', ϕ', can usually be determined with confidence. c and ϕ, however, vary widely and are determined on an *ad hoc* basis by means of triaxial tests on 'undisturbed' samples. This makes the effective stress method a powerful tool for assessing failure when the pore pressures are known as for example in the monitoring of dams. (This advantage is shared by effective stress limit analyses. Finite element methods offer the advantage that they can give information on the failure mechanism. They can show where slipping starts.)

When used to compute excess pore pressures, the uncertainty which, in the total stress approach, lies in the choice of pore pressure parameters is transferred to the $\mathbf{D}'$ matrix. This must incorporate the correct dilatancy characteristics of the soil. A model which predicts the correct volume change under drained loading will predict the correct pore pressure change under undrained loading.

REFERENCES

1. BISHOP, A. W. (1973). The influence of an undrained change in stress on the pore pressure in porous media of low compressibility, *Géotechnique*, 23(3), 435–442.
2. CHRISTIAN, J. T. (1968). Undrained stress analysis by numerical methods, *Proc. Am. Soc. Civ. Engrs.*, 94(SM6), 1333–1345.
3. CLOUGH, R. W. and PENZIEN, J. (1975). *Dynamics of Structures*, McGraw-Hill, New York.
4. CORMEAU, I. (1975). Numerical stability in quasi-static elasto/visco-plasticity, *Int. J. Num. Meth. Engng.*, 9, 109–127.
5. GALLAGHER, R. H. (1975). *Finite Element Analysis Fundamentals*, Prentice-Hall, Englewood Cliffs, N.J.
6. HINTON, E. H. and OWEN, D. R. J. O. (1977). *Finite Element Programming*, Academic Press, London, New York, San Francisco.
7. HINTON, E. H., ROCK, A. and ZIENKIEWICZ, O. C. (1976). A note on mass lumping and related processes in the finite element method, *Int. J. Earthquake Engng. and Struct. Dyn.*, 4, 245–249.
8. HUEBNER, K. H. (1975). *The Finite Element Method for Engineers*, Wiley, New York.
9. JAEGER, J. C. (1962). *Elasticity Fracture and Flow*, 2nd edition, Methuen, New York.
10. LEE, I. K. (1968). *Soil Mechanics, Selected Topics*, Butterworths, London.

11. NAYLOR, D. J. (1974). Stresses in nearly incompressible materials by finite elements with application to the calculation of excess pore pressures, *Int. J. Num. Meth. Engng.*, **8**, 443–460.
12. NAYLOR, D. J. and HOOPER, J. A. (1975). An effective stress finite element analysis to predict the short and long term behaviour of a piled raft foundation on London clay, In *Settlement of Structures*, Pentech Press, pp. 394–402.
13. SANDHU, R. S. and WILSON, E. L. (1969). Finite element analysis of seepage in elastic media, *Proc. Am. Soc. Civ. Engrs.*, **95**(EM3), 641–652.
14. SMITH, E. A. L. (1960). Pile driving analysis by the wave equation, *Proc. Am. Soc. Civ. Engrs.*, **86**(SM4), 35–61.
15. SMITH, I. M. (1976). Some time-dependent soil–structure interaction problems, *Numerical Methods in Soil and Rock Mechanics, Vol. I*, Institut für Bodenmechanik und Felsmechanik, Universität Karlsruhe.
16. TONG, P. and ROSSETTOS, J. N. (1977). *Finite-Element Method, Basic Technique and Implementation*, MIT Press, Cambridge, Mass., and London.
17. WILSON, E. L. (1978). Numerical methods for dynamic analysis, paper presented at *Int. Symp. on Num. Meth. for Offshore Engng.*, Swansea, January 1977.
18. ZIENKIEWICZ, O. C. (1971). *The Finite Element Method in Engineering Science*, 2nd edition, McGraw-Hill, London.
19. ZIENKIEWICZ, O. C. and CORMEAU, I. (1974). Visco-plasticity and creep in elastic soils—a unified numerical solution approach, *Int. J. Num. Meth. Engng.*, **8**, 821–845.
20. ZIENKIEWICZ, O. C. and LEWIS, R. W. (1973). An analysis of various time stepping schemes for initial value problems, *Int. J. Earthquake Engng. and Struct. Dyn.*, **1**, 407–408.
21. ZIENKIEWICZ, O. C. and PAREKH, C. J. (1970). Transient field problems: two-dimensional and three-dimensional analysis by isoparametric finite elements, *Int. J. Num. Meth. Engng.*, **2**(1), 62–72.

Chapter 2

STRESS–STRAIN LAWS FOR SOIL

D. J. NAYLOR

Department of Civil Engineering, University College of Swansea, UK

SUMMARY

This chapter describes the principal stress–strain laws used in the analysis of load-deformation boundary value problems in soil mechanics. The simple linear elastic law is first described, and the determination of suitable parameters is discussed. Several non-linear variable elastic and elastic-plastic laws (including the critical state model) are then described, and their advantages and disadvantages for the finite element method and for other purposes are discussed.

2.1 INTRODUCTION

The solution of any load-deformation boundary value problem requires a known relation between stress and strain of the form:

$$\Delta\boldsymbol{\sigma} = \mathbf{D}\Delta\boldsymbol{\varepsilon} \tag{1}$$

or

$$\Delta\boldsymbol{\varepsilon} = \mathbf{D}^{-1}\Delta\boldsymbol{\sigma} \tag{2}$$

where $\Delta\boldsymbol{\sigma}, \Delta\boldsymbol{\varepsilon}$ are vectors of the components of stress and strain increment respectively, and $\mathbf{D}$ is a modulus matrix. The increments may be large (as in linear analysis) or almost infinitely small (as in incremental plasticity). The finite element formulation for displacement usually requires the law in the form of eqn. (1). However, for certain non-linear applications, such as the analysis of elasto-plastic materials using the visco-plastic method (see Chapter 1), the inverse form of eqn. (2) is required.

Most of the laws described in this chapter may be defined in terms of either total or effective stress. Effective stress is usually more appropriate, and in the case of the critical state model it is essential. The advantages and disadvantages of total and effective stress approaches were discussed in Section 1.6 of the previous chapter.

No real soil can be accurately modelled by any stress–strain law, partly because of the complexity of its behaviour and partly because of its variability in the ground. Predictions of deformations that lie between half and twice the observed values must be considered to be good. There is therefore seldom any value in a complicated model: the fewer parameters which need evaluating the better. This is a major advantage of the isotropic linear elastic model which requires only two parameters. Non-linear models having up to six independent parameters are described in later sections of this chapter and can be justified in certain circumstances, especially if the results are not very sensitive to some of them. The critical state model is a good example of this, as only three of the six parameters are really important.

Soil is usually anisotropic. Even if such a material is modelled as linear and elastic with isotropy in the horizontal plane, five constants must be supplied. Such a law does seem to be justified for modelling stiff clays. It can, for instance, reproduce the high ratios of horizontal to vertical stress (K_0) which are encountered in London clay. A combination of anisotropy with non-linearity would, however, be hard to justify since the number of parameters would be excessive.

2.2 LINEAR ELASTIC LAWS—APPROPRIATE PARAMETERS

Elastic isotropic stress–strain laws are fully defined by two independent parameters. These are conventionally taken to be Young's modulus E and Poisson's ratio ν. In soils, however, there are advantages in using the bulk modulus K and the shear modulus G, as the behaviour of soil in the separate modes of volume change and shear are reasonably well understood (see Section 2.3). The moduli K and G may be expressed in terms of E and ν as follows:

$$K = \frac{E}{3(1 - 2\nu)} \tag{3}$$

$$G = \frac{E}{2(1 + \nu)} \tag{4}$$

Given the effective stress moduli, E' and ν' or K' and G', it is a simple matter to obtain the total stress equivalents for the analysis of a saturated soil. The pore water is almost incompressible compared with the soil skeleton (except in a few, very dense, well-graded soils) so that ν_u tends to 0·5 and K_u is nearly infinite. Also, since changes in mean stress cause no distortion in elastic isotropic materials, the shear moduli G_u and G' are equal. Then

$$G_u = \frac{E_u}{2(1 + \nu_u)} = G' = \frac{E'}{2(1 + \nu')} \tag{5}$$

and

$$E_u = \frac{1 \cdot 5}{1 + \nu'} E'$$

2.2.1 Parameter Values

The parameters needed to define stress–strain laws are most commonly based on the results of laboratory tests. Initial moduli (i.e. moduli at very small strain) can, however, be obtained from measurements of the speed of propagation of shock waves through the soil. Laboratory measurements of stiffness very often under-estimate the values *in situ*. This is particularly true of measurements in unconfined undrained triaxial shear tests. It has been suggested[14] that the modulus thus obtained should be increased five times. Oedometer tests, certainly when applied to rolled fills,[19,20] appear to give more realistic values of the vertical stiffness.

The models described here will require knowledge of the strength parameters c, ϕ or c', ϕ'. Also, a shear stress–strain curve (Fig. 1) will

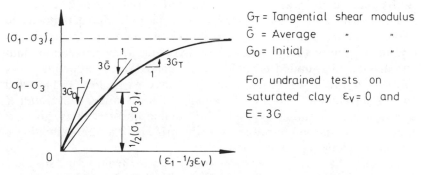

G_T = Tangential shear modulus
$\bar{G}$ = Average   ,,   ,,
G_0 = Initial   ,,   ,,

For undrained tests on saturated clay $\varepsilon_V = 0$ and $E = 3G$

FIG. 1. Deviator stress–strain curve for triaxial test showing alternative shear moduli.

be needed to obtain the shear stiffness, and a mean stress–volume change relation will be required to determine the bulk stiffness in terms of effective stress. For clay soils the latter will be measured by C_c or C_s, the slopes of the virgin consolidation and swelling curves respectively on a plot of void ratio against $\log_{10} \sigma'$ (either from oedometer or triaxial consolidation tests). A further requirement for clays is a knowledge of the preconsolidation pressure, σ_{pc}. It is usually assumed that C_c and C_s are constants, so that the bulk modulus increases linearly with stress:

$$K_T = \frac{p(1+e)}{0 \cdot 434 C} \qquad (6)$$

where K_T is the tangential bulk modulus at a mean effective stress p and void ratio e, and C stands for C_c or C_s as appropriate.

The undrained Young's modulus for saturated clays can be expected to lie in the range $100 < \bar{E}/c_u < 1000$, in which c_u is the undrained cohesion intercept. Note that for these clays $\nu \simeq 0 \cdot 5$, therefore by eqn. (4), $\bar{E} = 3\bar{G}$. *In situ* values will tend towards the upper end of this range and unconfined triaxial test values towards the lower end. Young's modulus increases with both the mean stress, p, and the pre-consolidation stress, σ_{pc}. Ladd[13] finds $\bar{E}/\sigma_{pc}$ to lie in the range 100–200 for normal and lightly overconsolidated clays. E_0/σ_{pc} is roughly twice as large.

Engineering operations which result in a reduction of stress—as in excavations—will require a larger modulus value than if the stresses were increased, and very large values may be appropriate. Thus values of Young's modulus in the range 100–200 MN/m^2 have been used in the analysis of large excavations in London clay (see Chapter 3).

The stiffness of granular soils is even more variable than that of clays. It depends principally on the relative density, the gradation, and the nature of the particles. A well-graded sandy gravel containing sub-angular or rounded particles (e.g. a glacial outwash deposit) may have $\bar{E}$ in excess of 100 MN/m^2. Or it may be as low as 14 MN/m^2 for an angular material with breakable particles under 100 KN/m^2 (1 atmosphere) confining pressure.[14] Doubling these values gives an indication of the initial value (E_0) which is also an approximation of the value under repeated loading.

The effective stress Poisson's ratio (ν') varies from near zero under initial loading to $0 \cdot 5$ as failure is approached. Under repeated loading of granular soils it will typically lie in the range $0 \cdot 3$ to $0 \cdot 4$.

2.3   VARIABLE ELASTIC LAWS

Most generally these laws express the elastic moduli as functions of stress or strain or both. They may be set up to define a secant or a tangential modulus.

2.3.1 Hyperbolic Model

A model for undrained saturated clays need only involve one variable modulus since Poisson's ratio is 0.5. The hyperbolic model originally attributed to Kondner[11] defines the stress–strain relation in the form

$$(\sigma_1 - \sigma_3) = \frac{\varepsilon_1}{a + b\varepsilon_1} \tag{7}$$

where a and b are constants when the equation is applied to conventional (constant σ_3) triaxial tests. $1/a$ is then the initial tangential Young's modulus and $1/b$ would be the failure value $(\sigma_1 - \sigma_3)_f$ were it not for a refinement. This is to allow $(\sigma_1 - \sigma_3)$ to become asymptotic to a stress in excess of $(\sigma_1 - \sigma_3)_f$, the curve being cut off at this value.

The model is not restricted to the undrained analysis of saturated soils. For general applications a and b vary with σ_3. When used to analyse Oroville dam[12] values had to be assigned to eight parameters. Despite this the model has had wide application, particularly in the western United States.

2.3.2 Differential Models

A differential model defines a tangential modulus. It is suited to incremental methods of analysis in which the steps are sufficiently small that the increment approximates a differential.

In general, the bulk modulus of soil increases with confining stress, and the shear modulus reduces with shear stress becoming zero at failure. It is therefore logical to define the tangential values of K and G separately. 'K–G' models can then be defined in which K and G are assumed to vary linearly with stress invariants. One possibility is

$$K = K_1 + \alpha_K p' \tag{8}$$

$$G = G_1 + \alpha_G p' + \beta_G q \tag{9}$$

in which

$$p' = \tfrac{1}{3}(\sigma_1' + \sigma_2' + \sigma_3')$$

and

$$q^2 = \sigma_1(\sigma_1 - \sigma_2) + \sigma_2(\sigma_2 - \sigma_3) + \sigma_3(\sigma_3 - \sigma_1)$$

q is related to the second deviatoric invariant, $J_2(= \frac{1}{3}q^2)$, by $q = \sqrt{3J_2}$, and to the octahedral shear stress, τ_{oct}, by $q = (3/\sqrt{2})\tau_{oct}$. The expression for q is the same whether the stress is effective or total.

An alternative is

$$K = K_1 + \alpha_K \sigma_s' \tag{10}$$

$$G = G_1 + \alpha_G \sigma_s' + \beta_G \sigma_d \tag{11}$$

in which

$$\sigma_s' = \tfrac{1}{2}(\sigma_1' + \sigma_3')$$

$$\sigma_d = |\sigma_1 - \sigma_3|$$

Either alternative requires values to be assigned to five parameters. The α_K and α_G are positive and β_G is negative. A suitable choice of G_1, α_G and β_G will make G zero when the stresses satisfy a failure criterion. Equations (8) and (9) can satisfy a yield criterion of the conical type described in Section 2.4.5, whereas eqns. (10) and (11) can satisfy a Mohr–Coulomb criterion. Since soils appear to adhere more closely to a Mohr–Coulomb criterion,[1] eqns. (10) and (11) are generally preferred to (8) and (9).

A different form of K–G model has been proposed by Nelson and Baron.[18] Their specification of G is essentially the same as eqn. (9), but they assume K to be a quadratic function of the volumetric strain.

Provided K and G always remain positive the elastic material defined by them will be theoretically admissible. The corresponding Poisson's ratio will lie between its bounds of -1 and 0.5.

A warning should be given about differential models which assume ν to be constant, and vary E (or G). If the tangential E is made to disappear when a failure stress state is reached not only does G disappear but the tangential bulk modulus does as well. The yielding material disappears into a 'black hole'. This is why, in practice, ν increases to 0.5 and K tends to infinity as failure is approached.

2.3.3 Evaluating Constants for K–G Models

Consider a normally consolidated clay for which K and G are to be defined by eqns. (10) and (11). It is assumed that a set of three consolidated drained triaxial tests (or three consolidated undrained tests with pore pressure measurement) is available.

From these data c' and ϕ' are obtained in the usual way. Since we are considering a normally consolidated clay c' will be small and can be neglected.

The Mohr–Coulomb criterion requires

$$\sigma_d = 2 \sin \phi' \cdot \sigma'_s \tag{12}$$

If we put $G = 0$ in eqn. (11), we find that $G_1 = 0$ and

$$\frac{\alpha_G}{\beta_G} = -2 \sin \phi' \tag{13}$$

satisfies the Mohr–Coulomb criterion.

G is next obtained from the slope of the triaxial test deviator stress–strain curves at a number of points on each curve. At each point the value of σ'_s and σ_d is recorded. A surface in a σ'_s, σ_d, G space is thus defined. Equation (11) defines a plane in this space, equal G contours on which are parallel to the failure line (Fig. 2). This plane can be made to best fit the data by adjusting α_G and β_G but keeping their ratio constant according to eqn. (13).

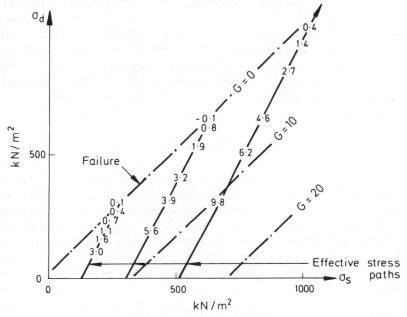

FIG. 2. Boulder clay shear modulus from triaxial tests. Contours are of best fit surface; $G = 0{\cdot}5 + 27\,\sigma_s - 27\,\sigma_d$ MN/m^2. Spot values are G in MN/m^2.

Figure 2 illustrates the procedure for a compacted boulder clay. The contours are of the best fit surface defined by $\alpha_G = 27$, $\beta_G = -27$ and $G_1 = 0 \cdot 5 \, \text{MN/m}^2$ (in this case the cohesion intercept was not neglected).

The bulk stiffness constants are obtained independently from the consolidation stage of the tests. If the $e - \log p'$ plot is a straight line, K_1 can be taken as zero, and by eqn. (6),

$$\alpha_K = \frac{1+e}{0 \cdot 434 C_c} \qquad (14)$$

Note that since we are considering isotropic consolidation ($\sigma_1 = \sigma_2 = \sigma_3$) $\sigma'_s = p'$. For the boulder clay the consolidation test data gives $K_0 = 0$ and $\alpha_K = 100$.

A good fit to test data is of course no guarantee that the parameters apply to the *in situ* soil. Consideration must also be given to whether the soil is normally or overconsolidated, or if the soil passes the preconsolidation pressure during loading. These factors are clearly important because a large reduction in stiffness can occur when the preconsolidation pressure is reached and this is not taken into account by the model.

A further limitation shared by all the variable elastic models is that abrupt increases in stiffness on unloading cannot automatically be brought about as they can be in elastic-plastic models. None the less they can provide a much better representation of soils than can a linear elastic (isotropic) model.

2.4   ELASTIC-PLASTIC LAWS

Elastic-plastic stress–strain laws relate *small increments* or *rates* of stress and strain. In soils applications it is usual to set aside (or ignore) true time effects so that the rate of loading has no effect. Thus $d(\ )/dt$ can be interpreted as $d(\ )$ when it appears to the same power on either side of the equation. Following common practice we shall denote the rate by a superimposed dot. Stresses are shown unprimed. They may, however, be effective or total, depending on the physical formulation.

2.4.1 Yield Surface
A first requirement is the specification of a yield surface. This is a function of stress which when evaluated for the stress components of

a typical point must have a value less than or equal to zero, i.e.

$$F(\sigma) \leqslant 0 \tag{15}$$

It is conveniently viewed as a surface in a stress space having as axes either the components of stress or some functions of the components. We shall see later that a stress space defined by two stress invariants will prove useful. The requirement of eqn. (15) in geometric terms is that the point representing the state of stress must lie on or within the yield surface.

If the point lies within the yield surface the stress–strain law is assumed to be elastic—usually linear and isotropic. If, during loading, the point reaches the yield surface and tries to cross it it is constrained to cling to the surface and plastic strains become superimposed on the elastic. These are governed by a *flow rule* and a *strain hardening* law.

Yield surfaces must be convex (they can have local flat areas) and must contain the stress origin.[8,21]

2.4.2 Strain Hardening

The yield surface may change in size when plastic yielding occurs. If it gets larger strain *hardening* occurs, if smaller, strain *softening*. It is usually assumed that the shape of the yield surface remains the same as it hardens (or softens). It expands (or shrinks) about the origin. This type of hardening is called *isotropic*. An alternative is *kinematic* hardening.[4,15] This has recently been incorporated in a sophisticated model for soil.[22]

Strain hardening is usually assumed to be controlled by a single parameter, although there is no theoretical objection to there being more than one. Let it be denoted by h. We shall see that for the critical state model h is the accumulated plastic volumetric strain. In soils where post-peak strength loss is important (*in situ* clays, dense sands) h may be a measure of accumulated plastic shear strain. Modelling post-peak strength loss, however, has problems as will be discussed in Section 2.6. Equation (15) may be generalised to include h, i.e.

$$F(\sigma, h) \leqslant 0 \tag{16}$$

Sometimes the yield function will be expressed in the form of eqn. (15) even when there is hardening. This will be done when variation of F with σ is considered so that h can be treated (temporarily) as a constant.

2.4.3 Flow Rule

The flow rule fixes the proportions of the components of the plastic strain rates. To do this the assumption is made that the directions of principal plastic strain rate coincide with the principal stress directions. This is called the assumption of coaxiality of the stress and plastic strain rate tensors. There is controversy about the extent to which it is valid in soils.

The flow rule is expressed as

$$\dot{\boldsymbol{\varepsilon}}^p = \dot{\lambda}\, \mathbf{a}_q \tag{17}$$

in which $\dot{\boldsymbol{\varepsilon}}^p$ is the vector of plastic strain rate components, $\dot{\lambda}$ is a parameter related to $\dot{\sigma}$ if there is strain hardening (or softening)—otherwise it is determined by the boundary conditions—and $\mathbf{a}_q$ is the gradient vector to a scalar function of stress, $Q(\boldsymbol{\sigma})$, known as the *plastic potential.* i.e.

$$\mathbf{a}_q = \frac{\partial Q}{\partial \boldsymbol{\sigma}}$$

The assumption of coaxiality allows the flow rule to be interpreted geometrically. A constant Q defines a surface in stress space which passes through the point representing the current state of stress. If the components of plastic strain rate are also assigned to the stress component axes, an outward pointing normal to the surface originating from the current stress point has the plastic strain rates as its components (Fig. 3).

If $Q = F$ the flow rule is associative. *Normality* is then said to apply. Otherwise the flow rule is *non-associative.* Stress–strain laws

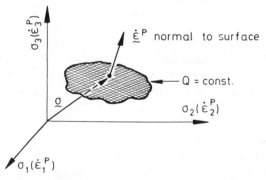

FIG. 3.   Flow rule in three-component space.

incorporating associative flow rules are on a firmer theoretical basis than those with non-associative rules.[8,21] Non-associative flow rules are commonly used for soils with the type of yield surface considered in this section because associative flow rules result in excessive dilatancy (see Section 2.4.5).

2.4.4 Formulation of Stress–Strain Relation

Consider a small increment of loading throughout which the stress point lies on the yield surface. The differential of F (eqn. (16)) is therefore zero, i.e.

$$F = \mathbf{a}_f^T \dot{\boldsymbol{\sigma}} + \frac{\partial F}{\partial h} \dot{h} = 0 \tag{18}$$

in which

$$\mathbf{a}_f = \frac{\partial F}{\partial \boldsymbol{\sigma}}$$

Defining a parameter A by

$$\dot{\lambda} = \frac{1}{A} \mathbf{a}_f^T \dot{\boldsymbol{\sigma}} \tag{19}$$

and writing $\dot{h} = f(\dot{\boldsymbol{\varepsilon}}^p)$ where $f(\ )$ has a form such that we can write, by eqn. (17),

$$\dot{h} = \dot{\lambda} f(\mathbf{a}_q) \tag{20}$$

We obtain, by combining eqns. (18), (19) and (20),

$$A = -\frac{\partial F}{\partial h} f(\mathbf{a}_q) \tag{21}$$

A is either obtained directly from test data or is expressed in terms of other experimental constants. It is related to the slope of the plastic deviator stress–strain curve if hardening or softening are determined by the deviatoric plastic strain. Alternatively it is related to the plastic volumetric strain if hardening is determined by plastic volume reduction. For metal plasticity, A becomes simply the slope of the uniaxial stress–plastic strain curve.[17]

The total strain rates are the sum of the elastic and plastic, i.e.

$$\dot{\boldsymbol{\varepsilon}} = \dot{\boldsymbol{\varepsilon}}^e + \dot{\boldsymbol{\varepsilon}}^p$$

Relating $\dot{\boldsymbol{\varepsilon}}^e$ to $\dot{\boldsymbol{\sigma}}$ by the elastic (tangential if variable) modulus $\mathbf{D}$ and introducing eqn. (17) with $\dot{\lambda}$ eliminated by eqn. (19) we have

$$\dot{\boldsymbol{\varepsilon}} = \mathbf{D}^{-1}\dot{\boldsymbol{\sigma}} + \frac{1}{A}\,\mathbf{a}_q\mathbf{a}_f^T\dot{\boldsymbol{\sigma}} \tag{22}$$

This can be inverted[17] to give

$$\dot{\boldsymbol{\sigma}} = \mathbf{D}\dot{\boldsymbol{\varepsilon}} - \frac{1}{\beta}\,\mathbf{b}_q\mathbf{b}_f^T\dot{\boldsymbol{\varepsilon}} \tag{23}$$

in which

$$\mathbf{b}_q = \mathbf{D}\mathbf{a}_q$$
$$\mathbf{b}_f = \mathbf{D}\mathbf{a}_f$$
$$\beta = A + \mathbf{a}_f^T\mathbf{b}_q$$

The explicit expression for $\dot{\boldsymbol{\sigma}}$ of eqn. (23) is needed in the equivalent load and tangential stiffness methods of non-linear finite element analysis. Equation (22) is needed in the visco-plastic method (see Chapter 1).

2.4.5 Specific Forms of Yield Surface
Yield surfaces relevant to soil mechanics may be divided into those which are open ended cones in principal stress space, and those which are capped cones (Fig. 4). In both classes the cones may or may not be right circular. Those which are right circular are fully defined by the stress invariants p and q. Of various possible non-

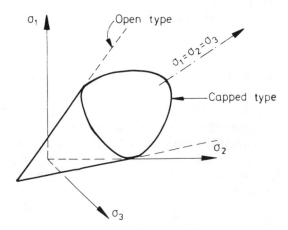

FIG. 4.   Yield surfaces in principal stress space.

circular cones (e.g. that proposed by Gudehus[7,31]) we shall restrict attention to that representing the Mohr–Coulomb criterion. It is a hexangular pyramid, and is fully defined by the stress invariants σ_s and σ_d. The critical state model has a yield surface which is a capped cone (although versions exist in which the cone is replaced by a continuation of the cap shape[30]). It will be described in the next section.

Figure 5 illustrates the Mohr–Coulomb yield surface and four of the right circular family of conical yield surfaces. They are shown as intersections of the three-dimensional surface with the pi plane—the plane in principal stress space at right angles to the line $\sigma_1 = \sigma_2 = \sigma_3$. There is a three-fold symmetry in the pi plane. Consequently the surfaces are fully defined in the 120° segment shown. Figures 5(b) and 5(c) show alternative plots for the surfaces in terms of the relevant pairs of invariants.

The open yield surfaces are generally assumed not to strain harden (that is, the cone angle does not increase) although they may strain soften. Consequently yielding implies local 'failure'. The yield surface becomes a failure criterion.

The equations defining the yield surface may be expressed in a variety of ways. That for Mohr–Coulomb may be written in terms of the shear and normal stress (τ_p, σ_p) on the plane of failure as

$$F(\tau_p, \sigma_p) = \tau_p - \sigma_p \tan \phi - c \leqslant 0 \qquad (24)$$

or in terms of σ_s and σ_d as

$$F(\sigma_s, \sigma_d) = \sigma_d - S\sigma_s - T \leqslant 0 \qquad (25)$$

in which $S = 2 \sin \phi$ and $T = 2 c \cos \phi$, or yet again as

$$F(p, q, \theta) = q - Mp - N \leqslant 0 \qquad (26)$$

in which M and N are functions of c, ϕ, and a third invariant, θ. θ is an angle in the pi plane (Fig. 5) whose tangent is $1/\sqrt{3}$ times Lode's parameter,[8] i.e.

$$\tan \theta = \frac{\sigma_1 - 2\sigma_2 + \sigma_3}{\sqrt{3}(\sigma_1 - \sigma_3)} \qquad (27)$$

θ measures the relative value of the intermediate principal stress. (An alternative expression for θ in terms of the stress components which is useful for three-dimensional finite element applications is given in reference 16.)

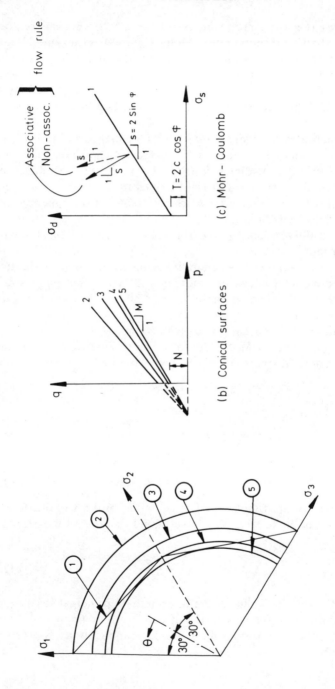

(a) Pi-plane plot

(b) Conical surfaces

(c) Mohr-Coulomb

Fig. 5.   Yield surfaces. 1 = Mohr–Coulomb, 2 = extended Von Mises, 3 = compromise cone, 4 = axial extension cone, 5 = Drucker–Prager.

TABLE 1
THE PARAMETERS M AND N IN EQN. (26)

Criterion	M	N
1. Mohr–Coulomb	$\dfrac{3 \sin \phi}{\sqrt{3} \cos \theta - \sin \theta \sin \phi}$	$\dfrac{3 c \cos \phi}{\sqrt{3} \cos \theta - \sin \theta \sin \phi}$
2. Extended Von Mises	$\dfrac{6 \sin \phi}{3 - \sin \phi}$	$\dfrac{6 c \cos \phi}{3 - \sin \phi}$
3. Compromise cone	$2 \sin \phi$	$2 c \cos \phi$
4. Axial extension cone	$\dfrac{6 \sin \phi}{3 + \sin \phi}$	$\dfrac{6 c \cos \phi}{3 + \sin \phi}$
5. Drucker–Prager	$\dfrac{3 \sin \phi}{\sqrt{3} + \sin \phi}$	$\dfrac{3 c \cos \phi}{\sqrt{3} + \sin \phi}$

The conical yield surfaces (2,3,4,5, in Fig. 5) may also be represented by eqn. (26) but with M and N containing c and ϕ only. The expressions for M and N are given in Table 1 for the five cases.

The more complicated form of eqn. (26) for the Mohr–Coulomb criterion has some advantage over eqn. (25) for three-dimensional applications. The main advantage is generality: a variety of yield criteria can be fitted into a single finite element program. For plane strain applications the simpler eqn. (25) may be preferred.

It has been mentioned that associative flow rules would cause excessive dilatancy with yield surfaces of the type considered here. This is illustrated in Fig. 5(c). The solid arrow represents an associative flow rule. Its direction is given by

$$\frac{\dot{\varepsilon}_d^p}{\dot{\varepsilon}_s^p} = - S = - 2 \sin \phi \tag{28}$$

in which $\dot{\varepsilon}_d^p = \frac{1}{2}(\dot{\varepsilon}_1^p - \dot{\varepsilon}_3^p)$ and $\dot{\varepsilon}_s^p = (\dot{\varepsilon}_1^p + \dot{\varepsilon}_3^p)$.* This means that the plastic deviator strain rate $\dot{\varepsilon}_d^p$ is associated with very significant negative plastic volumetric strain rate $(-\dot{\varepsilon}_s^p)$. This causes more dilatancy than actually occurs. A way out is to introduce a non-associative flow rule in which S in eqn. (28) is replaced by $\bar{S} = 2 \sin \Psi$. Ψ, the dilatancy angle, will be less than ϕ, typically in the range 0–20°.

*These have been defined so that for $\dot{\varepsilon}_2^p = 0$ the scalar product $\sigma_s \dot{\varepsilon}_s^p + \sigma_d \dot{\varepsilon}_d^p$ = rate of dissipation of plastic work, $\boldsymbol{\sigma}^T \dot{\boldsymbol{\varepsilon}}^p$.

2.5 CRITICAL STATE MODEL

The critical state model here is interpreted as an elastic-plastic cap model with a particular type of hardening law. The model may have a variety of different shaped yield surfaces. The conical part, for instance, may be circular or hexangular corresponding respectively to the (p, q) family described in the previous section and the Mohr–Coulomb yield criterion. The cap may have a variety of shapes. The two most common shapes for the curve formed by the intersection of the cap and the plane containing the $\sigma_1 = \sigma_2 = \sigma_3$ line are a log spiral (Cam clay)[26,28] and an ellipse (modified Cam clay).[24] Attention is restricted here to a yield surface incorporating the Mohr–Coulomb criterion in the conical part and which has an elliptical axial section through the cap. It is fully defined in terms of the stress-invariants σ_s, σ_d.

The yield surface is illustrated in Fig. 6. There is a discontinuity at the critical state line. The part to the left is called here the *super-critical* part and that to the right, i.e. the cap part, the *subcritical*. The supercritical part has a non-associative flow rule and the subcritical part has an associative flow rule. (The architects of the model assigned a flow rule to the subcritical part only.[25,26] Flow rules for the supercritical part have been added by the writer and others[17,30] to complete the stress–strain law for finite element applications). Yielding in the supercritical region is associated with failure. It is unstable

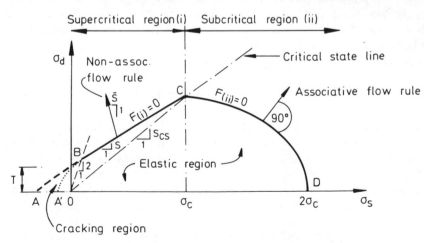

FIG. 6.   A Mohr–Coulomb critical state model.

in the sense defined by Drucker.[3] Yielding in the subcritical region is stable and is not associated with failure. This is an important distinguishing feature of the critical state model. Supercritical yielding results in strain softening, subcritical in strain hardening. How this occurs is explained by the hardening law.

2.5.1 Strain Hardening Law

The 'size' of the yield surface—as measured by σ_c (Fig. 6)—is related to the plastic component of volumetric strain according to

$$\sigma_c = \sigma_{c0} \exp\left(\frac{h}{\chi}\right) \tag{28}$$

in which σ_{c0} is the initial value of σ_c, h is the hardening parameter equal to the accumulated plastic volumetric strain, and χ is an empirical constant. Application is restricted here to plane strain for which it is assumed that $\dot{\varepsilon}_2^p = 0$ (where 2 denotes the out-of-plane direction). Therefore

$$h = \int d\varepsilon_s^p \tag{29}$$

The integral is taken from the start of plastic yielding. χ is related to the average voids ratio e and the compression indices λ and κ of the virgin and swelling consolidation curves* by

$$X = \frac{\lambda - \kappa}{1 + e} \tag{30}$$

It can be seen that this hardening law causes the yield surface to expand (strain harden) if volumetric plastic strain increments are compressive and to contract (strain soften) if they are expansive.

Interpreted geometrically on Fig. 6, the hardening law causes the yield surface to expand or contract about 0. Since the yield surface is assumed to retain its geometric proportions (isotropic hardening) the locus of point C—the critical state line—has a constant slope.

2.5.2 Flow Rules

These are illustrated in Fig. 6. The subcritical yield surface is so defined (see Section 2.5.3) that it has a horizontal tangent at the

---

*λ and κ are the slopes of these curves plotted to a natural log base. They are related to the log base 10 parameters C_c and C_s by $C_c = 2 \cdot 303\lambda$ and $C_s = 2 \cdot 303\kappa$.

critical state (point C). The associative flow rule therefore imposes zero plastic volume change at this point. It is reasonable that there should be no abrupt change in flow rule as the point moves into the supercritical region, and that the amount of dilatancy, as measured by plastic volume expansion, should progressively increase as the stress point moves to the left of C. The following specification for the inclination to the vertical $(\bar{S})$, of the flow rule vector satisfies this requirement.

$$\bar{S} = S_0 \left(1 - \frac{\sigma_s}{\sigma_c}\right) \tag{31}$$

S_0 is chosen empirically, and $\bar{S} = 2 \sin \psi$ where ψ is the dilatancy angle mentioned previously.

A plastic potential corresponding to eqn. (31) is

$$Q_{(i)} = \sigma_d + \frac{S_0}{2\sigma_c} (\sigma_c - \sigma_s)^2 \tag{32}$$

It can be readily verified that its gradient vector defines the flow rule, i.e. that

$$\frac{\partial Q_{(i)} / \partial \sigma_s}{\partial Q_{(i)} / \partial \sigma_d} = - \bar{S}$$

The subscript (i) relates to the supercritical region and (ii) will relate to the subcritical.

2.5.3 Equations of Yield Surface

The equation of the supercritical yield surface is simply that of the Mohr–Coulomb criterion, i.e.

$$F_{(i)} = \sigma_d - S\sigma_s - T = 0 \tag{33}$$

That for the subcritical, which is also the plastic potential, may be written

$$F_{(ii)} = Q_{(ii)} = \frac{\sigma_d^2 - S_{cs}^2 \sigma_s (2\sigma_c - \sigma_s)}{\sigma_d + S_{cs}\sigma_c} = 0 \tag{34}$$

in which S_{cs} is the slope of the critical state line. The denominator has been introduced so that $F_{(ii)} = F_{(i)}$ when $\sigma_s = \sigma_c$ for stress states that do *not* satisfy the yield criterion. This makes contours of $F_{(i)}$ and $F_{(ii)}$ connect up on the line $\sigma_s = \sigma_c$. This is desirable in certain computing schemes.

2.5.4 Tensile Zone

The supercritical yield surface has been shown dashed to the left of B in Fig. 6. In this region the minor principal stress is tensile. Various treatments are possible here. For example, the yield surface may be curved round as shown by the dotted line BA′. BA′ would be bounded by BO (the $\sigma_3 = 0$ line) and BA. Provided the intercept, T, is small—as it often is—it probably does not matter which course is adopted. The simplest procedure is to assume that the straight Mohr–Coulomb line continues to A.

2.5.5 Stress–Strain Relations

We are now in a position to derive explicit relations between stress and strain rate using the general expressions derived in the previous section. This will first be done to relate the invariants $\dot{\sigma}_s, \dot{\sigma}_d$ to $\dot{\varepsilon}_s, \dot{\varepsilon}_d$ and then to relate the three components of the plane strain stress and strain rate tensors.

The Parameter A

We first differentiate eqn. (33) to obtain $\partial F/\partial h$ for the supercritical region. Only T varies with h

so that
$$\frac{\partial}{\partial h} F_{(i)} = -\frac{\partial T}{\partial h}$$

From the geometry of Fig. 6 $T = (S_{cs} - S)\sigma_c$, and using eqn. (28) we obtain

$$\frac{\partial T}{\partial h} = (S_{cs} - S)\frac{\partial \sigma_c}{\partial h} = (S_{cs} - S)\frac{\sigma_{c0}}{\chi}\exp\left(\frac{h}{\chi}\right) = \frac{(S_{cs} - S)\sigma_c}{\chi}$$

i.e.

$$\frac{\partial F_{(i)}}{\partial h} = -\frac{(S_{cs} - S)\sigma_c}{\chi} \tag{35}$$

Now since $\dot{h} = \dot{\varepsilon}_s^p$, $f(a_q)$ in eqns. (20) or (21) is $\partial Q/\partial\sigma_s$, and

$$\frac{\partial Q_{(i)}}{\partial \sigma_s} = -S_0\left(1 - \frac{\sigma_s}{\sigma_c}\right) = -S$$

Equation (21) therefore gives

$$A_{(i)} = -\frac{\partial F}{\partial h} \cdot \frac{\partial Q}{\partial \sigma_s} = -S\frac{(S_{cs} - S)\sigma_c}{\chi} \tag{36}$$

D. J. NAYLOR

Proceeding similarly for the subcritical region we obtain

$$\frac{\partial F_{(ii)}}{\partial h} = -\frac{2S_{cs}^2\sigma_s\sigma_c}{\chi(\sigma_d + S_{cs}\sigma_c)} \tag{37}$$

and

$$\frac{\partial Q_{(ii)}}{\partial \sigma_s} = \frac{2S_{cs}^2(\sigma_s - \sigma_c)}{(\sigma_d + S_{cs}\sigma_c)} \tag{38}$$

whence

$$A_{(ii)} = \frac{4S_{cs}^4\sigma_s\sigma_c(\sigma_s - \sigma_c)}{\chi(\sigma_d + S_{cs}\sigma_c)^2} \tag{39}$$

Elastic Stress–Strain Relation
It is easily shown from the isotropic elastic stress–strain laws that

$$\dot{\sigma}_d = 4G\dot{\varepsilon}_d^e \tag{40}$$

and

$$\dot{\sigma}_s = \left(K + \frac{1}{3}G\right)\dot{\varepsilon}_s^e \tag{41}$$

Relation between Stress and Strain Rate Invariants
The gradient vectors $\mathbf{a}_q$ and $\mathbf{a}_f$ in the (σ_s, σ_d) space are respectively $[\partial Q/\partial\sigma_s, \partial Q/\partial\sigma_d]^T$ and $[\partial F/\partial\sigma_s, \partial F/\partial\sigma_d]^T$. Differentiating eqns. (32) and (33) with respect to σ_s and σ_d and introducing $A_{(i)}$ from eqn. (36) into the right-hand term of eqn. (22), we obtain the following expression for the plastic strain rate invariants in the supercritical region

$$\begin{Bmatrix} \dot{\varepsilon}_s^p \\ \dot{\varepsilon}_d^p \end{Bmatrix}_{(i)} = -\frac{\chi}{S(S_{cs} - S)\sigma_c} \begin{bmatrix} S\bar{S} & -\bar{S} \\ -S & 1 \end{bmatrix} \begin{Bmatrix} \dot{\sigma}_s \\ \dot{\sigma}_d \end{Bmatrix} \tag{42}$$

To this may be added the elastic strain rate components from the inverse of eqns. (40) and (41) to give the total strain rates.

Explicit expressions for $\dot{\sigma}_s$ and $\dot{\sigma}_d$ may similarly be obtained by substitution into eqn. (23). Note that **D** takes the form

$$\mathbf{D}^* = \begin{bmatrix} K + \dfrac{1}{3}G & 0 \\ 0 & 4G \end{bmatrix} \tag{43}$$

Proceeding similarly for the subcritical region we obtain

$$\begin{Bmatrix} \dot{\varepsilon}_{cs}^p \\ \dot{\varepsilon}_d^p \end{Bmatrix}_{(ii)} = \frac{\chi}{\sigma_s\sigma_c} \begin{bmatrix} S_{cs}^2(\sigma_s - \sigma_c) & \sigma_d \\ \sigma_d & t \end{bmatrix} \begin{Bmatrix} \dot{\sigma}_s \\ \dot{\sigma}_d \end{Bmatrix} \tag{44}$$

in which

$$t = \frac{\sigma_d^2}{S_{cs}^2(\sigma_s - \sigma_c)}$$

The symmetry of the matrix in eqn. (44) is in contrast to the asymmetry in eqn. (42). This arises from the use of a non-associative flow rule in the supercritical region. If it is made associative by making $\bar{S} = S$ the matrix in eqn. (42) becomes symmetric.

For brevity either eqns. (42) or (44) may be represented by the matrix equation

$$\dot{\varepsilon}_p^* = \mathbf{C}_p^* \dot{\boldsymbol{\sigma}}^* \qquad (45)$$

The asterisks are used to distinguish vectors and matrices defined in the (σ_s, σ_d) or $(\dot{\varepsilon}_s, \dot{\varepsilon}_d)$ space from those defined in a space containing all the stress or strain rate components.

Relation between Stress and Strain Rate Components
There are two ways of deriving the relation. The partial derivatives of Q and F with respect to the stress components σ_x, σ_y and τ can be obtained and used to define the gradient vectors $\mathbf{a}_q$ and $\mathbf{a}_f$ in a three-component stress space. The procedure then is the same as above. To obtain the partial derivatives it is necessary first to express σ_s, σ_d in terms of σ_x, σ_y and τ. The other way is first to derive the relation between invariants (eqn. (45)) and then proceed as follows.

Let the stress components be denoted by the vector $\boldsymbol{\sigma} = [\sigma_x, \sigma_y, \tau]^T$ and strains by $\boldsymbol{\varepsilon} = [\varepsilon_x, \varepsilon_y, \gamma]^T$.

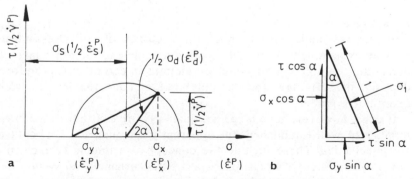

FIG. 7. Stress and plastic strain rate transformation. (a) Mohr circle of stress (plastic strain rate); (b) forces on elemental wedge.

Figure 7 shows a Mohr's circle which applies both to stress and plastic strain rate. This is possible from the assumption of coaxiality of the respective tensors. From the geometry of the diagram

$$\dot{\boldsymbol{\varepsilon}}^p = \mathbf{R}^T \dot{\boldsymbol{\varepsilon}}_p^* \tag{46}$$

in which

$$\mathbf{R} = \begin{bmatrix} \frac{1}{2} & \frac{1}{2} & 0 \\ \cos 2\alpha & -\cos 2\alpha & 2\sin 2\alpha \end{bmatrix}$$

Considering stresses,

$$\sigma_s = \tfrac{1}{2}(\sigma_x + \sigma_y)$$

so that

$$\dot{\sigma}_s = \tfrac{1}{2}(\dot{\sigma}_x + \dot{\sigma}_y)$$

$$\sigma_d^2 = (\sigma_x - \sigma_y)^2 + 4\,\tau^2$$

so that

$$2\,\sigma_d \dot{\sigma}_d = 2(\sigma_x - \sigma_y)(\dot{\sigma}_x - \dot{\sigma}_y) + 8\,\tau\dot{\tau}$$

Introducing the angle α from Fig. 7,

$$\dot{\sigma}_d = (\dot{\sigma}_x - \dot{\sigma}_y)\cos 2\alpha + 2\,\dot{\tau}\sin 2\alpha$$

whence

$$\dot{\boldsymbol{\sigma}}^* = \mathbf{R}\dot{\boldsymbol{\sigma}} \tag{47}$$

Substituting eqn. (47) in (45) to eliminate $\dot{\boldsymbol{\varepsilon}}_p^*$ from eqn. (46) gives

$$\dot{\boldsymbol{\varepsilon}}^p = \mathbf{R}^T \mathbf{C}_p^* \mathbf{R}\dot{\boldsymbol{\sigma}} = \mathbf{C}_p\dot{\boldsymbol{\sigma}} \tag{48}$$

The 3×3 plastic compliance matrix $\mathbf{C}_p$ is readily computed for the supercritical and subcritical regions by taking $\mathbf{C}_p^*$ from eqns. (42) and (44) respectively. The $\mathbf{D}$ matrix needed to relate $\dot{\boldsymbol{\varepsilon}}^e$ to $\dot{\boldsymbol{\sigma}}$ is the conventional 3×3 plane strain modulus matrix.

2.5.6 Comments
The remarks made earlier about the accuracy of soil stress–strain laws apply as much to the critical state model as to the other laws. It does offer the potential of better predictions in certain cases (uniform deposits of soft clay, for example) but is none the less a crude quantitative model.

It does, however, have a special qualitative role. It links together a number of well-established but apparently unconnected concepts in soil mechanics. These include the concept of a unique surface in a voids ratio effective stress space attributed originally to Rendulic,[23] the important criterion of Hvorslev that, for saturated clays, the shear strength depends on the moisture content at failure,[10] the Mohr–

Coulomb criterion, the irrecoverability of strains in soils consolidated to new effective stress levels, and the critical state concept itself. The critical state concept requires that continuous shearing is associated with no further volume change in the case of fully drained loading, and no further change in pore pressure in the case of undrained loading. The flow rule, by causing the ratio $\dot{\varepsilon}_s^p/\dot{\varepsilon}_d^p$ to tend to zero as the critical state is approached (point C in Fig. 6), automatically satisfies this requirement. Furthermore, the second requirement of the critical state concept that there should be no further changes in the stress components as the critical state is approached is also embodied in the model.

The critical state model was largely developed at Cambridge in the 1950s and 1960s under the later Professor Roscoe's leadership.[25–27] Some of the credit for the idea must, however, go to Drucker *et al.* who, in the 1950s, first suggested the application of plasticity theory to soil mechanics[6] and produced what was probably the first 'cap' model.[5]

The model performs best for roughly radial effective stress paths leading to subcritical yielding. In the supercritical region, deformation predictions are highly suspect. This is largely because the instability introduced on yielding causes the plastic strains to become concentrated into a narrow zone or zones. The material becomes discontinuous and, since the model assumes continuity, it ceases to apply. It is possible that the results may still be useful, but this requires further research. It is, however, necessary to have a stress–strain law defined in this area to cater for the limited regions which might enter into it in an analysis where yielding is predominantly subcritical.

The shape of the supercritical yield surface may not be important, provided the hardening law causes the yield surface to shrink so that the stress point eventually ends up on the critical state line. For this reason some workers extend the subcritical ellipse into the supercritical region and retain an associative flow rule throughout.[30]

The critical state model is suited to clays because of the hardening law which is based on a clear distinction between overconsolidated and normally consolidated states. This concept has little meaning for sands the stiffness properties of which depend more on the mechanics of its deposition than on its stress history. None the less the model may be useful for sands when yielding is supercritical, since it can incorporate the Mohr–Coulomb yield criterion and the hardening law will reproduce the post-peak strength loss exhibited by dense and medium-dense sands. Clays yielding supercritically, however, may fare less well, for, in addition to the errors due to the forming of discontinuities, a further error associated with the reduction in c' and ϕ' to residual values as shearing takes place will occur. This error can be compensated for by introducing a second strain softening parameter (h_2 in reference 17).

2.6 ASSESSMENT OF STRESS–STRAIN LAWS

Figure 8 compares four stress–strain laws for a drained triaxial test on a normally consolidated clay. The differences in the shapes of the $(\sigma_1 - \sigma_3) : \varepsilon_1$ curves are not significant, particularly if the curve for the hyperbolic model is adjusted upwards as mentioned in Section 2.3 to

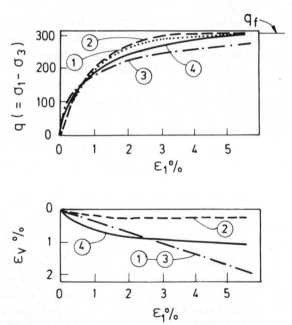

FIG. 8. Comparison of four stress–strain laws for drained triaxial tests. 1. $E - \nu$: $E = 100p - 130q$, $\nu = \frac{1}{3}$; 2. K–G: $K = 100p$, $G = 38p - 49q$; 3. Hyperbolic: a = 0·234, b = 0·320, $\nu = \frac{1}{3}$. 4. Critical state: $E = 250\,000$, $\nu = \frac{1}{3}$, $p_c = 150$, $\chi = 0·01$. (Units = kN and m). $\sigma_3 = 300$ kN/m^2.

bring it more in line with the others. All these curves can be made to approximate an experimental curve with sufficient accuracy. Of greater significance is the $\varepsilon_v : \varepsilon_1$ relation. The progressive decrease in volume exhibited by the $E - \nu$ model illustrates the 'black hole' phenomena mentioned earlier. If shearing continues for long enough a material obeying this law will disappear. Although both the K–G and critical state models incorporate the critical state concept, the latter will reproduce more closely the volume reduction on shearing experienced by soft clays. The critical state model also performed better in this respect for a compacted boulder clay. It can be concluded from this that for undrained loading the critical state model will give a better prediction of excess pore pressures than the K–G model. The K–G model, because it is isotropic and elastic, implies that Skempton's pore pressure parameter A is one third, and will underestimate the pore pressure response of soft clays.[29]

There is very little difference between the (p, q) and (σ_s, σ_d) versions of the variable elastic models when used to reproduce triaxial test curves. Thus the curves of Fig. 8 will be almost unchanged (the volume change curve is slightly affected) if p and q are replaced by σ_s and σ_d, and the constant β_G altered by an appropriate amount.

But this is only true for the special case of the triaxial test in which $\sigma_2 = \sigma_3$ and the extended Von Mises and Mohr–Coulomb failure criteria coincide ($\theta = 30°$ in Fig. 5(a)). In more general situations (p, q) and (σ_s, σ_d) versions can produce quite different predictions of the failure load. This difference becomes greatest (usually) when the (p, q) version incorporates the extended Von Mises criterion. It can be minimised by choosing a more appropriate criterion such as the compromise cone, axial extension cone, or Drucker–Prager (Fig. 5).* Figure 9 illustrates this difference and shows how it becomes greater as the effective stress path inclines to the right. This occurs in the drained analysis of footings where it is possible if the friction angle is fairly high for a Mohr–Coulomb (σ_s, σ_d) law to predict a failure load whereas with a (p, q) extended Von Mises law a failure load may never be reached. The matter is studied in reference 9.

---

*Of these, the axial extension cone is good for undrained plane strain analyses for which θ tends to be close to zero. The compromise cone provides a reasonable compromise for drained analyses for which θ varies more widely (see last chapter of reference 17).

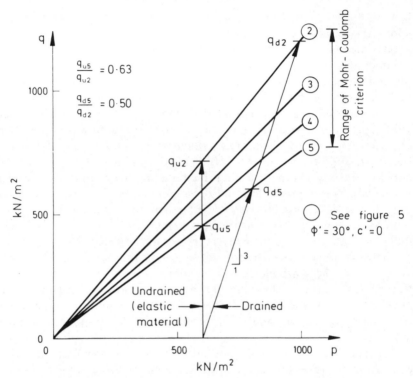

FIG. 9.   Effect of criterion and effective stress on failure stress.

The triaxial test is misleading in another respect. It has very little boundary constraint. This is not typical of practical situations. The differences between stress–strain laws become exaggerated when there is much boundary constraint. Figure 10 shows the difference between a (σ_s, σ_d) and a (p, q) version of the critical state model when used to reproduce two hypothetical shear box tests. Both clay samples are consolidated under an assumed σ_3'/σ_1' ratio of 0.5 to point A. One is then sheared under drained conditions with the normal effective stress constant at 480 KN/m² to reach the critical state at C_1. The other is unloaded to 280 KN/m² and then sheared to reach the critical state at C_2. The first therefore represents a normally consolidated clay and the second an overconsolidated clay. The (p, q) and (σ_s, σ_d) versions were matched to the same triaxial test data.

Boundary constraint has an important effect on elastic-plastic

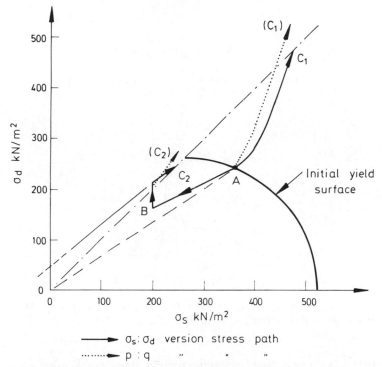

FIG. 10. Drained shear box test: critical state model.

stress–strain models, particularly when—as is nearly always the case—the elastic stiffness is relatively high. Figure 11 shows the stress path followed by a hypothetical lightly overconsolidated clay consolidated in an oedometer. The critical state (p, q) model is used. The 'dog's leg' arises from the abrupt change in the stress–strain law which occurs when the stable yield surface is encountered. The boundary controls the strains. Consequently the stresses must alter to accommodate the changed law. This must be considered a weakness of elastic-plastic laws for such a stress response is not likely to actually occur.

When loading is monotonic, variable elastic laws are attractive, particularly the K–G model, (σ_s, σ_d) version. When loading is reversed elastic-plastic laws offer the advantage that they automatically reproduce the increased stiffness which occurs on unloading. They

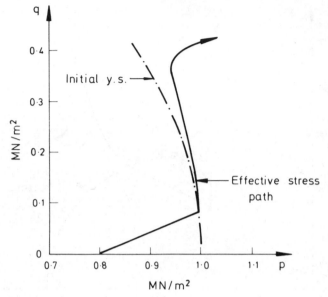

FIG. 11.   Oedometer test: stress path for critical state model. Material properties: $E = 2800$ MN/m^2, $\nu = 0\cdot4$, $M = 1\cdot2$, $\chi = 0\cdot011$, $p_{co} = 0\cdot5$ MN/m^2.

therefore have a potential for cyclic loading. The critical state class of models has the advantage over other elastic-plastic laws in that it can reproduce the reduction in stiffness in clay soils which occurs when they are stressed beyond the pre-consolidation pressure.

REFERENCES

1. BISHOP, A. W. (1966). The strength of soils as engineering materials, *Géotechnique*, **16**(2), 89–130.
2. CALLADINE, C. R. (1963). Correspondence, *Géotechnique*, **13**(3), 250–255.
3. DRUCKER, D. C. (1959). A definition of stable inelastic material, *Trans. Am. Soc. Mech. Engrs.*, **26**, 101–106.
4. DRUCKER, D. C. (1964). Concept of path independence and material stability for soils, *International Symposium on Rheology and Soil Mechanics*, Grenoble, 1964, 23–46.
5. DRUCKER, D. C., GIBSON, R. E. and HENKEL, D. J. (1957). Soil mechanics and work hardening theories of plasticity, *Trans. Am. Soc. Civ. Engrs.*, **122**, 338–346.

6. DRUCKER, D. C. and PRAGER, W. (1952). Soil mechanics and plastic analysis or limit design, *Qtly. Applied Maths*, **10**, 157–165.
7. GUDEHUS, G. (1973). Elastoplastische Stoffgleichungen für trockenen Sand, *Ingenieur-Archive*, **42**.
8. HILL, R. (1950). *The Mathematical Theory of Plasticity*, Clarendon Press, Oxford.
9. HUMPHESON, C. and NAYLOR, D. J. (1976). The importance of the form of the failure criterion, In *Numerical Methods in Soil and Rock Mechanics*, Vol. II, BORN, G. and MEISSNER, H. (Eds.), Institut für Bodenmechanik und Felsmechanik, Universität, Karlsruhe, pp. 17–30.
10. HVORSLEV, M. J. (1960). Physical components of the shear strength of saturated clays, *Am. Soc. Civ. Engrs. Conf. on Shear Strength of Cohesive Soils*, University of Colorado, 1960, 169–273.
11. KONDNER, R. L. (1963). Hyperbolic stress–strain response: cohesive soils, *Proc. Am. Soc. Civ. Engrs.*, **82**(SM1), 115–143.
12. KULHAWY, F. H. and DUNCAN, J. M. (1972). Stresses and movements in Oroville Dam, *Proc. Am. Soc. Civ. Engrs.*, **98**(SM7), 653–665.
13. LADD, C. C. (1964). Stress–strain modulus of clay from undrained triaxial tests, *Proc. Am. Soc. Civ. Engrs.*, **90**(SM5), 103–132.
14. LAMBE, T. and WHITMAN, R. V. (1969). *Soil Mechanics*, Wiley, New York.
15. MROZ, Z. (1967). On the description of anisotropic work-hardening, *J. Mech. Phys. Solids*, **15**, 163–175.
16. NAYAK, G. C. and ZIENKIEWICZ, O. C. (1972). Convenient form of stress invariants for plasticity, *Proc. Am. Soc. Civ. Engrs.*, **98**(ST4), 949–953.
17. NAYLOR, D. J. (1975). Non-linear finite elements for soils, Ph.D. Thesis, University of Wales.
18. NELSON, J. and BARON, M. L. (1971). Application of variable moduli to soil behaviour, *Int. J. Solids and Structures*, **7**, 399–417.
19. PENMAN, A. D. M., BURLAND, J. B. and CHARLES, J. A. (1971). Observed and predicted deformations in a large embankment dam during construction, *Proc. Inst. Civ. Engrs.*, **49**, 1–21.
20. PENMAN, A. D. M. and CHARLES, J. A. (1971). Constructional deformations in rockfill dam, *Proc. Am. Soc. Civ. Engrs.*, **99**(SM2), 139–163.
21. PRAGER, W. (1959). *An Introduction to Plasticity*, Addison–Wesley, Amsterdam and London.
22. PREVOST, J. and HOEG, K. (1975). Mathematical model for static and cyclic undrained clay behaviour, *Norwegian Geotechnical Inst. Report* 52412.
23. RENDULIC, L. (1938). A consideration of the question of plastic limiting states, *Bauingenieur*, **19**, 159–165.
24. ROSCOE, K. H. and BURLAND, J. B. (1968). On the generalised stress–strain behaviour of wet clay, In *Engineering Plasticity*, HEYMAN, J. and LECKIE, F. (Eds.), Cambridge University Press, pp. 539–609.
25. ROSCOE, K. H. and POOROOSHASB, H. B. (1963). A theoretical and experimental study of strains in triaxial compression tests on normally consolidated clays, *Géotechnique*, **13**(1), 12–38.

26. ROSCOE, K. H., SCHOFIELD, A. N. and THURAIRAJAH, A. (1963). Yielding of clays in states wetter than critical, *Géotechnique*, **13**(3), 211–240.
27. ROSCOE, K. H., SCHOFIELD, A. N. and WROTH, C. P. (1958). On the yielding of soils, *Géotechnique*, **8**(1), 22–53.
28. SCHOFIELD, A. N. and WROTH, C. P. (1968). *Critical State Soil Mechanics*, McGraw-Hill, New York.
29. SKEMPTON, A. W. (1954). The pore pressure coefficients A and B, *Géotechnique*, **4**(2), 143–147.
30. ZIENKIEWICZ, O. C., HUMPHESON, C. and LEWIS, R. W. (1975). Associated and non-associated visco-plasticity in soil mechanics, *Géotechnique*, **25**(4), 671–689.
31. ZIENKIEWICZ, O. C. and PANDE, G. N. (1976). Some useful forms of isotropic yield surfaces for soil and rock mechanics, *Numerical Methods in Soil Mechanics and Rock Mechanics*, Vol. II, BORN, G. and MEISSNER, H. (Eds.), 1976, Institut für Bodenmechanik und Felsmechanik, Universität, Karlsruhe, pp. 3–16.

Chapter 3

APPLICATION OF THE FINITE ELEMENT METHOD
TO PREDICTION OF GROUND MOVEMENTS

J. B. Burland

Geotechnics Division, Building Research Station, Garston, UK

SUMMARY

In this chapter a few examples are given of field studies of civil engineering projects for which ground deformations have been measured and compared with predictions made by the finite element method. All the studies were carried out by the Building Research Station and have been selected to illustrate the application of the finite element method to: (a) surface loading (b) embankment construction, and (c) excavations and retaining walls.

3.1 INTRODUCTION

The finite element method is an immensely powerful and versatile analytical technique which is easy to use. As such it is of great practical value to the engineer, but precisely because it is so easy to use it has to be handled with care, knowledge and judgement.

Given this analytical power there is a temptation to believe that geotechnics will become an exact science. Nothing could be further from the truth. It is instructive to consider for a moment an Utopian situation in which the engineer has unlimited analytical power. In such a situation the engineer would be in a position to solve any boundary value problem given: (a) *the geometry*, (b) *the loading*, and (c) *the material properties*. A moment's reflection reveals that despite such analytical power the engineer would not, in fact, be much better off than he is at present. This is because of the wide ranging idealisations and assumptions that have to be made before an analysis can be carried out. It is worth considering briefly some of the idealisations that have to be made under the above three headings.

(a) Geometry

The final geometry of a structure such as an embankment or a retaining wall is often well defined. However, the geometry at various stages during construction may be quite unpredictable. For example, the sequence in which the fill is placed in an embankment dam will depend on the contractor, the weather, the condition of various borrow pits and many other factors. Similarly, the precise sequence in which a deep basement is excavated will depend on a number of factors outside the control of the designer. Yet for many structures the behaviour is profoundly influenced by the construction sequence.

The geometry of the ground itself is, of course, very difficult to establish with any certainty. In the majority of situations it is not economic to put down enough boreholes to give an accurate three-dimensional picture of the various strata underlying the proposed structure. The engineer is usually compelled to make very sweeping assumptions.

(b) Loading

The resultant loads (as opposed to their distribution) acting on foundations are usually reasonably well defined. A major uncertainty is the initial stress distribution in the ground and this has a profound influence on the performance of subsurface structures and retaining walls. Difficulties also arise for structures subject to dynamic forces, e.g. waves, earthquakes, etc.

(c) Material Properties

It is here that the engineer is faced with the largest uncertainties. It is tempting to believe that by carrying out a number of careful laboratory tests it will be possible to define with some precision the stress–strain properties of the ground. Unfortunately it is just not possible to predict nor to reproduce in the laboratory the complexities of stress changes that occur in reality. Moreover, even after two decades of research on ideal reconstituted soils such as kaolin or pure sand soil, soil mechanics specialists cannot claim to be in a position to predict the stress–strain behaviour of such materials with confidence over a wide range of stress paths. Add to these factors the problems associated with 'undisturbed' sampling and it will be appreciated how formidable is the task of ascertaining realistic *in situ* stress–strain relations for natural soils and the variations with depth and plan. At best the engineer can only hope to represent the soil by means of highly idealised simple models.

It is evident from the foregoing that even if engineers were in possession of unlimited analytical power the uncertainties in the ground are so great that the precision in the prediction of behaviour would be unlikely to improve significantly on present capabilities. As in so many fields of engineering, analysis is only one of the many tools required in geotechnics. In most circumstances the real value of analysis will be in assisting the engineer to place bounds on likely overall behaviour and in assessing the influence of various assumptions. There is no doubt that the finite element method can be of great value here. However, it cannot be stressed too strongly that analytical sophistication will not compensate for the basic uncertainties about the ground. Moreover, over-sophistication can lead to serious difficulties in interpreting the results of an analysis and its value is then lost.

In this chapter a few examples are given of field studies of civil engineering structures for which ground deformations have been measured and compared with predictions made by the finite element method. All the studies have been carried out by the Building Research Station and have been carefully selected to illustrate the application of the method to (i) surface loading, (ii) embankment construction, and (iii) excavations and retaining walls. It is assumed that the reader is familiar with the basic principles of the finite element method as outlined in Chapter 1. For all the cases considered the ground is assumed to behave in accordance with simple linear elastic materials. By comparing the predictions with real behaviour the reader will be able to judge the relative importance of various assumptions.

3.2   A SETTLEMENT PROBLEM

In 1967 the Building Research Station was asked to undertake a site investigation for a large proton accelerator in chalk at Mundford, Norfolk. A central feature of this investigation was a detailed and precise study of the displacement of the ground beneath and around a large, circular, steel, water tank.

J. B. BURLAND

FIG. 1. Mundford test tank.

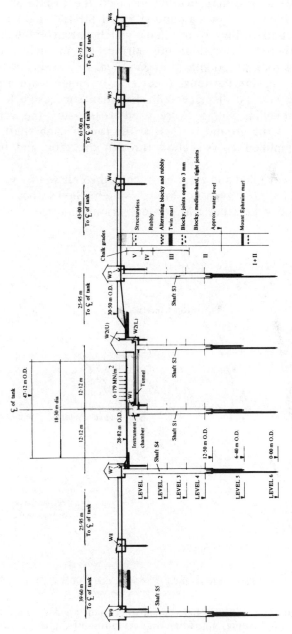

FIG. 2. Section through tank site showing the position of the instrument shafts and water levelling gauges.

Full details of the tank loading test and the results are given by Ward *et al.*[1] Figure 1 shows a photograph of the 18·3 m diameter, 18 m high tank founded directly on chalk which varied gradually from a highly weathered material at the surface to an intact completely unweathered rock at a depth of about 13 m. In a series of five shafts under and alongside the tank, precise instruments were arranged to measure the vertical displacements between successive levels. Precise water levelling gauges were used to measure the vertical displacements of the ground surface at the top of each shaft. Figure 2 shows a simplified cross-section through the tank and instrument shafts.

The tank was filled and emptied once in the course of a week. Some typical results are given in Fig. 3 which shows the relationship between applied pressure and vertical strain for three levels beneath

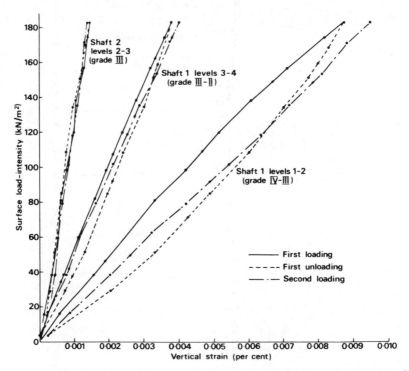

FIG. 3. Typical relationships between surface load-intensity and vertical strain during loading, unloading and reloading of the test tank.

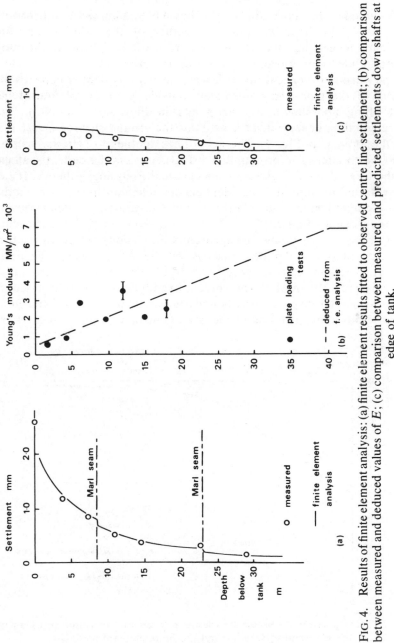

Fig. 4. Results of finite element analysis: (a) finite element results fitted to observed centre line settlement; (b) comparison between measured and deduced values of E; (c) comparison between measured and predicted settlements down shafts at edge of tank.

and outside the tank during loading, unloading and subsequent reloading. The linearity and reversibility of the relationships are evident, suggesting that the ground beneath and around the tank approximates very closely to an elastic material.

The observed vertical displacements at various depths beneath the centre of the tank and immediately outside it on completion of the first filling are shown by the points in Figs. 4(a) and 4(c). The corresponding observed settlements of the ground surface are shown by the points in Fig. 5. The solid lines in these diagrams will be referred to later. The broken line in Fig. 5 depicts the deflected shape of the ground surface given by the classical Boussinesq theory. It can be seen that in spite of the evident elastic behaviour of the ground the observed settlements outside the loaded area are very much less than the theoretical values.

In order to investigate this apparently anomalous result, Burland *et al.*[2] carried out a detailed analysis of the results using the finite element method and a rigorous analysis. The finite element mesh extended to a depth of 180 m where a fixed boundary was imposed. A zero horizontal displacement boundary was imposed at a distance of 100 m from the centre line of the tank. Constant strain triangles were

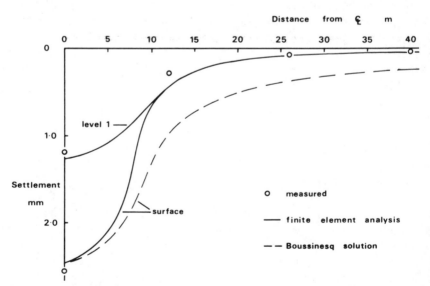

FIG. 5.   Comparison of observed settlement profile of the ground surface with the Boussinesq solution and the finite element analysis.

used and the mesh consisted of over 800 elements concentrated in the region of high stress gradients near the loaded area.

The approach adopted was to use the measured displacements beneath the centre of the tank to evaluate the variations of Young's modulus E with depth. This involved a trial and error procedure whereby a distribution of E was first assumed, an analysis was carried out and the centre-line displacements compared with the measured displacements. The value of E at each depth was then modified and another calculation made. The procedure was repeated until a satisfactory fit was obtained between the measured and calculated centre-line displacements. In carrying out the analysis a value of Poisson's ratio equal to 0·24 was used. This value was measured in the laboratory on intact samples of chalk and was subsequently confirmed by full-scale measurements of horizontal displacement beneath the tank.

In Fig. 4(a) the solid line represents the centre-line settlements given by the finite element analysis. A close fit has been achieved between the measured settlements and the theoretical values obtained by trial and error. The small discontinuities in the theoretical relationships are caused by the presence of thin compressible marl seams, the influence of which was included in the analysis. In Fig. 4(b) the theoretical distribution of E with depth required to give the fit is given by the broken line. It can be seen that the deduced stiffness increases approximately linearly with depth from a very low value at the surface. The points representing the results of plate loading tests will be referred to later.

Making use of the distribution of deduced E with depth given in Fig. 4(b), the displacement of a large number of points beneath and around the tank has been evaluated. In Fig. 4(c) the theoretical displacements down the shafts immediately adjacent to the test tank are compared with the observed values. It can be seen that the agreement is very good, the theoretical values being slightly larger than the observed values at all depths.

In Fig. 5 the theoretical deflected shape of the ground surface is compared with the measured displacements. Once again the agreement is excellent. It is evident that the unexpected observation that the settlement of the ground surface is localised around the loaded area much more than the simple isotropic elastic theory predicts, is accounted for almost entirely by the influence of non-homogeneity. Burland et al.[2] confirmed the accuracy of the finite element analysis by comparing the numerical and measured results with a closed form solution for a circular load on a non-homogeneous elastic half-space.

In Fig. 4(b) reference is made to the results of plate loading tests. The tests were carried out at various depths at the bottom of a 900 mm diameter unlined shaft alongside the test tank.[3,4] The pressure settlement curves from the tests were found to be nearly linear and reversible over the range of stresses imposed by the test tank. The value of E can be determined from the well-known expression for a rigid punch of diameter D:

$$E = \frac{q}{\rho} \cdot \frac{\pi}{4} \cdot D(1 - \nu^2) \cdot I_D$$

where q is the average pressure, ρ is the settlement, ν is Poisson's ratio, and I_D is the depth correction factor.

A difficulty arises over the evaluation of the depth correction factor I_D. It has been customary to make use of values derived by Fox[5] for a uniformly loaded area at depth within an elastic half-space. However, the actual situation in which the load is applied at the base of an unlined shaft departs considerably from this ideal case. No analytical solution exists and the opinions of various experts as to the likely influence of the unlined shaft differ significantly. Some have felt that the load could be assumed to be acting at the surface of a half-space ($I_D = 1$), while others have felt sure that the values given by Fox could be used. A series of finite element analyses have been carried out. The same mesh was used each time with the load applied at successively greater depths and the overlying elements given zero stiffness. In this way an unlined shaft of varying depth was simulated. The results of the analysis are given in Fig. 6 and are compared with Fox's solution. It is evident that the presence of the unlined shaft considerably reduces the correction for depth.

It may be worth noting in passing that the mesh used to study the unlined shaft problem can also be used to study the behaviour of a pile under load. Instead of giving the elements low stiffness they can just as easily be given a high stiffness. Although analytically the two problems are very different, the same finite element mesh may be used to study each. This demonstrates the versatility of the method.

This study has drawn attention to the following important aspects in relation to the application of the finite element analysis to the prediction of settlements:

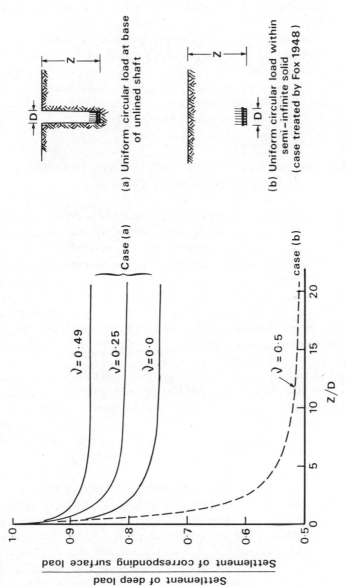

FIG. 6.   Influence of an unlined shaft on the depth correction factor.

(a)   The method is a valuable tool for the back analysis of field
      measurements of settlement to give *in situ* properties of the
      ground.
(b)   Non-homogeneity has a profound influence on the deformations
      beneath and around loaded areas and is easily handled by the
      finite element method.
(c)   The method can be used to analyse complicated boundary
      conditions with comparative ease, as for a load at the base of an
      unlined shaft or a vertically loaded pile.

3.3   DEFORMATIONS OF EMBANKMENTS DURING CONSTRUCTION

The previous section dealt with the displacements resulting from the
application of pressure to the ground surface. In the present section
we will consider the displacements within an embankment or fill
during construction. For this problem the displacements result from
the weight of material as it is added. However, the additional material
also contributes to the stiffness of the embankment as a whole. Hence
the problem is one in which the geometry of the structure to be
analysed changes during construction and the analysis has to be
carried out in a step-by-step manner following the actual construction
history.

The importance of following closely the construction history can be
conveniently illustrated by considering the internal vertical displacements within a layer of compressible material of thickness H
possessing self-weight as it is constructed (see Fig. 7). A simple
analysis could be carried out neglecting the construction history of
the layer and simply applying gravity to the whole layer. This method
is often termed 'gravity switch-on'. The broken curve (1) in Fig. 7
gives the form of vertical displacement with height that would result
from such an analysis. Note that the surface of the layer settles the
most. In fact, curve (1) is grossly in error as can be quickly demonstrated by considering the settlement of a point embedded at the top
of the layer. This top, infinitely thin element will induce virtually no
increase in stress in the underlying material and hence a point
embedded in it will experience no settlement. Yet the 'gravity switch-
on' method predicts maximum settlement for the topmost element. A
moment's reflection will reveal that a point embedded in the layer at a

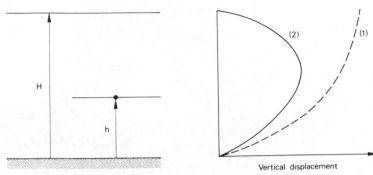

FIG. 7. Vertical displacements during construction of a thick compressible layer. Curve 1 due to 'gravity switch-on'. Curve 2 due to progressive construction.

height h (say) can only begin to settle once the layer has reached a height equal to h. Prior to that the point does not exist! The correct form of relationship between settlement and height is given by the solid line (2) in Fig. 7 and it can be seen that the maximum settlement is at about half the height of the layer. Penman et al.[6] give a detailed analysis of this one-dimensional problem using various stress–strain relationships for the fill. It is important to note that although the displacements are dependent on the construction history the final stresses are not.

The foregoing discussion helps to demonstrate the importance of simulating the construction history if the construction displacements of an embankment are to be correctly analysed. The finite element method is well suited for such an analysis and its application will be illustrated by describing its use in predicting the construction deformations of Scammonden Dam.

The 70-m high dam is of rockfill construction with an upstream clay core and at the time of construction was the highest dam in Great Britain. An aerial photograph of the dam is shown in Fig. 8. The dam has been fitted with very comprehensive instrumentation.[6,7] In particular the downstream rockfill shoulders contain horizontal plate gauges at four elevations consisting of rigid PVC pipes fitted with steel plates at about every 15-m length as indicated in Fig. 9. Sensing units passed through the pipes detect the steel plates and enable their horizontal and vertical positions to be measured relative to terminal plates at the end of each gauge in small instrument houses on the

FIG. 8.   Aerial view of Scammonden Dam.

downstream slope of the dam (these are visible in Fig. 8). The displacements of each gauge house are measured relative to a reference pillar founded on hard rock well downstream of the dam.

The finite element idealisation for Scammonden Dam is shown in Fig. 10. Initially the stiffness of every element was set equal to zero. The analysis involved evaluating the increments in displacement (and

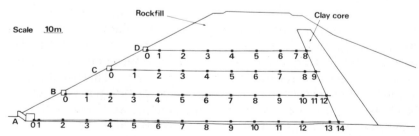

FIG. 9.   Simplified cross-section of Scammonden Dam showing positions where displacements were measured.

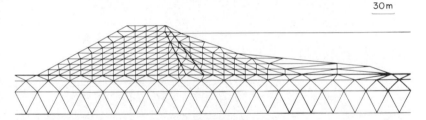

30m

FIG. 10.   Finite element idealisation of Scammonden Dam.

stress) as each successive layer was given the appropriate stiffness and its dead weight was applied. The total displacements and stresses at any given stage of construction were obtained by summing the respective increments. During the construction of the dam the fill was not always placed in horizontal layers and the actual construction sequence was modelled as far as possible in the finite element analysis.

Values of Young's modulus and Poisson's ratio had to be selected for the rockfill, the clay core and the shale on which the dam rests. Although these materials seldom approach the theoretical requirements of linearity and reversibility, experience has shown that, provided the elastic parameters are chosen as representative for the stresses encountered, satisfactory predictions can often be made by the elastic theory. Penman *et al.*[6] have outlined a simple method of selecting the appropriate elastic parameters from confined compression tests on samples of the rockfill compacted to the appropriate *in situ* density. They emphasised that the method, termed the 'equivalent compressibility' method, applies only to the analysis of a dam during construction.

A selection of the predicted and observed displacements are plotted as vectors in Fig. 11, each plotted point corresponding to the movement associated with the placement of a higher layer, except at the base of the dam where only the net final displacement is given.

Considering the simplifications and assumptions made in the analysis it can be seen that the overall agreement is very satisfactory. In the centre of the dam the agreement is excellent. Very near the downstream slope the agreement is less good. This is hardly surprising since near the downstream slope the material is subjected to the greatest changes in principal stress ratio and principal stress direction and is evidently far removed from the condition of confined

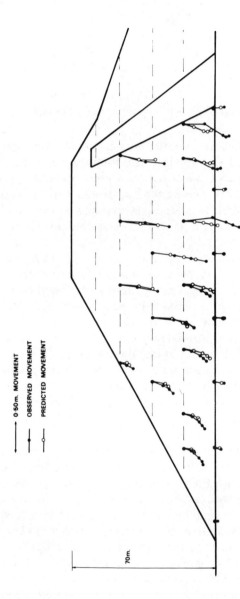

FIG. 11. Comparison of observed and predicted displacements of Scammonden Dam during construction.

compression assumed in deriving the material properties. The fact that the deformations are not symmetrical about the centre-line of the dam is due to the presence of the clay core of weaker material.

More recently Penman and Charles[8] used the 'equivalent compressibility' method to analyse the constructional deformations of the 90-m high Llyn Brianne Dam which is constructed of rockfill with a central clay core. Once again the agreement between the predicted and observed displacements was remarkably good.

The elastic constants used in the 'equivalent compressibility' method are derived from simple laboratory oedometer tests together with simplifying assumptions about the stress conditions in the field. Recently Charles[9] investigated the reasons for the apparent success of this very simple approach. A more sophisticated analysis would require stiffness properties based on laboratory tests which accurately follow the stress paths imposed on the embankment material *in situ*. Moreover, such a method would involve successive alterations to the stiffness of every element after each construction increment. Such testing and analysis is extremely expensive and time consuming. There is no evidence as yet that the accuracy of the simple method just outlined can be significantly improved upon by increased sophistication.

3.4 DEFORMATIONS AROUND EXCAVATIONS

In this section we will consider some problems in which the finite element method has been used to investigate ground movements around excavations. Once again the ground is assumed to behave elastically.

As for the embankment problem it is the self-weight of the material which induces the movements. In this case, however, it is the *removal* of material rather than the *addition* of it which induces the movement. In order to carry out an analysis it is necessary to know the stresses in the ground which existed around the boundaries of the excavation prior to removal of the material.

Figure 12(a) shows the distribution of vertical and horizontal stress around the boundaries of a proposed excavation. After excavation these stresses will be zero. Hence the displacements can be calculated by applying stresses of the same magnitude, but of opposite sign, to the boundaries of the final excavation as shown in Fig. 12(b). Unlike

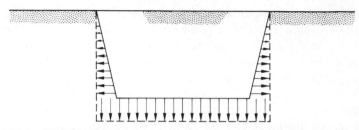

FIG. 12(a).   Distribution of initial *in situ* stress around the boundaries of proposed excavation.

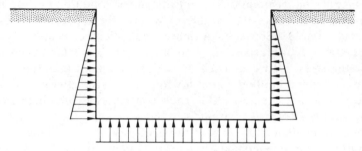

FIG. 12(b).   Stresses of opposite sign applied to the boundaries of completed excavation.

the embankment construction described in Section 3.3 the deformations resulting from the removal of material are *not* dependent on the excavation history. This is because the material outside the excavation boundary only responds to stress changes applied across that boundary. Since an elastic material is stress path-independent the precise manner in which the boundary stresses change from the initial conditions to zero has no influence on the result and the analysis can be carried out in a single step. This result does not hold for plastic materials. Even for elastic materials the excavation history becomes significant when supports are installed during excavation since the stage at which the supports are added determines how much load they carry and this determines the magnitude of the displacements. An example of a propped excavation will be described later in this section.

Cole and Burland[10] describe the use of the finite element method to back analyse the observed retaining wall movements associated with a deep excavation in London clay. The excavation, in 1963, was for

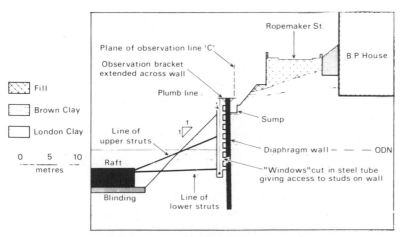

FIG. 13. Section through north wall of the basement excavation at Britannic House.

FIG. 14. Photograph of the north end of the excavation for the basement of Britannic House taken on 21 June 1963.

the basement of Britannic House in the City of London and a site monitoring system was carried out to ensure that the deformations of the diaphragm walls remained within tolerable limits during excavation. In particular the horizontal and vertical movements of the top of the north wall were measured by means of theodolite alignment and the inward tilt of the wall was measured from a man-sized plumbing tube just in front of the wall. Figure 13 is a section through the north wall at a stage of excavation when it was temporarily supported by struts. Figure 14 is a photograph of the north wall prior to strutting and the top of the plumbing tube is visible on the left-hand side.

The inward movement of the wall and the settlement of the neighbouring street responded slowly to early stages of excavation, accelerated rapidly as excavation approached full depth, and slowed down before the line of upper struts was placed. The wall displacements when excavation first reached full depth on 28 June, and just before the struts were placed on 21 July, are given on the left-hand side of Fig. 15.

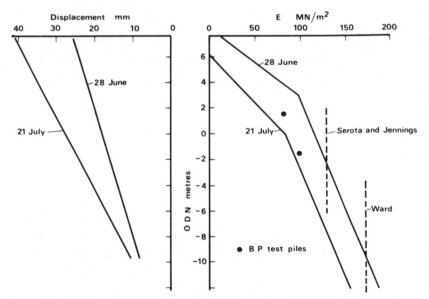

FIG. 15. Horizontal displacements of the north wall of the basement excavation at Britannic House and the corresponding distribution of Young's modulus with depth. Other field values of modulus are also shown.

A finite element analysis was carried out assuming the ground to behave as a linear elastic material having an undrained Young's modulus E_u varying with depth, account being taken of the stiffer Woolwich and Reading beds underlying the London clay. The initial vertical and horizontal total stresses in the ground were obtained from the following expressions:

$$\sigma_v = \gamma \cdot z$$

and

$$\sigma_h = K_0(\sigma_v - u) + u$$

where γ is the unit weight, u is the pore water pressure, K_0 is the coefficient of earth pressure at rest, and z is the depth.

The variations of u with depth were measured on the site. No direct measurements of K_0 are available, but values were deduced from laboratory tests on London clay by Skempton[11] and Bishop et al.[12] as shown in Fig. 16. For the analysis a distribution approximately mid-way between the two results was adopted.

The analysis involved a trial and error procedure, similar to that outlined in Section 2.2, in which the values of E_u were adjusted at each step so as to successively improve the fit between the observed and predicted wall displacements. The variations of E_u with depth corresponding to the wall movements immediately on completion of excavation (28 June) and a month later are shown on the right-hand side of Fig. 15. Once again the strong increase in stiffness with depth will be noted. The continued inward movement of the wall with time is associated with a general decrease in stiffness of the ground. Figure 15 also shows other values of stiffness of the London clay derived from field measurements. These include values derived by Ward[13] and Serota and Jennings[14] and some values deduced from pile loading tests carried out on the site.

Figure 17 shows the predicted displacement vectors outside and beneath the north wall of Britannic House. It can be seen that the horizontal displacements of the ground surface outside the excavation are generally two to three times as large as the vertical displacements—a feature which could be very significant in determining the performance of structures adjacent to open excavations in stiff, overconsolidated clays. The extent of the movement away from the excavation is also of significance and raises the question of how far ground anchors have to be inserted if they are to be effective in reducing movements. Recent observations made by the Building

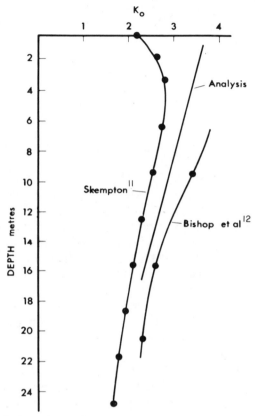

FIG. 16.   Variation of K_0 with depth below the surface of London clay.

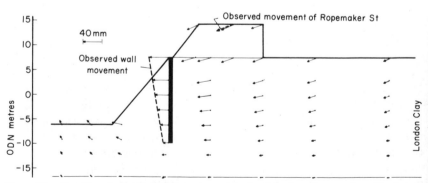

FIG. 17.   Predicted displacements associated with the observed wall movements on 21 July 1963. Note comparison between observed and predicted
movement of Ropemaker Street.

Research Station on a tied-back retaining wall in London clay show substantial horizontal movements extending back at least three times the depth of excavation.[15]

The results of the analysis of the Britannic House observations have been used in the design and analysis of other major excavations in the London area. These projects have provided opportunities for testing the accuracy of predictions and for further refining knowledge of the behaviour of the ground around excavations. The underground car park for Members of Parliament at the Palace of Westminster is a notable example of one of the above projects for which the finite element method of analysis has proved of considerable value in design. Burland and Hancock[16] have described the geotechnical aspects of this project in some detail.

The car park is 18 m deep and the diaphragm retaining walls come within a few metres of the foundations of both the Clock Tower ('Big Ben') and Westminster Hall. The stability of these important and historic structures was clearly a matter of prime importance in planning the construction.

A model depicting the construction of the car park is shown in Fig. 18. The diaphragm walls and piles were installed first. The steel columns supporting the floors were lowered into lined boreholes above each pile and grouted into position. The ground floor slab was then cast on the ground surface. Excavation took place beneath the ground floor slab and the next floor was then cast. The process was repeated until the lowest slab had been cast.

A complete stage-by-stage plane strain finite element analysis of the whole excavation and construction sequence was undertaken during the design stage. The variation of Young's modulus with depth used in the analysis is given by the solid line in Fig. 19. The stiffness of the overlying sand and gravel was set at an arbitrary low value of $10 \, MN/m^2$ since the material has a low stiffness in extension. The distribution of E_u with depth in the clay was based on the values derived from the Britannic House measurements. An upper limit of $E_u = 200 \, MN/m^2$ was adopted for depths in excess of 30 m as a conservative assumption. Much higher values had been deduced for the Woolwich and Reading beds elsewhere in London. Even so it is worth noting that the values of E_u used for the analysis were three to five times larger than the values derived from laboratory oedometer tests.

The initial horizontal stresses adopted for the analysis were based on the values used in the Britannic House analysis. However, for the

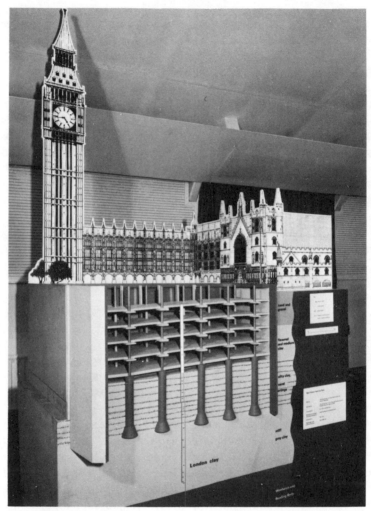

FIG. 18. Model showing the construction of the House of Commons car park.

car park it was necessary to take account of the effects of reloading resulting from the deposition of the sand and gravel over the surface of the clay. It was assumed that during reloading the clay had a Poisson's ratio of 0·15 so that the increase of horizontal effective stress was 0·18 times the increase of vertical effective stress induced by the deposition of the sand and gravel. The distributions of vertical and horizontal stresses used in the analysis are shown in Fig. 20.

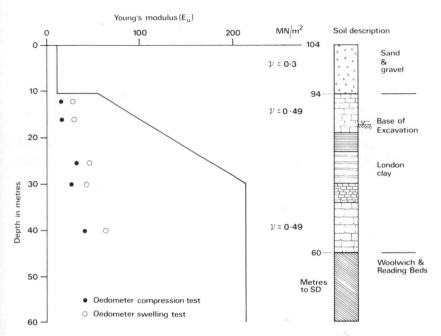

FIG. 19. Elastic parameters used in the finite element analysis.

The analysis of the ground movements was carried out as a step-by-step process and is best described by referring to the insert diagram in Fig. 21. For the analysis of the first stage, only prop number 1 (represented as a spring) was installed and the material to be removed during Stage 1 was given zero stiffness. Stresses were then applied at the base of the current excavation and on the retaining wall which were equal but of opposite sign to the initial stresses. The ground displacements were calculated and also the new stresses around *all* the subsequent excavation boundaries.

For the next step, the material to be removed during Stage 2 was given zero stiffness and prop number 2 was inserted. The current stresses around the Stage 2 excavation boundary were then released. The resulting ground displacements were added to the first set and the new stresses were again evaluated for the remaining excavation boundaries. The procedure was repeated for each excavation stage accumulating the displacements and stresses at each step.

Figure 21 shows the predicted wall displacements as the excavation reached various levels. It is important to note the deep seated inward movements which result more from the release of vertical stress within the excavation than from the release of horizontal stress on the inner face of the wall. Thus, even though stiff props are inserted at each stage of excavation some lateral yield of diaphragm retaining walls must still be anticipated. The predicted short-term horizontal and vertical displacements of the ground surface outside the excavation are given by the broken lines in Fig. 22. It should be emphasised that the predictions given in Figs. 21 and 22 were published by Ward and Burland[17] prior to the construction of the car park and they therefore constitute a Class A1 prediction in the terminology proposed by Lambe.[18]

A comprehensive programme of monitoring was undertaken by the Building Research Station during excavation and construction of the car park. This included inclinometers in the retaining walls and a large number of precision surveying stations around the Palace of Westminster and Westminster Hall. Figure 23(a) shows the observed displacements of a typical wall panel at each stage of excavation. The agreement between the predicted and observed behaviour is very

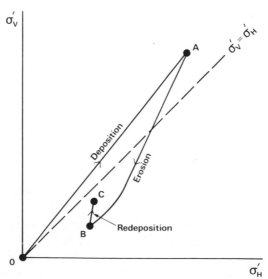

FIG. 20.   Effect of reloading on the London clay: effective stress history of the London clay.

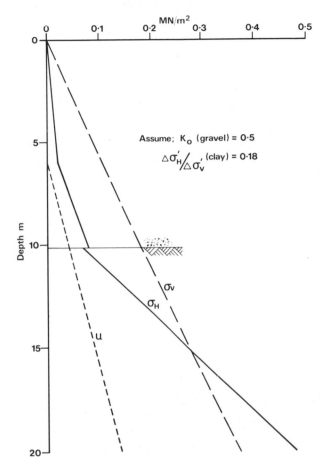

FIG. 20—*cont.* Initial *in situ* stress distribution used in the finite element
analysis.

satisfactory. It is evident that the movements beneath the final
excavation level were overestimated as a direct result of the deli-
berate choice of low values of E_u at depth. A wide range in the
deflected shapes of the various wall panels was observed (see Fig.
23(b)). Hence, even under relatively uniform ground conditions,
widely differing deflected shapes must be anticipated and account
should be taken of this when designing the reinforcing steel for such
walls.

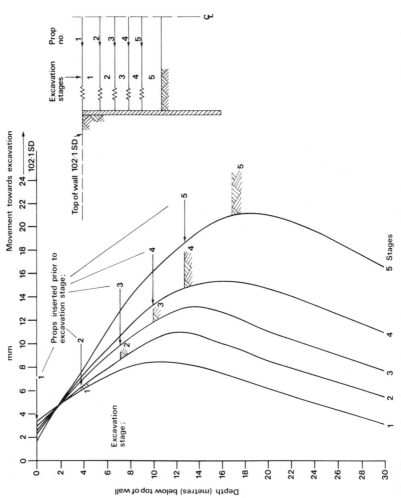

FIG. 21. Prediction of horizontal wall movements during excavation of the House of Commons car park.

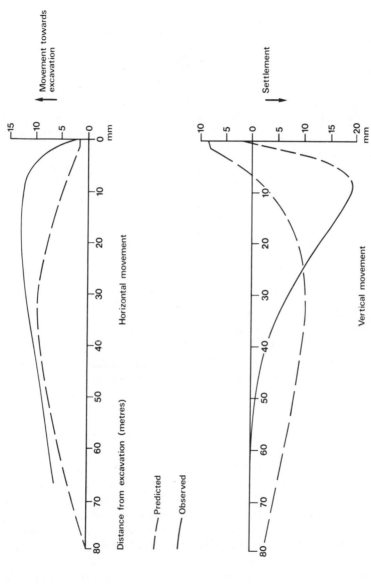

Fig. 22.   Comparison of observed and predicted ground surface movements around the House of Commons car park.

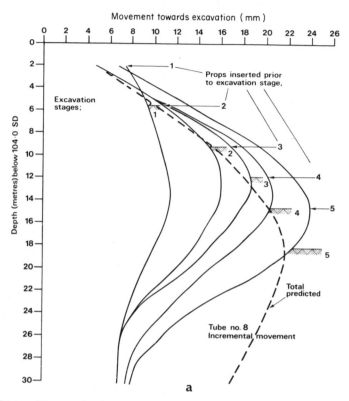

FIG. 23(a).   Observed movements of the diaphragm walls: inclinometer tube 8
at various stages of excavation.

Agreement between the predicted and observed ground surface
movements is less good, as can be seen from Fig. 22. The predicted
settlements are approximately of the correct magnitude, but the
observed settlement profile differs markedly from the predicted one in
that the settlement is concentrated close to the edge of the excavation. The difference between the predicted and observed vertical
movements can be attributed in part to the assumption of plane strain
in the analysis. However, the difficulty of correctly modelling the top
10 m of loose granular materials is thought to be the main reason for
the discrepancy. The predicted horizontal movements are in reasonable agreement with the observations.

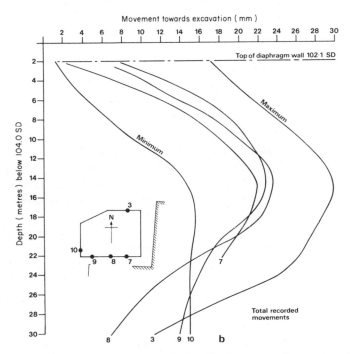

FIG. 23(b). Observed movements of the diaphragm walls: range of total
movements.

3.5 CONCLUDING REMARKS

In the introduction to this chapter, considerable emphasis was placed
on the inherent assumptions that the geotechnical engineer has to
make, irrespective of the degree of sophistication of the analytical
techniques he proposes to use. This was done deliberately in an effort
to maintain the balance between theory and practice at a time when
analytical techniques are developing rapidly and becoming widely
available to practising engineers.

It is tempting in a discussion on the application of finite element
methods to illustrate a large number of problems which have been
solved analytically. Such an approach does little to help the engineer
to tackle his own problems, as the extent of the underlying assumptions is seldom discussed and the real behaviour is not known. In this
chapter the author has confined himself to a discussion of a few

practical cases for which field measurements of movements have been made. In all cases the analysis has been kept simple yet the accuracy and insight it has provided are very apparent.

The examples have emphasised the importance of taking account of such factors as non-homogeneity, initial *in situ* stresses, and construction history. In many problems these factors may well prove to be much more sensitive than the detailed choice of stress–strain law. For example, there is little point in carrying out a highly sophisticated non-linear deformation analysis if the initial *in situ* stress conditions have not been accurately determined or if the variation of the initial tangent modulus with depth is neglected. Moreover, the incorporation of non-linearity or other complex features does not necessarily lead to improved predictions. Whereas the use of a certain non-linear relationship may appear reasonable when applied to the simple stress paths available in the laboratory, the same relationship may be quite inappropriate under the complex stress changes that occur in the field.

In analysing any problem, the complexity of the geometry and the material properties should be kept to a minimum. The influence of each variable should be tested separately in an effort to assess its importance. For many deformation problems the use of linear elasticity with appropriate elastic constants may be adequate. Where a non-linear analysis is felt to be more appropriate it is very useful to obtain a first solution using linear elastic properties as this can serve as a frame of reference against which to test and evaluate later solutions. Used in this way, with care and discrimination, the finite element method is a valuable tool for the engineer. However, when used blindly it can actually inhibit an engineer's 'feel' for and physical appreciation of a problem.

REFERENCES

1. WARD, W. H., BURLAND, J. B. and GALLOIS, R. E. (1968). Geotechnical assessment of a site at Mundford, Norfolk, for a large proton accelerator, *Géotechnique*, **18**, 399.
2. BURLAND, J. B., SILLS, G. C. and GIBSON, R. E. (1973). A field and theoretical study of the influence of non-homogeneity on settlement, *Proc. 8th Int. Conf. SMFE*, Moscow, 1973, **1.3**, 39.
3. BURLAND, J. B. and LORD, J. A. (1969). The load-deformation behaviour of Middle Chalk at Mundford, Norfolk. A comparison between full-scale performance and *in situ* and laboratory measurements. *Proc. Conf. on In*

Situ *Investigations in Soils and Rock*, Institution of Civil Engineers, London, 1969, 3.

4. BURLAND, J. B. (1969). Reply to discussion. *Proc. Conf. on In Situ Investigations in Soils and Rock*, Institution of Civil Engineers, London, 1969, 62.

5. FOX, E. N. (1948). The mean elastic settlement of uniformly loaded area at a depth below ground surface, *Proc. 2nd Int. Conf. SMFE*, Rotterdam, 1948, **1**, 129.

6. PENMAN, A. D. M., BURLAND, J. B. and CHARLES, J. A. (1971). Observed and predicted deformations in a large embankment dam during construction, *Proc. Inst. Civ. Engrs.*, **49**, 1–21.

7. PENMAN, A. D. M. and MITCHELL, P. B. (1970). Initial behaviour of Scammonden Dam, *Proc. 10th Congress on Large Dams*, Montreal, 1970, **1**, 723–747.

8. PENMAN, A. D. M. and CHARLES, J. A. (1973). Constructional deformations in a rockfill dam, *Proc. ASCE*, **99**(SM2), 139–163.

9. CHARLES, J. A. (1976). The use of one-dimensional compression tests and elastic theory in predicting deformations of rockfill embankments, *Canad. Geotech. J.*, **13**, 189–200.

10. COLE, K. W. and BURLAND, J. B. (1972). Observation of retaining wall movements associated with a large excavation, *Proc. 5th Eur. Conf. SMFE*, Madrid, 1972, **1**, 445–453.

11. SKEMPTON, A. W. (1961). Horizontal stresses in an overconsolidated Eocene clay, *Proc. 5th Int. Conf. SMFE*, Paris, 1961, **1**, 351–387.

12. BISHOP, A. W., WEBB, D. L. and LEWIN, P. I. (1965). Undisturbed samples of London clay from the Ashford Common shaft: strength–effective stress relationships, *Géotechnique*, **15**, 1–31.

13. WARD, W. H. (1961). Displacements and strain in tunnels beneath a large excavation in London, *Proc. 5th Int. Conf. SMFE*, Paris, 1961, **2**, 749–753.

14. SEROTA, S. and JENNINGS, R. A. J. (1959). The elastic heave of the bottom of excavations, *Géotechnique*, **9**, 62–70.

15. SILLS, G. C., BURLAND, J. B. and CZECHOWSKI, M. K. (1977). Behaviour of an anchored diaphragm wall in stiff clay, *Proc. 9th Int. Conf. SMFE*, Tokyo, 1977, **2**, 147–154.

16. BURLAND, J. B. and HANCOCK, R. J. R. (1977). Underground car park at the House of Commons, London: geotechnical aspects, *The Structural Engineer*, **55**(2), 87–100.

17. WARD, W. H. and BURLAND, J. B. (1973). The use of ground strain measurements in civil engineering, *Phil. Trans. R. Soc.*, London, **A274**, 421–428.

18. LAMBE, T. W. (1973). Prediction in soil engineering, *Géotechnique*, **23**, 151–202.

Chapter 4

DEVELOPMENTS IN TWO- AND THREE-DIMENSIONAL CONSOLIDATION THEORY

R. T. Murray

Transport and Road Research Laboratory, Crowthorne, Berkshire, UK

SUMMARY

This chapter reviews recent developments in two- and three-dimensional theories of consolidation, and discusses the relevance of such developments to engineering design.

Most of the consolidation problems found in practice can now be solved, although all but the simplest require techniques of numerical analysis. The finite difference methods which have been most widely used for this purpose are described.

Much recent research has been directed towards assessing the importance of the various factors involved in the analysis. Several such research papers are reviewed in this chapter, and the authors' results are presented—often in the form of design charts—to allow the influence of the various parameters to be studied.

In addition, comparisons are presented of the behaviour predicted both by the Terzaghi–Rendulic theory and by Biot's more rigorous analysis. These should be of value in assessing when the simpler Terzaghi–Rendulic theory is adequate.

4.1 INTRODUCTION

Consolidation theory has reached an advanced stage of development, and solutions are now available for most practical problems, either in closed analytical form or by numerical techniques. However, the

103

predicted consolidation behaviour may be fundamentally affected by the accuracy of the presumed soil profile, and by the reliability of the sampling and testing procedures used to obtain the soil data. Thus, a simple analytical procedure, using reliable soil parameters, will often provide the designer with an adequate prediction of consolidation behaviour. There is often a feeling that complex theory will somehow compensate for inadequate data. In fact, the reverse is true, and the use of sophisticated theory can only be justified by data of high quality.

Figure 1 demonstrates the very significant effect that incorrect testing procedures may have on the predicted behaviour. The figure compares the predicted and observed relations between settlement and time for an embankment on the A40 trunk road near Gloucester. The calculations were based on two-dimensional consolidation theory, taking account of the multi-layer soil profile and of the non-linear soil behaviour. For curve 2, the soil parameters were obtained from small scale consolidation tests. This prediction seriously underestimated the rate of settlement, and might have led the design engineer to adopt a more expensive form of construction. The figure also shows a second prediction (curve 3). This was obtained using the same analytical procedures, but the coefficients of consolidation were based on the results of *in situ* permeability tests in each of the main subsoil strata. This second prediction agrees well with the observed behaviour.[1]

The variations which result from the use of different consolidation theories will not in general be as large as those discussed previously. None the less, the differences may be important, and the use of an insufficiently accurate procedure could significantly affect the cost of a structure. An example of this is the common use of one-dimensional theory in cases where the geometry clearly shows that additional drainage boundaries are present, and that a multi-dimensional theory is more appropriate.

Thus the engineer needs to know when a simple theory will be adequate, and when he must use a more rigorous analysis. Much recent work on consolidation theory has aimed to determine this, and some of the findings are outlined in this chapter. Moreover, parametric studies of the more complex theories have allowed some measure of simplification, by isolating those parameters to which the results are least sensitive.

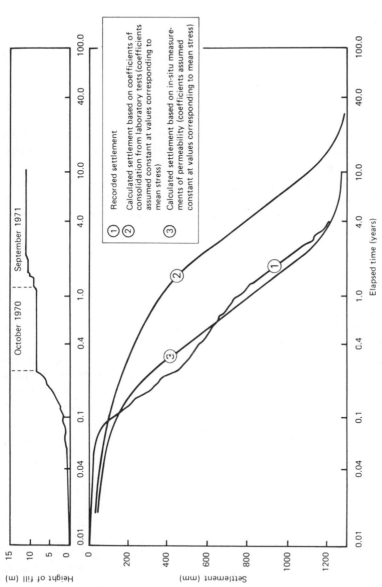

FIG. 1.   Recorded settlement compared with the calculated relations between settlement and time, using coefficients of consolidation from both laboratory and *in situ* tests.

4.2 CONSOLIDATION THEORY

Two basic approaches are commonly used for analysing consolidation problems in two and three dimensions. The first was developed from diffusion theory by Terzaghi[2] and Rendulic.[3] The second was developed directly from elastic theory by Biot.[4]

The diffusion theory has been widely used, largely because it is mathematically much simpler to apply, but also because a considerable range of solutions has been obtained in the field of heat conduction. It is, however, less rigorous than Biot's theory and is generally referred to as pseudo two- and three-dimensional consolidation theory.

Although published in 1941, Biot's method has been less frequently used, mainly because of its mathematical complexity. However, with the increasing availability of computers, solutions to previously intractable problems can be obtained by techniques of numerical analysis. The equations for both theories were originally derived for a homogeneous isotropic and fully saturated soil medium. Many recent developments have extended the theories well beyond the original concepts, particularly as regards restrictions to homogeneity and isotropy. The basic equations for the two methods for both two and three space dimensions in Cartesian co-ordinates are given in Table 1. The one-dimensional equation is also included for comparison.

In employing Rendulic's equations it is common to use the appropriate one-dimensional coefficient of consolidation for each space direction or alternatively to assume that a single value of the coefficient applies for one-, two-, or three-dimensional conditions, as shown in Table 1. Elastic theory shows, however, that the coefficient of consolidation depends on the conditions of strain, and thus the values generally differ in one-, two-, and three-dimensional consolidation.[5]

A comparison of the three coefficients given in Table 1 shows that:

$$c_1 = 2(1 - \nu)c_2 = 3\left(\frac{1 - \nu}{1 + \nu}\right)c_3$$

Only for undrained conditions, when ν is 0·5, are these three coefficients equal. In this case, Biot's and Rendulic's equations are identical, as the terms

$$\frac{1}{2}\frac{\partial}{\partial t}(\sigma_x + \sigma_y)$$

TABLE 1

EQUATIONS OF CONSOLIDATION IN CARTESIAN CO-ORDINATES

Number of space dimensions	Rendulic system of equations (pseudo-consolidation)	Biot system of equations	Coefficient of consolidation
1	$\dfrac{\partial u}{\partial t} = c_1 \dfrac{\partial^2 u}{\partial z^2}$	$\dfrac{\partial u}{\partial t} = c_1 \dfrac{\partial^2 u}{\partial z^2}$	$c_1 = \dfrac{kE(1-\nu)}{\gamma_w(1+\nu)(1-2\nu)}$
2	$\dfrac{\partial u}{\partial t} = c_1\left(\dfrac{\partial^2 u}{\partial x^2} + \dfrac{\partial^2 u}{\partial z^2}\right)$ or: $\dfrac{\partial u}{\partial t} = c_x \dfrac{\partial^2 u}{\partial x^2} + c_z \dfrac{\partial^2 u}{\partial z^2}$	$\dfrac{\partial u}{\partial t} = c_2\left(\dfrac{\partial^2 u}{\partial z^2} + \dfrac{\partial^2 u}{\partial x^2}\right)$ $+ \dfrac{1}{2}\dfrac{\partial(\sigma_x + \sigma_z)}{\partial t}$	$c_2 = \dfrac{kE}{2\gamma_w(1-2\nu)(1+\nu)}$
3	$\dfrac{\partial u}{\partial t} = c_2\left(\dfrac{\partial^2 u}{\partial x^2} + \dfrac{\partial^2 u}{\partial y^2} + \dfrac{\partial^2 u}{\partial z^2}\right)$ or: $\dfrac{\partial u}{\partial t} = c_x \dfrac{\partial^2 u}{\partial x^2} + c_y \dfrac{\partial^2 u}{\partial y^2} + c_z \dfrac{\partial^2 u}{\partial z^2}$	$\dfrac{\partial u}{\partial t} = c_3\left(\dfrac{\partial^2 u}{\partial x^2} + \dfrac{\partial^2 u}{\partial y^2} + \dfrac{\partial^2 u}{\partial z^2}\right)$ $+ \dfrac{1}{3}\partial\left(\dfrac{\sigma_x + \sigma_y + \sigma_z}{\partial t}\right)$	$c_3 = \dfrac{kE}{3\gamma_w(1-2\nu)}$

u = excess pore-water pressure
k = coefficient of permeability
ν = Poisson's ratio of soil skeleton
E = elastic modulus of soil skeleton
t = elapsed time
γ_w = unit weight of water
$\sigma_x, \sigma_y, \sigma_z$ = total stress increment in x, y and z directions, respectively
c_x, c_y, c_z = one-dimensional coefficient of consolidation in x, y and z directions, respectively.

and

$$\frac{1}{3}\frac{\partial}{\partial t}(\sigma_x + \sigma_y + \sigma_z)$$

in Biot's equations—which represent the rates of change of mean total stress with time—are zero. These terms were excluded from Rendulic's equations because it was incorrectly assumed that the total stress remained constant during consolidation.

The bulk modulus of water is many times greater than that of the soil skeleton. An undrained saturated soil is therefore nearly incompressible, and the pore water initially supports any increment of the external load. However, as consolidation proceeds the load is gradually transferred to the soil skeleton. The equivalent elastic moduli, E_u and v_u, for the undrained state differ from the values, E' and v', for the fully consolidated, or drained state. The soil adjacent to a free draining boundary consolidates relatively quickly, while regions remote from the boundary remain nearly undrained, so that differential strains develop through the soil. These require changes in total stress to satisfy the stress–strain law. The effect of these changes in total stress on the excess pore pressure are provided for by the additional terms in Biot's equations.

An interesting feature of Biot's theory is the *Mandel–Cryer* effect.[6,7] Since the excess pore-water pressure is affected by changes in the mean total stress, it may—at some places within the soil mass—continue to increase for some time after the application of a load increment. It may finally attain values much larger than the applied pressure. This phenomenon has been confirmed experimentally on a laboratory scale but has not been reported from field observations. This is not altogether surprising, as very precise measurements of applied load and pore pressure are needed. In practice, the rate of construction in the field is not usually constant, while field piezometers do not always respond immediately to changes in pore pressure. It is therefore difficult to separate the effects of increases in applied pressure, the response time of the piezometers, and the dissipation of excess pore-water pressure due to consolidation.

Nevertheless, the Mandel–Cryer effect is now considered to be firmly established, and can have important consequences. In some cases of multi-dimensional consolidation, the extra time taken to build up the pore-water pressure can actually increase the time required for completion of a given degree of consolidation. Thus, a one-dimensional study may predict more rapid consolidation than is indicated by two- or three-dimensional analyses. This is contrary to former opinion, since it has generally been assumed that one-dimensional theory provides a lower bound to the rate of consolidation, and that the inclusion of additional drainage boundaries in a multi-dimensional analysis must necessarily increase the rate.

Although Rendulic's method does not account for the effects of changes in mean total stress, these effects are often small. When this is so, the simpler mathematical basis of the theory allows it to be used more conveniently, to study, for example, non-linear behaviour. However, with the development of such powerful numerical techniques as the finite element method, there is likely to be more general use of Biot's theory, especially as non-linear effects can be conveniently included in such numerical techniques.

4.3   RENDULIC EQUATIONS OF CONSOLIDATION (PSEUDO-CONSOLIDATION THEORY)

Solutions to most types of consolidation problem likely to arise in engineering practice in terms of small-strain theory have been obtained by this approach and are published in the literature on soil mechanics, heat conduction and diffusion theory.

A number of solutions have been obtained in closed, analytical form. These provide valuable insight into consolidation behaviour, and also present standards against which the solutions based on numerical analysis can be checked. Examples of such results, in addition to the rectilinear and axi-symmetric solutions for a homogeneous layer, are the sand drain problem solved by Barron,[8] the multi-layer solution by Horne,[9] and the extension of this theory by Rowe[10] to include the presence of a sand drain included in a multi-layer system.

4.3.1 Finite Difference Solutions

The vast majority of solutions to the more complex problems of pseudo-consolidation theory have been obtained by the method of finite differences. This method involves the replacement of the derivatives in the partial differential equations by finite difference approximations in terms of values at adjacent node or mesh points.

The region is first subdivided by a rectilinear mesh having sides (in the case of a two-dimensional problem) of dx and dy. Known initial values of the pore-water pressure are assigned to each node on the mesh, and the appropriate boundary conditions are inserted. From these initial values, the value of the pore-water pressure at each node point is determined after successive time intervals dt.

There are two basic approaches to determining the unknown pore-water pressure, described as *explicit* and *implicit* methods. In the former method, the unknown values of the pore-pressure u after a time interval dt are stated explicitly for each node in turn in terms of the known values at adjacent nodes at the beginning of the time interval. Implicit methods, on the other hand, require the solution of a set of simultaneous equations (one for each node) at each time step. Although the explicit method is simple to use, it is only stable if the time intervals are very small. (Otherwise errors of increasing magnitude and alternating sign appear at each time step). It is not therefore generally suitable for two- and three-dimensional problems, as the volume of computation is prohibitive.

4.3.2 Crank–Nicholson Implicit Methods

The implicit method most commonly used is that due to Crank and Nicholson,[11] in which the first order partial derivative $\partial u/\partial t$ is replaced by the mean values of the second order derivatives $\partial^2 u/\partial x^2$ and $\partial^2 u/\partial z^2$ over a time interval dt. Thus, for the solution of a two-dimensional problem in a rectangular region, the co-ordinates of the node points in the solution are defined as

$$x = i\,dx; \quad z = j\,dz; \quad t = n\,dt$$

and the value of excess pore-water pressure after a time t at the co-ordinate locations x, z is given in the following symbolic form:

$$u(x, z, t) = u(i\,dx, j\,dz, n\,dt) = u_{i,j,n}$$

The finite difference representation of the two-dimensional pseudo-consolidation equation

$$\frac{\partial u}{\partial t} = c_v \left(\frac{\partial^2 u}{\partial x^2} + \frac{\partial^2 u}{\partial z^2} \right)$$

in Crank–Nicholson form is as follows:

$$\frac{u_{i,j,n+1} - u_{i,j,n}}{dt} = c_v \left[\frac{1}{2} \left(\frac{u_{i+1,j,n+1} - 2u_{i,j,n+1} + u_{i-1,j,n+1}}{dx^2} \right) \right.$$

$$+ \frac{u_{i+1,j,n} - 2u_{i,j,n} + u_{i-1,j,n}}{dx^2} \right) + \frac{1}{2} \left(\frac{u_{i,j+1,n+1} - 2u_{i,j,n+1} + u_{i,j-1,n+1}}{dz^2} \right.$$

$$\left. + \frac{u_{i,j+1,n} - 2u_{i,j,n} + u_{i,j-1,n}}{dz^2} \right)$$

The solution of the above equation would then be achieved by equating (in matrix form) the unknown excess pore-water pressures containing the subscript term $(n + 1)$ to the known values with subscript n. If the overall size of the region has length L and depth Z then, with the known boundary values, this would involve the solution of $(P - 1) \times (Q - 1)$ simultaneous algebraic equations where $P = L/dx$ and $Q = Z/dz$. Thus even with problems of relatively small size, fairly large systems of equations may be involved. To improve the efficiency in solving two- and three-dimensional problems, additional methods have been developed—in particular, the *alternating direction implicit* (ADI) method[12] and the *locally one-dimensional* (LOD) method.[13] In both methods the Crank–Nicholson implicit finite difference equations are generally employed. The ADI method has been the most commonly used of the two. In principle, the method consists of replacing only one second-order derivative, say $\partial^2 u/\partial x^2$, by the implicit finite difference approximation at the $(n\text{th} + 1)$ time step. The remaining second-order derivative, $\partial^2 u/\partial z^2$, is replaced by an explicit form of finite difference approximation.

Thus, each unknown row of excess pore-water pressure values at the $(n\text{th} + 1)$ time level is solved independently and this procedure is carried out for the $(Q - 1)$ sets of equations. The solution for the next time step $(t = (n + 2) dt)$ is achieved by reversing the order of the finite difference approximation to solve $(P - 1)$ sets of equations relating to the excess pore-water pressures arranged in columns. The method is stable provided the alternate solution of rows and columns is maintained.

The implicit representation of $\partial^2 u/\partial x^2$ to be employed at time $(n + 1) dt$ is

$$\frac{(u_{i,j,n+1} - u_{i,j,n})}{c_v dt} = \frac{(u_{i-1,j,n+1} - 2u_{i,j,n+1} + u_{i+1,j,n+1})}{dx^2}$$

$$+ \frac{(u_{i,j-1,n} - 2u_{i,j,n} + u_{i,j+1,n})}{dz^2}$$

The implicit representation of $\partial^2 u / \partial z^2$ to be employed at time $(n + 2)\,\mathrm{d}t$ is:

$$\frac{(u_{i,j,n+2} - u_{i,j,n+1})}{c_v\,\mathrm{d}t} = \frac{(u_{i-1,j,n+1} - 2u_{i,j,n+1} + u_{i+1,j,n+1})}{\mathrm{d}x^2}$$

$$+ \frac{(u_{i,j-1,n+2} - 2u_{i,j,n+2} + u_{i,j+1,n+2})}{\mathrm{d}z^2}$$

Putting

$$r_x = \frac{c_v\,\mathrm{d}t}{\mathrm{d}x^2} \qquad r_z = \frac{c_v\,\mathrm{d}t}{\mathrm{d}z^2}$$

and re-arranging the above equations in a form suitable for matrix operation, we obtain

$$-r_x u_{i-1,j,n+1} + (1 + 2r_x)u_{i,j,n+1} - r_x u_{i-1,j,n+1}$$
$$= r_z u_{i,j-1,n} + (1 - 2r_z)u_{i,j,n} + r_z u_{i,j-1,n}$$

and

$$-r_z u_{i,j-1,n+2} + (1 + 2r_z)u_{i,j,n+2} - r_z u_{i,j+1,n+2}$$
$$= r_x u_{i-1,j,n+1} + (1 - 2r_x)u_{i,j,n+1} + r_x u_{i+1,j,n+1}$$

The finite difference formulation given above relates to the condition of a homogeneous, isotropic layer. In practice, there are often several layers present in the subsoil with different properties, while the vertical and horizontal coefficients of consolidation may differ. It may also be necessary to take account of an impermeable boundary at the base of a compressible layer. In addition, considerable economy in the volume of computation can be achieved by taking into account the symmetry of a problem. All of the above conditions require modifications to the operational forms of the expressions given for a homogeneous system. These will now be provided.

4.3.3 Axes of Symmetry

On such a boundary, the symmetry of the problem is recognised by taking account of the condition that the values of u at points on either side of the axis of symmetry, and adjacent to it, are equal. Thus, if we define the ith grid point as occurring on the axis of symmetry, the value of u at points on either side of this axis could be equated

$$u_{i-1,j,n+1} = u_{i+1,j,n+1}$$

Now, as the $(i\text{th} - 1)$ point occurs outside the region of integration, the operational expressions are formed by replacing the value of the pore-water pressure subscripted $(i + 1)$ with the value subscripted

$(i - 1)$, to produce the following forms

$$(1 + 2r_x)u_{i,j,n+1} - 2r_x u_{i+1,j,n+1} = r_z u_{i,j-1,n} + (1 - 2r_z)u_{i,j,n} + r_z u_{i,j+1,n}$$

and

$$- r_z u_{i,j-1,n+2} + (1 + 2r_z)u_{i,j,n+2} - r_z u_{i,j+1,n+2} = (1 - 2r_x)u_{i,j,n+1} + 2r_x u_{i+1,j,n+1}$$

Precisely the same forms of equation are used for impermeable boundaries.

4.3.4 Internal Boundaries between Layers with Different Properties

Consider two soil strata with an interface parallel to the x axis, their properties being distinguished by subscripts r and s respectively. For flow along the interface (in the x direction) the equivalent coefficient of consolidation is the mean of the coefficients in the two strata. For flow across the interface there must be continuity of flow. These conditions lead to the following operational expressions for pore pressure at nodes adjacent to the interface:[14]

$$- \bar{r}_{rs}u_{i-1,j,n+1} + (1 + 2\bar{r}_{rs})u_{i,j,n+1} - \bar{r}_{rs}u_{i+1,j,n+1}$$
$$= r'_r u_{i,j-1,n} + (1 - r'_r - r'_s)u_{i,j,n} + r'_s u_{i,j+1,n}$$

$$- r'_r u_{i,j-1,n+2} + (1 + r'_r + r'_s)u_{i,j,n+2} - r'_s u_{i,j+1,n+2}$$
$$= \bar{r}_{rs}u_{i-1,j,n+1} + (1 - 2\bar{r}_{rs})u_{i,j,n+1} + \bar{r}_{rs}u_{i+1,j,n+1}$$

where

$$\bar{r}_{rs} = \frac{(c_r + c_s)\,dt}{2\,dx^2}$$

$$r'_r = \frac{2c_r}{dz_r^2} \cdot \frac{dz_r m_r\,dt}{(m_r\,dz_r + m_s\,dz_s)}$$

$$r'_s = \frac{2c_s}{dz_s^2} \cdot \frac{dz_s m_s\,dt}{(m_s\,dz_s + m_r\,dz_r)}$$

where c_r = coefficient of consolidation of the r layer
c_s = coefficient of consolidation of the s layer
m_r = coefficient of volume compressibility of the r layer
m_s = coefficient of volume compressibility of the s layer
dz_r = vertical distance between node points in the r layer
dz_s = vertical distance between node points in the s layer

It is worth noting that in a situation where a reduction in vertical stress occurs, the coefficients referred to above would be obtained from consolidation-unloading tests.

Similar expressions may be derived for boundaries parallel to the z axis.

4.3.5 Variable Soil Properties and Rates of Loading

Where the coefficient of consolidation varies with the effective stress, a simple procedure is to approximate the relations between the soil parameters and effective stress by step functions. Thus, it is assumed that the particular parameter remains constant during a finite range of effective stress. Following completion of this stage, the next value is selected, the operational matrices are reconstructed, and the solution is continued.[14]

A similar step function approach can be employed for analysing problems influenced by variable loading rates. The actual loading rate can be approximated by a series of incremental loads applied to the node points at the commencement of each time step. Then,

$$u'_{i,j,n} = u_{i,j,n} + \Delta u_{i,j}$$

and

$$\Delta u_{i,j} = \Delta\sigma_{3,i,j} + A'(\Delta\sigma_{1,i,j} - \Delta\sigma_{3,i,j})$$

where $u'_{i,j,n}$ = the known value of excess pore-water pressure at node point (i, j) just prior to evaluating the unknown pore-water pressures at the $(n + 1)$ time level

$\Delta u_{i,j}$ = the increment of excess pore-water pressure at the node point (i, j) induced by the loading over the time interval dt (at free-draining boundaries this value will be zero)

$\Delta\sigma_{3,i,j}$ = increment of the minor principal stress at node (i, j) induced by the increment of applied load

$\Delta\sigma_{1,i,j}$ = increment of the major principal stress at node (i, j) induced by the increment of applied load

$A' = 0\cdot866A + 0\cdot211$

A = the pore-water pressure parameter obtained from tri-axial tests; A' is the equivalent pore-pressure parameter for plane–strain conditions.

4.3.6 Applications to Three-dimensional Problems

The application of finite difference methods to pseudo three-dimensional consolidation problems in Cartesian co-ordinates could be achieved by extending the ADI procedure outlined above by using two intermediate explicit expressions in conjunction with the implicit form applied to each co-ordinate direction at every third time step. Unfortunately the procedure is no longer unconditionally stable, and variations on the method have been put forward by Douglas[15] and also by Brian.[16] According to the scheme proposed by Brian for an anisotropic soil, the finite difference expressions to be employed in sequence are as follows:

$$\frac{r_x}{2}\{-u_{i+1} + 2u_i(1+1/r_x) - u_{i-1}\}_{j,k,n+1} = \frac{r_y}{2}\{u_{j+1} - 2u_j(1-1/r_y) + u_{j-1}\}_{i,k,n}$$

$$+ \frac{r_z}{2}\{u_{k+1} - 2u_k + u_{k+1}\}_{i,j,n}$$

$$\frac{r_y}{2}\{-u_{j+1} + 2u_j(1+1/r_y) - u_{j-1}\}_{i,k,n+2} = \frac{r_x}{2}\{u_{i+1} - 2u_i + u_{i-1}\}_{j,k,n+1}$$

$$+ \frac{r_z}{2}\{u_{k+1} - 2u_k(1-1/r_z) + u_{k-1}\}_{j,k,n+1}$$

$$\frac{r_z}{2}\{-u_{k+1} + 2u_k(1+1/r_z) - u_{k-1}\}_{i,j,n+3} = \frac{r_x}{2}\{u_{i+1} - 2u_i + u_{i-1}\}_{j,k,n+2}$$

$$+ \frac{r_y}{2}\{u_{j+1} - 2u_j(1-1/r_y) + u_{j-1}\}_{i,k,n+2}$$

where

$$r_x = \frac{c_x\,dt}{dx^2} \quad r_y = \frac{c_y\,dt}{dy^2} \quad r_z = \frac{c_z\,dt}{dz^2}$$

and $i\,dx = x$ co-ordinate; $j\,dy = y$ co-ordinate; $k\,dz = z$ co-ordinate.

The procedure outlined for dealing with internal boundaries, etc. in two space dimensions can also be employed for three-dimensional consolidation.

4.3.7 Problems with Axial Symmetry

Many problems in three space dimensions are axi-symmetric. The diffusion equation in cylindrical co-ordinates for an anisotropic soil is then of the form:

$$\frac{\partial u}{\partial t} = c_R\left(\frac{\partial^2 u}{\partial R^2} + \frac{1}{R}\frac{\partial u}{\partial R}\right) + c_z\frac{\partial^2 u}{\partial z^2}$$

where c_R, c_z are coefficients of consolidation in the radial and axial directions respectively. The corresponding finite difference equations in ADI form are as follows:

$$\frac{r_R}{2}\{-u_{i+1}(1+1/2i)+2u_i(1+1/r_R)-u_{i-1}(1-1/2i)\}_{j,n+1}$$

$$=\frac{r_R}{2}\{u_{i+1}(1+1/2i)+u_{i-1}(1-1/2i)\}_{j,n}$$

$$+u_{i,j,n}(1-r_R-2r_z)+r_z(u_{j+1}-u_{j-1})_{i,n}$$

$$\frac{r_z}{2}\{-u_{j+1}+2u_j(1+1/r_z)-u_{j-1}\}_{i,n+2}$$

$$=r_R\{u_{i+1}(1+1/2i)+u_{i-1}(1-1/2i)\}_{j,n+1}$$

$$+\frac{r_z}{2}\{u_{j+1}-u_{j-1}\}_{i,n+1}+u_{i,j,n+1}(1-r_z-2r_R)$$

where

$$r_R=\frac{c_R\,dt}{dR^2}\qquad r_z=\frac{c_z\,dt}{dz^2}$$

and $i\,dR=R$ co-ordinate (radial direction); $j\,dz=z$ co-ordinate.

The term $(1/R)(\partial u/\partial R)$ is indeterminate at the axis of symmetry, since both R and $\partial u/\partial R$ are zero. However, it may be shown that, as R tends to zero,

$$\left(\frac{\partial^2 u}{\partial R^2}+\frac{1}{R}\frac{\partial u}{\partial R}\right)\Rightarrow 2\frac{\partial^2 u}{\partial R^2}$$

At the axis, therefore, the diffusion equation has the form

$$\frac{\partial u}{\partial t}=2c_R\frac{\partial^2 u}{\partial R^2}+c_z\frac{\partial^2 u}{\partial z^2}$$

and the corresponding finite difference expressions to use alternately are:

$$r_R\{-u_{i+1}+u_i(2+1/r_R)-u_{i-1}\}_{j,n+1}$$

$$=r_R\{u_{i+1}+u_{i-1}\}_{j,n}+r_z\{u_{j+1}+u_{j-1}\}_{i,n}+u_{i,j,n}(1-2r_R-2r_z)$$

and

$$\frac{r_z}{2}\{-u_{j+1}+2u_j(1+1/r_z)-u_{j-1}\}_{i,n+2}$$

$$=2r_R\{u_{i+1}+u_{i-1}\}_{j,n+1}+\frac{r_z}{2}\{u_{j+1}+u_{j-1}\}_{i,n+1}+u_{i,j,n+1}(1-4r_R-r_z)$$

4.4 EXAMPLES OF PSEUDO-CONSOLIDATION SOLUTIONS

Koppula and Morgenstern[17] have described the use of ADI methods for studying the efficiency of drains in earth dams. The study considered a range of geometries of the clay core, the ratio of the coefficient of permeability of the drain to that of the core, and the relative lengths of the flow paths in the two regions. They expressed their results in terms of an *impedance factor*

$$\lambda = (k_2/k_1)(H/d)$$

where k_1 = permeability coefficient of the clay core
k_2 = permeability coefficient of the side drains
H = height of the clay core
d = width of a side drain

Large values of λ denote efficient filter drains. Thus, as the value approaches infinity, the drain approaches perfect efficiency and the excess pore-water pressure at the interface between the clay core and the drain is always zero. Conversely, if λ is zero, the filter drain completely inhibits flow and all dissipation of the excess pore-water pressure occurs vertically upwards.

The influence of the geometry of the clay core and impedance of the side drains is given for specified degrees of consolidation in Fig. 2.

Notice that, for small values of the ratio $H/2w$, the impedance of the drains does not significantly influence the results, as the horizontal drainage path is much longer than the vertical, so that vertical drainage predominates. However, as the authors pointed out, the impedance could result in a significant retardation of the dissipation of excess pore-water pressures for λ less than 10 in conjunction with $H/2w$ greater than about $0 \cdot 25$.

The application of two-dimensional pseudo-consolidation theory to the design of road embankments has also received considerable attention. The influence of embankment geometry, depth of compressible strata, and anisotropy of the soil has been extensively studied by Dunn and Razouki.[18] Their results are presented in a form suitable for use as design charts.

To show the influence of these factors on the progress of consolidation more clearly, some of their results for an isotropic soil are shown in Fig. 3. The ratio H/b_u is zero for conditions of one-dimensional consolidation and in that case, the results are

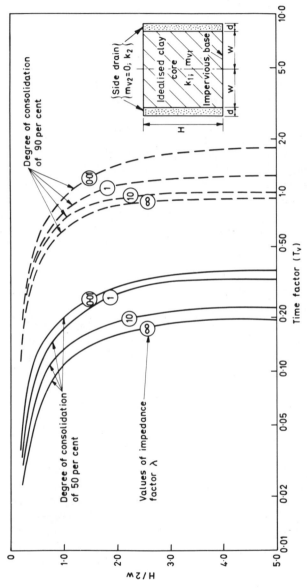

FIG. 2. Influence of geometry and side-drain impedance on time to specified degrees of consolidation of a clay core for an earth dam.

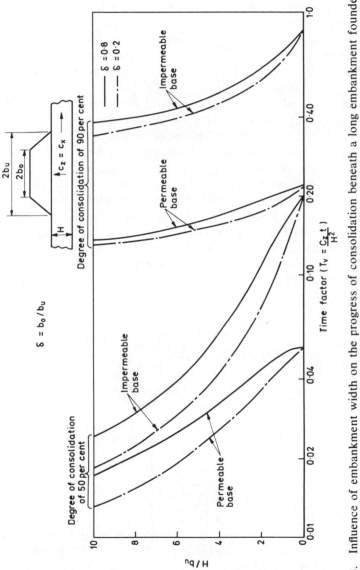

FIG. 3. Influence of embankment width on the progress of consolidation beneath a long embankment founded on a homogeneous, isotropic soil.

uninfluenced by embankment shape, (defined as the ratio (δ) of the width at the top of the embankment to that of the base). This is demonstrated in the figure by convergence of the pairs of curves. In any case the effect of embankment shape is not particularly large at any stage. Since a road engineer often assesses his design according to the time required to achieve a degree of consolidation of 90%, the influence of the shape of embankment could be ignored for the conditions described.

This influence of problem geometry on the progress of consolidation beneath the centre of an embankment on anisotropic soil is shown for a range of extreme situations in Fig. 4. The data used in this figure was also abstracted from the paper by Dunn and Razouki.[18] The figure would allow a design engineer to make a rapid assessment of the time to achieve 50% and 90% consolidation for shallow or deep deposits of anisotropic clay, and for embankments with either relatively flat or steep sides. Although the information is based on the results for a single layer of soil, an approximate analysis for multi-layered deposits could be obtained by determining equivalent values for the coefficients of consolidation in the horizontal ($\bar{c}_x$) and vertical directions ($\bar{c}_z$). The horizontal value may be simply achieved by taking a weighted average as follows

$$\bar{c}_x = \frac{\sum\limits_{i=1}^{i=N} c_{xi}H_i}{\sum\limits_{i=1}^{i=N} H_i}$$

where H_i = thickness of the ith layer
c_{xi} = horizontal coefficient of consolidation of the ith layer
N = total number of layers

The equivalent coefficient of consolidation in the vertical direction may be obtained as follows:

$$\bar{c}_z = \frac{\left[\sum\limits_{i=1}^{i=N} H_i\right]^2}{\left[\sum\limits_{i=1}^{i=N} H_i/\sqrt{c_{zi}}\right]^2}$$

where c_{zi} = vertical coefficient of consolidation of ith layer.

An extensive series of comparisons between the predicted and observed settlements of embankment is given in a paper published by Lewis et al.[19] Figure 5 shows a typical result for a site at Tickton in Yorkshire. Notwithstanding the many criticisms of the use of pseudo-

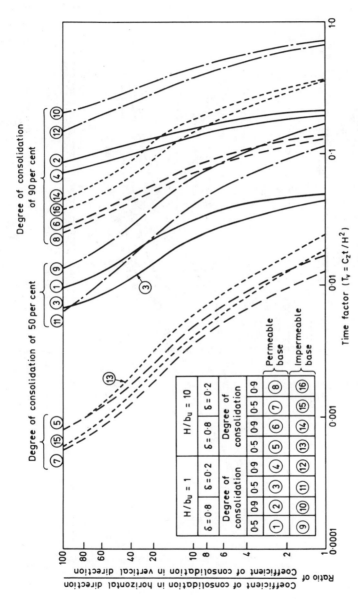

FIG. 4.   Influence of problem geometry on progress of consolidation for anisotropic soils.

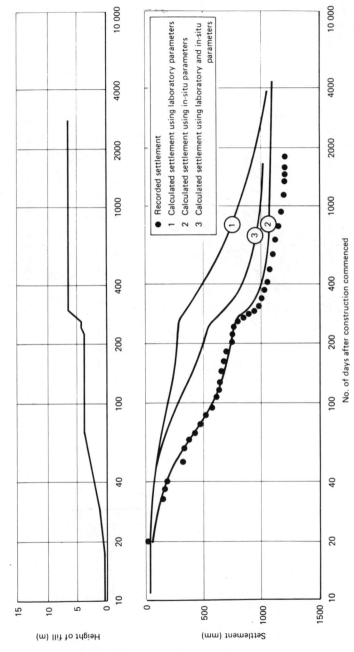

FIG. 5.   Recorded settlement compared with the calculated relations between settlement and time, using coefficients of consolidation from both laboratory and *in situ* tests.

consolidation theory, the results from this series of studies generally showed very good agreement between predicted and observed performance and it was concluded that sufficient accuracy was obtained by this approach for purposes of road engineering.

An interesting paper by Poskitt[20] provides a solution to the pseudo two-dimensional theory for the case of a layer in which the coefficients of permeability and compressibility varied with the effective stress. It was assumed for purposes of the analysis that linear relations were maintained between void ratio and the logarithm of effective stress and also between void ratio and the logarithm of the coefficient of permeability. The results were obtained in non-dimensional form (Fig. 6) in terms of the average relative strain against the time factor for a range of values of the grouped terms:

$$\frac{a_{v0}}{a_{vf}} \cdot \left(\frac{k_f}{k_0}\right)_v ; \quad \frac{a_{v0}}{a_{vf}} \cdot \left(\frac{k_f}{k_0}\right)_h ; \quad \frac{k_{0h}}{k_{0v}} \cdot \left(\frac{H_v}{H_h}\right)^2$$

where a_v = the tangent to the void ratio—effective stress curve
H_v = vertical length of drainage path
H_h = horizontal length of drainage path
k = coefficient of permeability
0 = subscript relating to initial values
f = subscript relating to final values
v = subscript relating to vertical direction
h = subscript relating to horizontal direction

An example will be given to demonstrate the application of the charts. Determine the time factor corresponding to a degree of consolidation of 90% for a soil in which the compressibility (a_{v0}) is halved during consolidation and in which both the vertical and horizontal coefficients of permeability are reduced to one-sixth of their original value. The horizontal length of drainage path (H_h) is equal to 9 m and the vertical length (H_v) is equal to 3 m. The ratio of the initial coefficients of horizontal and vertical permeability (k_{0h}/k_{0v}) is equal to 4·5.

Then

$$a_{vf} = 0·5\, a_{v0}; \quad k_{fv} = \tfrac{1}{6} k_{0v}; \quad k_{fh} = \tfrac{1}{6} k_{0h}$$

and

$$\frac{a_{v0}}{a_{vf}} \left(\frac{k_f}{k_0}\right)_v = \frac{a_{v0}}{a_{vf}} \left(\frac{k_f}{k_0}\right)_h = 2\left(\frac{1}{6}\right) = \frac{1}{3}$$

$$\frac{k_{0h}}{k_{0v}} \left(\frac{H_v}{H_h}\right)^2 = 4 \cdot 5 \left(\frac{3}{9}\right)^2 = 0 \cdot 5$$

Examining the appropriate chart given in Fig. 6 with these groups of terms a time factor of 1·4 is obtained. It is instructive to compare this result with that obtained assuming the compressibility and permeability coefficients were unchanged from their initial values. Thus we would refer to the same chart but use values of unity for the first two terms. A time factor of about 0·6 is obtained for this case, showing the considerable difference between the two results.

Extensive use of finite difference methods for solving the pseudo-consolidation theory has been made by Davis and Poulos.[21] Their

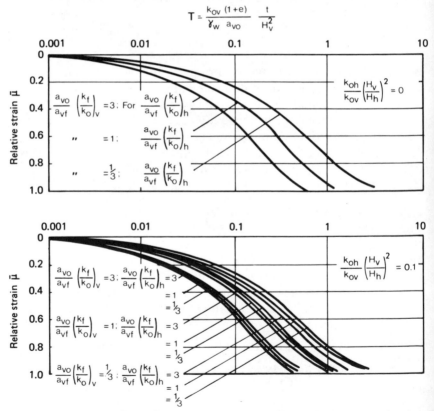

FIG. 6.   Influence of varying permeability and compressibility on strain–time relations. (After Poskitt, reference 20.)

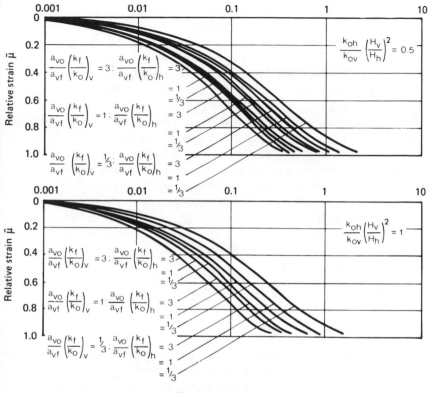

FIG. 6.—*cont.*

results have provided valuable insight into those parameters which contribute most in the disparities between one-dimensional and multi-dimensional consolidation analyses and also between pseudo and true three-dimensional behaviour.

They have carried out a comprehensive study of the rate of settlement of strip and circular footings subject to both two- and three-dimensional consolidation. The influence of various factors were investigated, including the shape of the footing, anisotropic permeability, and the effect of a sand layer overlying the clay. They also made a limited investigation into the progress of consolidation of rectangular footings. They point out that the results for circular footings are very close to those for square footings of identical area, particularly if the width of the footing is much less than the depth of the compressible strata. Thus, for design purposes, it is appropriate to use the available results for circular footings in determining the duration of consolidation of square footings of the same area. The non-dimensional time factor $(T_v = c_1 t/H^2)$ for degrees of consolidation of both 50% and 90% have been extracted from their results and are presented in Figs. 7 and 8 for the axi-symmetric case and strip case respectively.

The application of elastic theory for predicting immediate and final settlement under three-dimensional conditions has also been considered by Davis and Poulos. They point out that, although the use of elastic theory for predicting the deformation of soil appears questionable, provided the correct stress range is adopted in the soil testing procedures for assessing the equivalent elastic parameters, more reliable and consistent results will be obtained in three-dimensional situations than is given by the use of one-dimensional theory. Thus, they propose the use of the well-known elastic equation for predicting total settlement:

$$, \rho_{TF} = \Sigma \frac{1}{E'} (\Delta\sigma_z - \nu'(\Delta\sigma_x + \Delta\sigma_y)) \, \mathrm{d}z$$

where

E' and ν' = soil modulus and Poisson's ratio respectively, corresponding to the anticipated range of stress:

$\Delta\sigma_z, \Delta\sigma_x$ and $\Delta\sigma_y$ = increments of stress in vertical and horizontal directions respectively, obtained from elastic stress distribution theory corresponding to sub-layer $\mathrm{d}z$.

In assessing the settlement arising from undrained conditions, the value of soil modulus (E_u) is obtained from undrained triaxial tests and Poisson's ratio is taken to be 0.5. Thus, the equation for estimating undrained settlement is as follows.

$$\rho_u = \Sigma \frac{1}{E_u} (\Delta\sigma_z - 0.5(\Delta\sigma_x + \Delta\sigma_y)) \, \mathrm{d}z$$

In applying the above equations for predicting total and immediate settlements beneath footings, it is assumed that conditions of uniform stress prevail beneath the footing. For settlement of rigid footings, where conditions correspond more closely with uniform deformation,

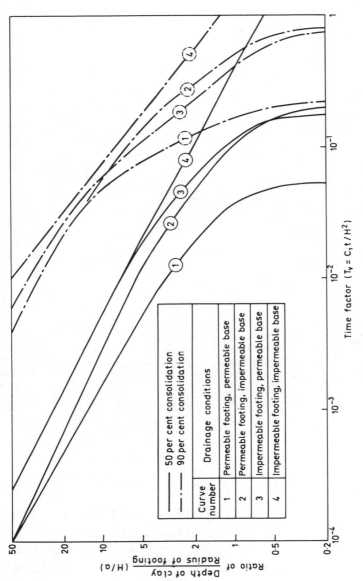

FIG. 7.   Influence of depth of clay layer on the progress of axi-symmetric consolidation beneath centre of circular footing according to Terzaghi–Rendulic theory.

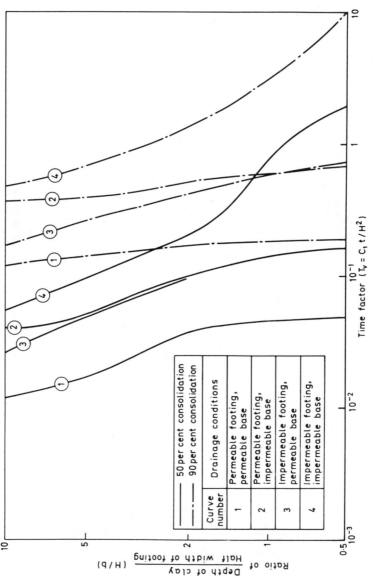

FIG. 8. Influence of depth of clay layer on the progress of plane–strain consolidation beneath centre of strip footing according to Terzaghi–Rendulic theory.

Davis and Poulos suggest taking $\frac{1}{2}$(settlement of the centre + settlement of the edge) for a strip or circular footing; and $\frac{1}{3}$ (2 × settlement of the centre + settlement of a corner) for a rectangular footing. They found that, except on very shallow layers, this gave results which were accurate within 5% where v' is zero and within 10% where v' is 0·5.

The authors also point out that the use of the one-dimensional approach results in relatively small error, for Poisson's ratio less than 0·25.

4.5   BIOT'S CONSOLIDATION THEORY

Solutions to this equation in closed analytical form are rather limited and are generally confined to flexible footings on a homogeneous, isotropic semi-infinite half-space, although recently some results have been obtained for a layer of finite thickness. However, solutions are now becoming more readily available for other situations, obtained by the application of numerical techniques such as the finite element method. Some of these results are presented below to allow comparisons with the more readily available pseudo-consolidation theory, and to allow the reader to judge when Biot's equations are essential.

A major difference between the Terzaghi–Rendulic theory and the Biot theory is the Mandel–Cryer effect on the excess pore-water pressures. The theoretical relation between excess pore-water pressure and adjusted time factor for various locations beneath the centre of a strip load on an infinite half-space subject to plane–strain consolidation, was obtained by Schiffman et al.,[22] and is reproduced in Fig. 9. Points of interest concerning the figure are the proportionate increase of excess pore-water pressure with depth and the increase in time to reach a peak value as the depth increases. As was also shown by Schiffman et al., the Mandel–Cryer effect is strongly influenced by surface drainage conditions, with the effect being almost completely absent when the surface is impervious. For comparison, the results obtained by Schiffman et al. for a point at depth $z/a = 0·5$ beneath the centre are reproduced in Fig. 10 on the basis of the time factor determined from the two-dimensional coefficient of consolidation according to the procedure proposed by Davis and Poulos.[21]

$$T_v = \frac{c_2 t}{H^2}$$

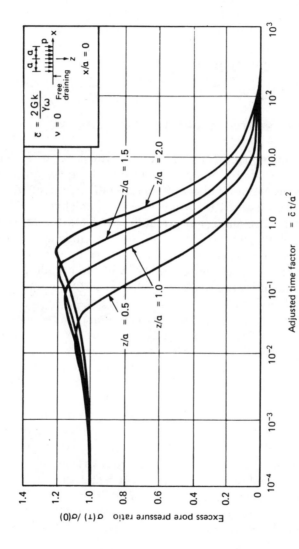

Adjusted time factor $= \bar{c} \, t/a^2$

Excess pore pressure ratio $\sigma(\tau)/\sigma(0)$

FIG. 9.   Effect of depth on excess pore pressure: plane–strain consolidation: normally loaded half-plane. (After Schiffman *et al.*, reference 22.)

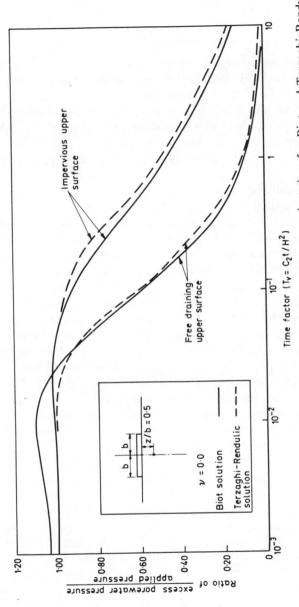

FIG. 10. Comparison of relations between excess pore-water pressure against time for Biot and Terzaghi–Rendulic theories.

where

$$c_2 = \frac{kE'}{2\gamma_w(1 - 2\nu')(1 + \nu')}$$

Also shown on the figure are the relations for a point at the same location obtained by the authors using the pseudo-consolidation theory. The differences between the two theories are never very large, once the initial increase due to the Mandel–Cryer effect has been overcome. Such results give support to the view held by some research workers that the simple theory may provide adequate accuracy for predictive purposes, particularly at later stages of consolidation.

As to the validity of the Mandel–Cryer effect, there now appears ample experimental evidence to justify the existence of this phenomenon. Results are reproduced in Fig. 11 from an experimental study carried out in 1955 by Aboshi.[23] Several other research workers have also published experimental data confirming the behaviour.[24,25]

In Fig. 12 are shown the results of a plane–strain analysis for a flexible and permeable strip footing overlying a clay layer with an impermeable base. The results are given in terms of degree of settlement against time factor $(T_v = c_1 t/H^2)$ for different values of Poisson's ratio and for a range of values of the ratio of the half width of footing to the depth of the clay layer. These data were originally published in the paper by Yamaguchi and Murukami.[26] It is of interest to note the considerable retarding effect on the progress of settlement induced by the Mandel–Cryer effect at low values of ν. This is clearly demonstrated in Fig. 12 where the two-dimensional theory has predicted a slower rate of settlement than given by the one-dimensional theory for the higher range of values of b/H.

Whereas in pseudo-consolidation theory, the degree of settlement and degree of consolidation are identical, this is not the case with Biot's theory, and for comparison between the two theories, the degree of settlement of the latter theory must be employed. It has been pointed out by Davis and Poulos[21] and separately by Christian et al.[27] that the one-dimensional value is the most suitable coefficient of consolidation to use in determining the time factor for these comparisons.

In their paper, Christian et al. have also considered the effects of width of load in the plane–strain consolidation of a strip footing. In addition they have studied both the influence of anisotropy and the

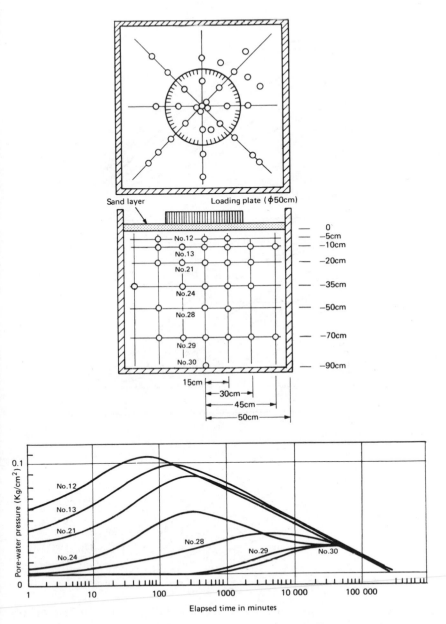

FIG. 11.   Experimental result by Aboshi.[23]

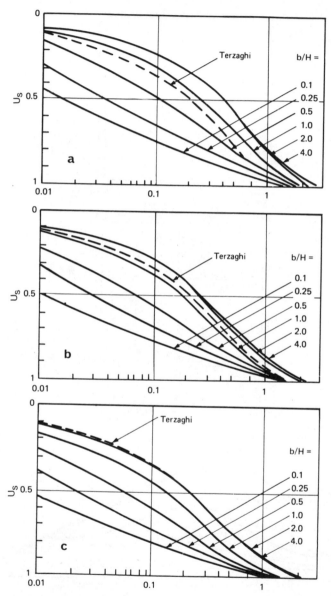

FIG. 12. Relationships between time factor and degree of settlement (under strip load). (a) $\nu' = 0$; (b) $\nu' = 0\cdot3$; (c) $\nu' = 0\cdot4$. (After Yamaguchi and Muruk-ami, reference 26.)

contribution made by the elastic parameters. In their investigation into the effect of load width, the authors point out that, for ratios of depth of clay to half width of strip (δ) of unity and less, the times for specified degrees of settlement arising from consolidation are virtually identical. This finding agrees with earlier results obtained from pseudo-consolidation theory.

They have also shown that for a range of values of Poisson's ratio between 0·25 and 0·4 and δ of unity, the time factor for 90% consolidation is 0·75. As this value is relatively close to that given by one-dimensional theory, they have proposed that conventional one-dimensional theory be used for predicting the range of consolidation settlement of strip loads with rough bases on uniform soils for values of δ less than unity. For values of δ equal to or greater than unity they recommend that the times calculated by conventional theory be multiplied by the factors given in Table 2.

In considering anisotropic effects, Christian *et al.* made use of a parameter α defined as the ratio of equivalent isotropic permeability (k_{eff}) to the equivalent permeability coefficient conventionally employed in steady state flow ($\sqrt{k_v k_h}$)

that is:
$$\alpha = \frac{k_{\text{eff}}}{\sqrt{k_v k_h}}$$

The α parameter is a useful measure of the influence of the anisotropy. For example, a value of unity indicates that flow is equally effective in both vertical and horizontal directions. Their results are reproduced here and show that during the early stages of consolidation (Fig. 13) the larger values of the permeability ratio

TABLE 2
MULTIPLYING FACTORS FOR DIFFERENT RATIOS OF DEPTH OF CLAY/HALF WIDTH OF STRIP LOAD (H/b)

$H/b = \delta$	Multiplying factor for degree of consolidation settlement of:			
	10%	50%	75%	90%
4	0·875	0·457	0·625	0·708
2	1·375	0·711	0·729	0·802
1	2	1·015	0·938	0·943
0·5	2	1·218	1·104	1·061

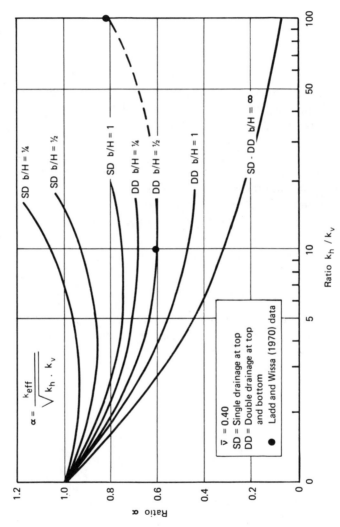

FIG. 13.   α versus k_h/k_v at 50% consolidation. (After Christian *et al.*, reference 27.)

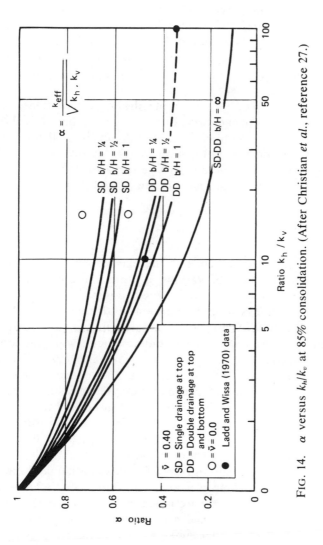

FIG. 14.   α versus k_h/k_v at 85% consolidation. (After Christian *et al.*, reference 27.)

produce a fairly significant effect. However, at later stages (Fig. 14) the effects were much smaller. The authors propose that the α parameter can be employed to 'correct' conventional one-dimensional theory for anisotropic effects by selecting the value from Figs. 13 and 14 corresponding to 50% and 90% (strictly, 85%) degrees of consolidation settlement for the particular values of k_h/k_v and then dividing the appropriate values given in Table 2 by $\alpha \sqrt{k_h/k_v}$. An example will make the procedure clear.

Given $H/b = 1$; $k_h/k_v = 10$; double drainage conditions, determine the time to 90% consolidation. First derive the time to achieve 90% consolidation (t_1) by conventional one-dimensional theory. Referring to Table 2 we find the multiplying factor for isotropic conditions is 0·943. From Fig. 14 we obtain a value for α of 0·45.

Thus the corrected time (t_2) is obtained as follows:

$$t_2 = \frac{0·943 t_1}{0·45\sqrt{10}} = 0·66 t_1$$

In considering the influence of the elastic parameters, Christian *et al.* evaluated the effect of varying ν, while the constrained modulus (D', defined below) was maintained constant. For the example chosen, the progress of consolidation settlement was influenced to only a small extent for a range of ν between 0 and 0·4 corresponding to a value of δ of unity. These findings agree with the conclusions reached by Davis and Poulos[21] when comparing the predicted time relations beneath the centre of a circular footing based on both Biot's and Rendulic's theories of consolidation. Their results for a circular footing on a semi-infinite half-space are reproduced in Fig. 15. It can be seen from this figure that the prediction based on pseudo-consolidation theory always lies between the curves based on the Biot theory using the extreme values of ν of 0 and 0·5.

The relations between degree of consolidation and time factor beneath the centres of circular and strip footings on a finite layer and for different Poisson's ratio are given in Figs. 16 and 17 respectively. These results were obtained by Gibson *et al.*[28] and relate to a free draining upper surface and an impervious base. Note that the results were presented by the authors in terms of an adjusted time factor, θ.

$$\theta = \frac{2Gk}{\gamma_w} \cdot \frac{t}{H^2} = \frac{\bar{c}t}{H^2}$$

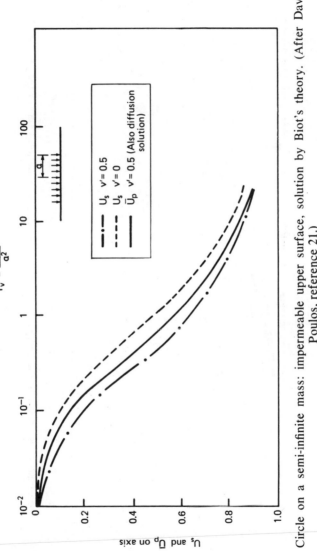

FIG. 15.   Circle on a semi-infinite mass: impermeable upper surface, solution by Biot's theory. (After Davis and Poulos, reference 21.)

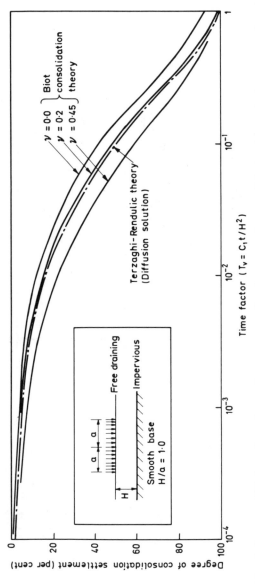

FIG. 16. Influence of Poisson's ratio on the progress of axi-symmetric consolidation beneath the centre of a circular loaded area.

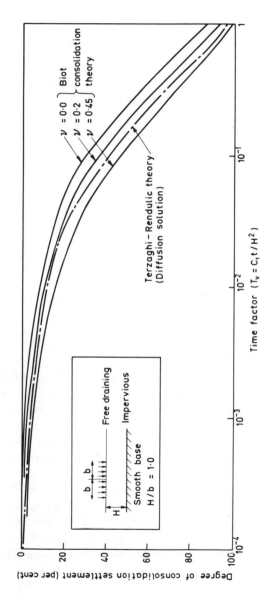

FIG. 17.   Influence of Poisson's ratio on the progress of plane-strain consolidation beneath the centre of a strip load.

where G = shear modulus = $E'/2(1 + v')$

γ_w = unit weight of water

k = coefficient of permeability.

The conventional time factor ($T_v = c_1 t/H^2$) which has been used in the figures is obtained by the relation between c_1 and $\bar{c}$ on the basis of the following well-known equations from elastic theory:

$$c_1 = \frac{k}{\gamma_w} D'$$

where:

$$D' = \text{constrained modulus} = \frac{E'(1 - v')}{(1 + v')(1 - 2v')}$$

$\therefore$

$$c_1 = \frac{k}{\gamma_w} \frac{E'(1 - v')}{(1 + v')(1 - 2v')}$$

$$\bar{c} = \frac{2Gk}{\gamma_w} = \frac{2k}{\gamma_w} \frac{E'}{2(1 + v')}$$

$\therefore$

$$\frac{c_1}{\bar{c}} = \frac{(1 - v')}{(1 - 2v')}$$

or:

$$c_1 = \frac{(1 - v')}{(1 - 2v')} \bar{c}$$

Also shown in Figs. 16 and 17 are the degree of consolidation against time relations obtained by Davis and Poulos on the basis of the simpler, pseudo-consolidation theory. For this particular geometry ($H/a = 1$), the differences between the curves are not very great, although an extreme range of Poisson's ratio has been employed. The curve based on the simpler theory lies generally near the centre of the range and agrees best with the curves based on v' of 0·2. To investigate further if the reasonable agreement obtained between the curves for the two theories is associated with the particular geometry, the relations between H/a and time factor for both 50% and 90% degrees of consolidation for the two theories are given in Fig. 18. The Biot solution is based on a Poisson's ratio of 0. The curves indicate that the choice of H/a of unity provides slightly better agreement than is obtained on average throughout the full range, but in any case the differences are never very great, there being at most a ratio of 2 between the time factors given by the two theories.

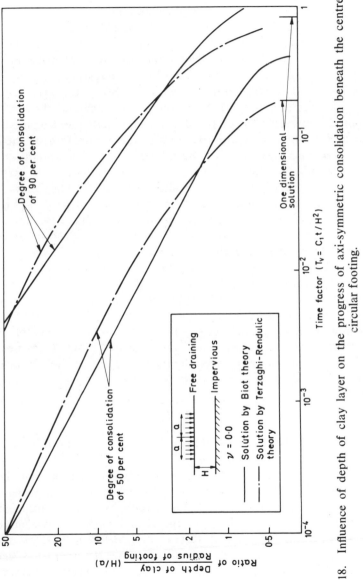

Fig. 18. Influence of depth of clay layer on the progress of axi-symmetric consolidation beneath the centre of a circular footing.

In contrast to the many case histories available in which comparisons are given between the recorded settlement behaviour and that predicted by the Terzaghi–Rendulic theory, relatively few such comparisons have been reported for the Biot theory.

Two recent examples, however, relate to comparisons between observed and calculated settlement performance of road embankments founded in compressible soils. In an interesting paper by Smith and Hobbs,[29] the application of the Biot theory for predicting the settlement performance of two road embankments showed good agreement between recorded and calculated settlement against time. Of particular interest in this paper was the application of the finite element method to a study of the influence of embankment stiffness on the consolidation behaviour. The authors' results for the particular embankment under study are reproduced in Fig. 19. They pointed out that increased bank stiffness led to higher values of predicted excess pore-water pressures beneath the centre and reduced values at the toe. Using a bank stiffness of 2000 tf/m^2 gave best overall agreement with the observed pore-water pressures, this value of bank stiffness corresponded approximately with the estimated value of between 5 and 10 times the subsoil stiffness of 200 tf/m^2.

The other recent paper by Shoji and Matsumoto[30] discussed the choice of elastic parameters and the effects of anisotropy. The authors propose that for purposes of finite element analysis the elastic modulus (E) of the soil may be best estimated from m_v rather than from the tangent modulus over the early portion of the stress–strain curve as advocated by some workers.

Thus:

$$E = \frac{(1+\nu)(1-2\nu)}{(1-\nu)} \frac{1}{m_v}$$

The authors also propose that a value of ν of one-third may be generally most satisfactory to use.

In considering the influence of anisotropy, they found that the use of the vertical coefficients of permeability from one-dimensional tests for both vertical and horizontal directions significantly underestimated the rates of movement. Increasing the horizontal values by a factor of 10 still produced serious underestimates of the rate of settlement. Back analysis of the observed results indicated that increases in the coefficients of permeability in the upper layers of the order of between 10 and 50 times was necessary to establish the rate of settlement to the current order of magnitude.

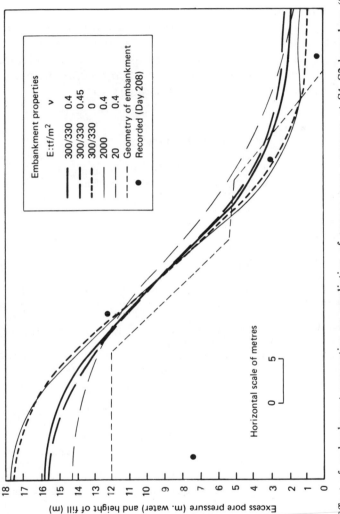

FIG. 19. Effect of embankment properties on prediction of excess pore pressure at S1–S2 boundary (Sale). (After Smith and Hobbs, reference 29.)

The particular point concerning the unreliability of data from the small-scale one-dimensional consolidation tests was also made in the paper by Smith and Hobbs[29] and is consistent with the conclusions based on many more case histories in which the Terzaghi–Rendulic equations were employed.

Thus as was pointed out in the introductory remarks to this chapter, the most important factor in controlling the accuracy of predictions of consolidation behaviour is the procedure adopted for assessing the soil parameters. This is best achieved by methods of *in situ* and large scale laboratory consolidation testing.

REFERENCES

1. SYMONS, I. F. and MURRAY, R. T. (1975). Embankments on soft foundations: settlement and stability study at Over causeway bypass, *TRRL Report 675*.
2. TERZAGHI, K. V. (1943). *Theoretical Soil Mechanics*, John Wiley & Sons, New York.
3. RENDULIC, L. (1936). Porenziffer und Porenwasser Drück in Tonen, *Der Bauingenieur*, 17(51/53), 559–564.
4. BIOT, M. A. (1941). General theory of three-dimensional consolidation, *J. Appl. Phys.*, 12, 155–164.
5. DAVIS, E. H. and POULOS, H. G. (1968). The use of elastic theory for settlement prediction under three-dimensional conditions, *Géotechnique*, 18(1), 67–91.
6. MANDEL, J. (1957). Consolidation des couches d'argiles, *Proc. 4th Int. Conf. Soil Mech. Found. Engng.*, 1, 360–367.
7. CRYER, C. W. (1963). A comparison of three-dimensional theories of Biot and Terzaghi, *Quart. Journal Mech. and Appl. Math.*, 16, 401–412.
8. BARRON, R. (1948). Consolidation of fine grained soils by drain wells. *Trans Amer. Soc. Civ. Engrs.*, 113, 718–742.
9. HORNE, M. R. (1964). The consolidation of a stratified soil with vertical and horizontal drainage, *Int. J. Mech. Sciences*, 6, 187–197.
10. ROWE, P. W. (1964). The calculation of the consolidation rates of laminated, varved or layered clays with particular reference to sand drains, *Géotechnique*, 14(4), 321–340.
11. CRANK, J. and NICHOLSON, P. (1947). A practical method for numerical solutions of partial differential equations of the heat conduction type, *Proc. Cambridge Phil. Soc.*, 43, 50–67.
12. PEACEMAN, D. W. and RACHFORD, H. H. (1955). The numerical solution of parabolic and elliptic differential equations, *Journal Soc. Indust. Appl. Math.*, 3, 28–41.
13. MITCHELL, A. R. (1969). *Computational Methods in Partial Differential Equations*, John Wiley & Sons, New York.

14. MURRAY, R. T. (1971). Embankments constructed on soft foundations: settlement study at Avonmouth, *TRRL Report LR 419*.
15. DOUGLAS, J. (1962). Alternating direction methods for three space variables, *Numerische Mathematik*, **4**, 41–63.
16. BRIAN, P. L. T. (1961). A finite difference method of high order accuracy for the solution of three-dimensional transient heat conduction problems, *A.I.Ch.E. Journal*, **7**, 367–370.
17. KOPPULA, S. D. and MORGENSTERN, N. R. (1972). Consolidation of clay layer in two dimensions, *J. ASCE Soil Mech. Found. Div.*, **SM1**, 79–93.
18. DUNN, C. S. and RAZOUKI, S. S. (1974). Two-dimensional consolidation under embankments, *J. Instn. Highway Engineers*, **21**(10), 12–24.
19. LEWIS, W. A., MURRAY, R. T. and SYMONS, I. F. (1976). Settlement and stability of embankments constructed on soft alluvial soils, *Inst. of Civ. Engrs.*, **59**, 571–593.
20. POSKITT, T. J. (1970). Settlement charts for anisotropic soils, *Géotechnique*, **20**(3), 325–330.
21. DAVIS, E. H. and POULOS, H. G. (1972). Rate of settlement under two- and three-dimensional conditions, *Géotechnique*, **22**(1), 95–114.
22. SCHIFFMAN, R. L., CHEN, T. F. A. and JORDAN, J. C. (1969). An analysis of consolidation theories, *J. ASCE Soil Mech. Found. Div.*, **SM1**, 285–312.
23. ABOSHI, H. (1955). Measurement of pore-water pressure during consolidation of fine-grained soils, *Bulletin of the Faculty of Engineers, Hiroshima University*, **4**(1), 1–12.
24. GIBSON, R. E., KNIGHT, K. and TAYLOR, P. W. (1963). A critical experiment to examine theories of three-dimensional consolidation, *Proc. European Conf. Soil Mech. Found. Engineering*, Weisbaden 1963, **1**, 69–76.
25. VERRUIJT, A. (1965). Discussion, *Proc. 6th Int. Conf. Soil Mech. Found. Engng.*, **3**, 401–402.
26. YAMAGUCHI, H. and MURUKAMI, T. (1976). Plane–strain consolidation of a clay layer with finite thickness, *Soils and Foundations*, **16**(3), 67–79.
27. CHRISTIAN, J. T., BOEHMER, J. W. and MARTIN, P. P. (1972). Consolidation of a layer under a strip load, *Journal ASCE Soil Mech. and Found. Div.*, **SM7**, 693–707.
28. GIBSON, R. E., SCHIFFMAN, R. L. and PU, S. L. (1968). Plane–strain and axially symmetric consolidation of a clay layer of limited thickness, *Soil Mechanics Report 3, Dept. Materials Engng., University of Illinois*, **SM3**.
29. SMITH, I. M. and HOBBS, R. (1976). Biot analysis of consolidation beneath embankments, *Géotechnique*, **26**(1), 149–171.
30. SHOJI, M. and MATSUMOTO, T. (1976). Consolidation of embankment foundation, *Soils and Foundations*, **16**(1), 59–74.
31. LADD, C. C. and WISSA, A. E. Z. (1970). Geology and engineering properties of Connecticut Valley varved clays with special reference to embankment construction, *Mass. Inst. Tech. Dept. of Civil Eng. Report R70-56*.

Chapter 5

FOUNDATION INTERACTION ANALYSIS

J. A. HOOPER

Ove Arup & Partners, London, UK

SUMMARY

Methods of analysing the interaction that occurs between foundations and the supporting soil are reviewed, with particular reference to building structures. The work mostly relates to the linear elastic analysis of foundations under static loading, although mention is made of various forms of non-linear analysis. Modelling of the soil strata is discussed, and a summary is given of a wide range of solutions for plain rafts, pavement slabs and strip and pad foundations. The analysis of piled foundations is also discussed, with the emphasis on pile groups rather than single piles. Finally, attention is focused on the influence of superstructure stiffness on the degree of foundation interaction.

5.1 INTRODUCTION

This chapter deals with problems of foundation analysis in which the stiffness of the foundation—and sometimes that of the superstructure—is taken into account, in addition to the stiffness of the soil. The emphasis is strongly directed towards the foundations of building structures, although much of the work referred to is relevant to the foundation analysis of dams, bridges and other types of structure. Aspects of soil–structure interaction relating specifically to retaining walls, tunnels and other buried structures are not discussed. The same applies to the assessment of allowable settlements and the response of foundations to dynamic loading.

In the past, it has been common practice to consider the analysis of the soil and the structure quite separately. Typically the structural engineer calculates the load distribution within a building on the assumption that the base of the structure is fixed. This loading is then used by the soil mechanics specialist to calculate foundation settlements based on a completely flexible structure. In reality, of course, the vertical deformations at the soil–structure interface are nearly always compatible, thereby modifying the calculated settlement pattern and the distribution of structural load. In recent years, however, methods of analysis have been developed which can often form the basis of a more rational approach to foundation design.

It is the aim of this chapter to present—if only in a somewhat condensed form—a summary of the work that has been carried out on foundation interaction analysis and, where appropriate, to comment on the efficacy of alternative solution methods. Although most of the work referred to is based on the assumption of linear elastic behaviour of the foundation, soil and superstructure, it is considered that this does provide a valuable yardstick against which to judge and compare actual building performance. It is also noteworthy that linear elasticity forms the basis of most structural analysis, and that the response of most soils is sensibly linear within the usual range of working stresses.

The references given in this chapter are by no means exhaustive either in number or in scope; in general, only those that are both readily accessible and are written in one of the principal European languages have been included. Furthermore, solutions for the limiting case of zero foundation stiffness are not mentioned in detail because they are reasonably well-documented elsewhere. In contrast, references concerning foundations of finite stiffness are spread far and wide, and for this reason it is considered useful in the present context to draw attention to the very extensive range of analytical solutions and numerical techniques currently available.

5.2 ANALYTICAL MODELS

5.2.1 Basic Material Model
In most practical cases the analysis of foundation interaction is concerned only with the assessment of structural behaviour at working loads. Under these circumstances the magnitudes of the applied

loads are relatively low and both the soil and the structural components can be realistically considered as linear elastic materials. Hence virtually the whole of the present discussion on interaction analysis is restricted to the domain of linear elasticity.

Of course, numerous refinements are possible. The soil, for example, could be considered as an elasto-plastic, visco-elastic or critical state material with time-dependent properties. Similar models could be used for the structural components, including one which allowed for the cracking of concrete or brickwork. However, it is usually possible to take account of these non-linear characteristics by modifying the basic material parameters and treating the problem as a linear elastic one; this is standard procedure, for example, in most structural design. Even then, a comparatively rigorous analysis of all but the simplest soil–structure interaction problems requires the use of a large computer.

5.2.2 Modelling of Soil Strata

There are numerous ways of modelling soil strata within the framework of linear elasticity, but the three most common models are:

(a) Winkler springs
(b) Half-space continuum
(c) Layered continuum

These models are represented diagrammatically in Fig. 1 for the case of a single layer, and it is useful to outline the principal features of each.

(a) Winkler Springs

Here the soil mass is represented simply by a series of independent linear springs (Fig. 1(a)). Hence the shear resistance of the soil is not taken into account and the reaction force acting on the foundation at any point is directly proportional to the foundation settlement at that point.

Very many papers have been published on the subject of beams and plates on Winkler springs, and much of the work has been summarised by Hetényi[1] and Timoshenko and Woinowsky-Krieger.[2] More refined spring models that incorporate cross-coupling members have also been devised, and details of many of these have been discussed by Kerr.[3]

Although the Winkler model has the apparent advantage that it can

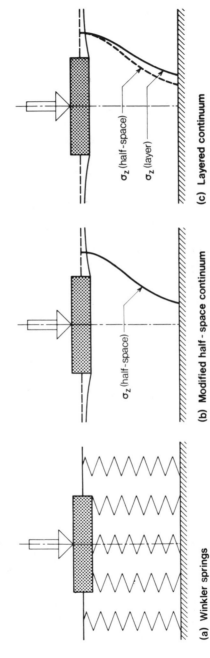

FIG. 1.   Analytical models for single soil layer underlain by rigid base.

be readily incorporated into standard computer programs for structural analysis, it generally constitutes a poor physical model of the soil mass and can give rise to grossly erroneous results. Consider, for example, a stiff raft or footing in contact with a homogeneous soil stratum and subjected to a uniformly distributed applied load. If the soil is represented by Winkler springs, then an almost uniform contact pressure distribution will be deduced from the near-uniform settlement profile, and the computed bending moments will be virtually zero. In reality, of course, the contact pressure distribution is likely to be distinctly non-uniform, giving rise to relatively high bending moments. Only in the idealised case of an incompressible elastic half-space, whose moduli increase linearly with depth from zero at the surface, does the Winkler settlement profile match that for a continuum; this correspondence has been demonstrated by Gibson[4] for an isotropic medium and by Gibson and Kalsi[5] and Gibson and Sills[6] for an orthotropic medium.

In this particular example, therefore, the Winkler spring model gives entirely the wrong answers. An improved correlation could be achieved by varying the spring stiffness values across the foundation, but the basis on which such variations are established is not clear. Better results would also be obtained for more flexible foundation structures and for non-uniform distributions of applied load, but the complete inability of the model to deal adequately with uniformly loaded stiff foundations on homogeneous soils strongly suggests that its general use in foundation analysis and design should not be encouraged. Further disadvantages of the Winkler model stem from the difficulty of relating the necessary spring constants to soil parameters, and from the total lack of information provided by a Winkler-type analysis on the stresses and deformations within the soil mass.

(b) Half-space Continuum

In the strictest sense, this model relates only to the case where both the stresses and displacements within the soil mass are assumed to be those corresponding to an elastic half-space. For a homogeneous half-space, several exact solutions are available for both isotropic and transversely isotropic media loaded at or below the plane surface. There are also a few closed-form solutions for the surface loading of a heterogeneous isotropic half-space whose elastic modulus increases or decreases monotonically with depth.

Direct application of this model is severely limited in practice because no account is taken of soil layering or the variation of elastic properties with depth within a given layer. However, a very useful extension of this model can be achieved by assuming a half-space stress distribution and then calculating the strains at any given depth using the relevant elastic parameters for the soil at that depth. Settlements can then be obtained simply by summating vertical strains.

This model, referred to here as a modified half-space continuum (Fig. 1(b)), is analogous to that proposed by Steinbrenner[7] for a single layer underlain by a rigid base. In addition, the elastic method of settlement analysis is directly related to the classical one-dimensional method of analysis for saturated clays developed by Terzaghi.[8] In this latter method the fully drained settlement is given by

$$\rho' = \Sigma \sigma_z m_v \delta h \tag{1}$$

where σ_z denotes the average increase in vertical stress within a layer of thickness δh, calculated from the foundation loading using half-space theory, m_v denotes the coefficient of volume compressibility, and the summation is carried out over an appropriate depth of soil beneath the foundation. The corresponding expression relating to the modified half-space continuum model is

$$\rho' = \Sigma \frac{1}{E'}[\sigma_z - \nu'(\sigma_x + \sigma_y)]\delta h \tag{2}$$

where $\sigma_x, \sigma_y, \sigma_z$ denote the average changes in applied normal stress and E', ν' denote the drained elastic soil parameters.

If ν' is known with reasonable accuracy, then E' can be deduced from measured values of m_v using the relationship

$$m_v = \frac{(1 + \nu')(1 - 2\nu')}{E'(1 - \nu')} \tag{3}$$

These m_v values are usually obtained from laboratory oedometer tests, but may also be obtained from K_0 drained triaxial tests. Alternatively, both elastic parameters may be obtained directly from drained triaxial tests.

Clearly eqns. (1) and (2) become identical when $\nu' = 0$, and it is reasonable to suppose that in many cases they would give similar computed settlements in view of the relatively low values of ν'

appropriate to most clay soils; this has been confirmed by Davis and Poulos[9] for the particular case of a circular load on a finite layer. In contrast to the classical method, however, the three-dimensional elastic method can also give direct values of undrained settlement— and thereby values of the consolidation component of settlement— simply by replacing E' by $E'(1 + \nu)/(1 + \nu')$ and ν' by ν in eqn. (2), where ν denotes the undrained Poisson's ratio. Similar procedures for dealing with a transversely isotropic soil medium have been described by Hooper.[10]

Hence, by using the appropriate soil parameters in conjunction with a half-space stress distribution, settlement values relating to both undrained and fully drained soil conditions may be obtained by means of a single unified approach. In contrast to the Winkler model, the modified half-space continuum model can provide much useful information on the stresses and deformations within the soil mass, and there is a much closer relationship between the model parameters and measured soil properties. It also has the important advantage of simplicity; in view of the acknowledged difficulties and uncertainties in determining *in situ* soil deformation parameters, there is little to be gained in carrying out over-sophisticated settlement analyses.

(c) Layered Continuum

In this model, the exact distribution of stress and strain within a layered elastic continuum is established in order to calculate surface settlements (Fig. 1(c)). However, the theoretical formulation implicit in such problems is difficult, and only a few closed-form solutions are available. Moreover, these solutions relate to relatively simple layer configurations, such as a single layer underlain by a rigid base.

Solutions for the more general case of a multi-layered continuum loaded at its horizontal plane surface can be obtained using numerical methods—see, for example, Peutz *et al.*[11] and Gerrard and Harrison[12]—but the necessary computations are very much more complicated than those for a homogeneous half-space. Even then, these rigorous solutions are only applicable to horizontal layers of uniform thickness. For the completely general case of inclined layers of variable thickness, there is little alternative but to resort to approximate calculations based on the modified half-space continuum model.

However, it has been shown by Poulos,[13] Ueshita and Meyerhof[14] and Hooper and Wood[15] that this latter model gives very satisfactory results for a wide range of soil layering and heterogeneity. The layered continuum model is undoubtedly useful in certain cases—for example, in pavement design or when assessing the behaviour of a raft foundation on a thin clay layer—but its general application to the solution of practical foundation problems is bound to be somewhat limited.

5.3 CIRCULAR RAFT FOUNDATIONS

A number of analytical solutions to the problem of an elastic circular plate or raft of finite flexibility supported by Winkler springs have been referred to by Timoshenko and Woinowsky-Krieger[2] and Galletly.[16] These solutions have been augmented by those of Frederick[17] for thick plates, Sinha[18] and Bolton[19] for large plate deflections, and Leonards and Harr[20] for conditions of partial support. The case of an annular foundation has been considered by Volterra[21] and Reismann.[22] Details of the more refined Winkler-type models, such as the two-parameter Pasternak model and the three-parameter Kerr model, have been discussed by Vlasov and Leont'ev,[23] Rades[24] and others.

For the reasons given in the previous section, a raft analysis based on a continuum soil model is generally preferable to a Winkler-type analysis. Moreover, the relatively simple case of a circular raft founded on an elastic continuum is particularly useful in making preliminary assessments of foundation interaction for a wide range of structures.

There are many analytical solutions to the problem of circular loading applied directly to the surface of an elastic continuum, including those which take account of soil layering and anisotropy. These solutions, many of which have been recorded by Giroud[25] and Poulos and Davis,[26] relate to the limiting case of a completely flexible raft, and therefore can be used to determine upper bound values of total and differential settlement.

Likewise there are available many solutions to the problem of a rigid circular punch indenting an elastic solid. These solutions relate to the opposite limit of a completely rigid raft, and are useful in giving lower bound values of total settlement, contact pressure and bending moment. The case of an axially loaded rigid raft in frictionless contact with a homogeneous isotropic elastic half-space has been considered by Sneddon,[27] Schiffman and Aggarwala[28] and Chan and Cheung,[29] and the corresponding case of completely adhesive contact has been studied by Mossakovski,[30] Keer,[31] Spence,[32] Popova[33] and Egorov et al.[34] Results for these two cases become identical for incompressible materials as no interfacial shear stresses are developed. The problem of finite interfacial friction has been solved by Spence,[35] and the effect of the elasticity of the indentor has been investigated by Khadem and O'Connor[36] and Zlatin and Ufliand.[37]

A few isolated solutions exist for eccentric loading as well as lateral and torsional loading, but a comprehensive set of results for several types of loading applied to a transversely isotropic half-space has been obtained by Gerrard and Harrison[38] and Gerrard and Wardle.[39] The effect of a linearly increasing modulus with depth has been discussed by Carrier and Christian,[40] and the problem of a homogeneous elastic layer has been considered by Egorov and Nichiporo-vich,[41] England,[42] Egorov and Serebrjanyi,[43] Poulos,[44] Milovic,[45] Dhal-iwal[46] and Shelest.[47] The indentation of an elastic medium comprising two parallel layers bonded to a homogeneous half-space has been studied by Chen and Engel.[48] The effect of depth of embedment has been investigated by Butterfield and Banerjee[49] for a half-space and by Banerjee[50] for a finite layer. Various solutions for a rigid annular foundation on an elastic continuum have been obtained by Egorov,[51] Valov,[52] Shibuya et al.[53] and Dhaliwal and Singh.[54]

Detailed results for the case of a uniformly loaded circular raft of finite flexibility in frictionless contact with the surface of an elastic isotropic half-space or layer have been obtained by Brown[55-57] using a collocation method. Here the soil modulus is assumed to be either constant or increasing linearly with depth. Results corresponding to the case of adhesive contact between the raft and the soil have been given by Hooper.[10,58] These results, which also include the effects of non-uniform applied loading and transverse isotropy of the soil, were obtained using isoparametric finite elements in their axisymmetric form; see, for example, Zienkiewicz.[59]

Results for differential settlement $\Delta\rho$ and maximum bending moment M^* are most conveniently given in terms of the relative raft stiffness K_c, which may be defined as

$$K_c = \frac{E(1 - \nu_s^2)}{E_s}\left(\frac{h}{R}\right)^3 \tag{4}$$

for the case of a homogeneous isotropic stratum (elastic parameters E_s, ν_s) and a raft of radius R, thickness h, and Young's modulus E.

Where the soil is anisotropic or heterogeneous, the definition of K_c will depend on how E_s is specified. Examples of the moment–curvature relations for a uniformly loaded circular raft on the surface of a deep homogeneous isotropic stratum are shown in Fig. 2, where q denotes the intensity of applied loading. Thus, uniformly loaded rafts with $K_c < 0.1$ may be termed flexible and those with $K_c > 10$ considered as being stiff. However, in the zone $0.1 < K_c < 10$ the moment–curvature relations change quite rapidly.

It follows that a useful first step in any analysis is to determine the relative raft stiffness and thereby to assess the possible effects of soil–structure interaction. At the same time it is essential to take account of the stiffness of the superstructure and the approximate distribution of applied loading. Where this loading is not even approximately uniform, a set of curves of the type shown in Fig. 2 would have to be constructed. In the analysis of storage tanks or silos, for example, radial edge moments and peripheral line loads could be added. Where horizontal as well as vertical forces are applied, the possibility of loss of contact between the raft and the soil should be considered.

It is apparent from Fig. 2 that the boundary condition at the interface between the raft and the soil can have a very marked effect on computed values of differential settlement and bending moment. This effect can be explained by considering an axially loaded stiff circular raft on the surface of a homogeneous elastic half-space. As load is applied to the raft, vertical settlement occurs and horizontal compressive stresses are generated in the soil immediately below the raft. This soil tends to move inwards towards the raft centre, and is free to do so if there is no friction between the raft and soil (Fig. 3(a)). In the case of adhesive contact, however, this inward movement is resisted by the raft and interfacial shear stresses are developed. Assuming the compressional stiffness of the raft to be greater than that of the soil, then the direction of the shear stresses is as indicated in Fig. 3(b).

It is clear that these interfacial shear stresses will tend to reduce raft bending moments and differential settlements, and also the horizontal compressive stresses at the soil surface beneath the raft. For design purposes the case of frictionless contact is generally to be preferred because it tends to overestimate raft bending moments and differential settlements. This also applies to non-circular rafts and strip footings as the effects of interfacial restraint will be similar. In

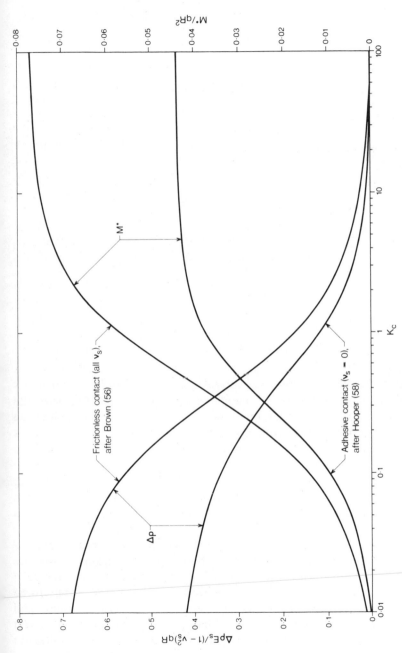

FIG. 2. Maximum values of differential settlement and bending moment for uniformly loaded circular raft on deep elastic layer.

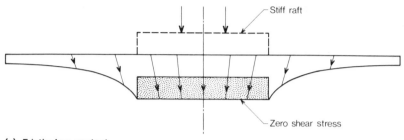

(a) **Frictionless contact**

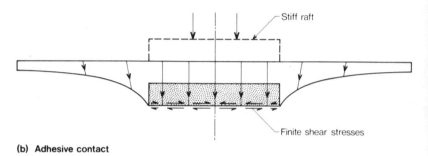

(b) **Adhesive contact**

FIG. 3.   Influence of interfacial boundary conditions.

practice, the adhesion between the raft and the soil will be reduced by interfacial slip and local yielding of the soil. Further discussion of the effect of interfacial contact, including the case of partial slip, has been given by Hooper,[60] Buragohain and Shah[61] and Varma et al.[62]

Although the majority of solutions referred to above relate to axisymmetric loading, Wilson[63] has suggested a method of solving asymmetric problems by expressing the applied loading as a Fourier series. However, this harmonic formulation is restricted to linear elastic analyses where there is no separation between the raft and soil.

Other work associated with the elastic analysis of circular rafts of finite flexibility includes the analytical half-space solutions of Borowicka[64–66] for both uniform loading and asymmetric concentrated loading, as well as those of Gladwell and Iyer[67] for the case where centrally applied loading leads to loss of contact between raft and half-space. The analytical approach has also been adopted by Pickett and McCormick[68] to study the problem of a circular raft on an elastic layer underlain by a rigid base. The finite element method has been used by Smith[69] and Boswell and Scott[70] to analyse symmetrically loaded rafts, and an iterative solution method has been developed by Zbirohowski-Koscia and Gunasekera,[71] based on a modified half-space continuum. Finite difference solutions for asymmetrically loaded rafts have been given by Chakravorty and Ghosh[72] and Banerjee and Jankov;[73] in the latter, raft separation is also considered.

The above-mentioned elastic solutions can be usefully employed in estimating total and differential settlements for the limiting cases of undrained and fully drained soil conditions. However, the variations in settlement occurring during the intervening period can only be determined by solving the consolidation equations. Almost all the available analytical solutions to these equations relate to the case of a completely flexible raft under uniform circular loading on the surface of a homogeneous isotropic elastic half-space or layer. Thus solutions have been obtained by Gibson and Lumb[74] and Davis and Poulos[75] based on simple diffusion theory, whilst De Jong,[76] Mandel,[77,78] Paria,[79] McNamee and Gibson[80] and Gibson et al.[81] have presented solutions based on the more rigorous theory developed by Biot[82] in which the pore-water drainage conditions are coupled to the soil displacement field. These two theories can yield substantially different results, except in the case of incompressible materials when the respective governing differential equations become identical. This is discussed more fully in Chapter 4.

There appear to be no analytical solutions to the corresponding problem of a raft of finite flexibility on a consolidating elastic continuum, although the time-settlement behaviour of a completely rigid circular raft has been investigated by Agbezuge and Deresiewicz.[83] The raft is considered to be either fully permeable or completely impermeable, and to be in frictionless contact with a homogeneous isotropic poro-elastic half-space having a free-draining surface. In developing the solution, Hankel–Laplace double transforms were used to derive a pair of dual integral equations, which were then solved numerically. Computed results for the time-variation of contact pressure $p(t)$ are shown in Fig. 4 for the case $v_s = 0$, the applied load being taken as constant and the coefficient of consolidation being denoted by c_v. The maximum changes in contact pressure occur at the raft centre during the initial stages of consolidation (Fig. 4(a)); they amount to an increase of 5% for the permeable raft and a decrease of 8% for the impermeable raft. In order to

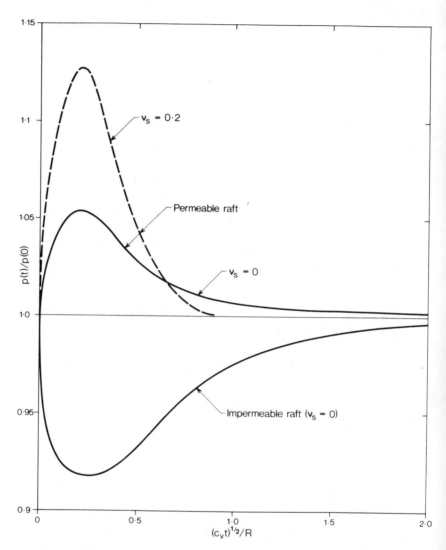

FIG. 4(a). Time–variation of contact pressure beneath rigid circular raft, after Agbezuge and Deresiewicz[83] and Chiarella and Booker.[84] Contact pressure at raft centre.

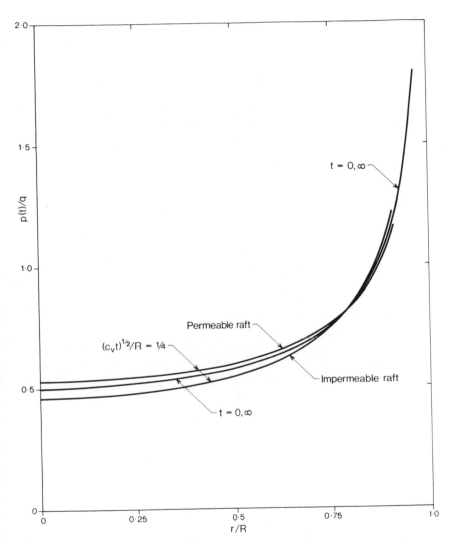

FIG. 4(b). Contact pressure distribution ($\nu_s = 0$).

maintain vertical equilibrium, these variations are accompanied by pressure changes of opposite sign within the edge region of the raft; this is shown in Fig. 4(b), in which the distribution of contact pressure is expressed as a proportion of the average applied pressure q. In accordance with simple elastic theory, the initial and steady-state contact pressures turn out to be identical.

In this connection, an independent solution to the same problem for a fully permeable raft has been obtained by Chiarella and Booker[84] using integral transform methods. In this solution, the dual integral equations were reduced to a single Fredholm equation prior to quadrature, and the resulting values of contact pressure for the case $v_s = 0.2$ are included in Fig. 4(a). The maximum pressure increase is more than double that given by Agbezuge and Deresiewicz[83] for $v_s = 0$, and it would be interesting to know whether the contact pressures are really so sensitive to v_s, or whether the computed pressure differences simply reflect the different numerical procedures used in the two analyses. From a more general viewpoint, however, these time-variations in contact pressure are relatively small, and probably would be even smaller in practice as a result of finite raft permeability and the gradual—as opposed to instantaneous—application of structural load.

Consolidation problems for rafts of finite flexibility can be solved using finite difference or finite element methods, along the lines indicated by Sandhu and Wilson,[85] Christian and Boehmer,[86] Hwang et al.,[87,88] Yokoo et al.,[89–91] Booker,[92] Ghaboussi and Wilson,[93] Valliappan and Lee[94] and Smith and Hobbs.[95] These numerical methods have the added advantage of being able to readily accommodate such parameters as soil layering, anisotropy and non-uniform applied loading.

5.4   NON-CIRCULAR RAFT FOUNDATIONS

Rafts or ground slabs having a large plan area, or those subjected to localised applied loading, can often be adequately treated as structural members of infinite lateral extent, thereby simplifying the analytical formulation of the problem. Solutions for an infinite elastic raft of uniform thickness on a Winkler foundation include those by Westergaard,[96–100] Wyman,[101] Naghdi and Rowley,[102] Livesley,[103] Campbell and Heaps[104] and Reissner,[105] and the reinforcing effect of a finite length beam rigidly connected to the raft has been investigated by Brown et al.[106] Solutions relating to an infinite raft on an elastic half-space instead of a Winkler foundation have been obtained by Hogg[107] and Holl[108] for frictionless contact at the raft–soil interface, and numerical results for both types of foundation have been presented by Pickett and Ray.[109] The conditions whereby loss of contact occurs between the raft and the half-space have been investigated by Richart and Zia,[110] Weitsman,[111] Pu and Hussain[112] and Keer et al.[113] The effect of interfacial shear stresses has been considered by Parkes,[114] and further results for the case of adhesive contact can be obtained from the various solutions for two-layer elastic systems; see, for example, Poulos and Davis.[26] The problem of an infinite raft on an elastic layer of finite depth has been solved by Hogg[115] for the case of frictionless raft contact, whilst the effect of adhesive contact can be assessed from the general solutions to the three-layer problem.

Various solutions for rectangular rafts on a Winkler foundation have been obtained by Vint and Elgood,[116] Murphy,[117] Fletcher and Thorne,[118] Timoshenko and Woinowsky-Krieger[2] and Mackey and Chung,[119] using classical methods of analysis. Approximate methods based on a grillage analogy have been outlined by Ewell and Okubo,[120] Hudson and Matlock[121] and Hudson and Stelzer,[122] and practical applications have been discussed by Panak et al.[123] and Panak and Rauhut.[124] These methods can be used for rafts of any plan shape; this also applies to the relaxation method proposed by Allen and Severn,[125] and to the finite element methods described by Cheung and Zienkiewicz[126] and Severn.[127] Applications of the latter to raft design have been reported by Winter[128] and Wegmuller et al.,[129] and applications to the design of jointed pavements have been described by Huang and Wang.[130] The bending of thick rectangular rafts has been analysed by Frederick,[131] and the effect of large deflections studied by Sinha[18] and Yang.[132] There are also solutions based on more sophisticated Winkler-type foundation models, such as those described by Vlasov and Leont'ev,[23] Yang[133] and Jones and Xenophontos.[134]

Relatively few analytical solutions have been obtained for rectangular rafts on an elastic continuum. A number of solutions for the limiting case of zero raft stiffness have been summarised by Poulos and Davis[26] for semi-infinite and layered media, but these are more applicable to the calculation of total raft settlements rather than differential settlements. At the other extreme, results for axial and eccentric loading of rigid rectangular rafts in frictionless contact with a half-space have been given by Gorbunov–Possadov and Sere-brjanyi,[135] Absi,[136] Brown,[137] Chan and Cheung[29] and Absi et al.,[138] whilst approximate results for rigid non-rectangular rafts have been given by Conway and Farnham.[139] It appears that the corresponding problems of adhesive raft contact have not yet been solved. The effect of finite layer thickness has been considered by Sovinc[140] and Milovic and Tournier,[141] while Butterfield and Banerjee[49] have examined the effect of depth of embedment; moreover, the combination of both effects has been analysed by Banerjee[50] using the boundary element method.

Results for rectangular rafts of arbitrary bending stiffness founded on an elastic continuum have been obtained by Pickett et al.,[142] Janes[143] and Valantagul et al.[144] using the finite difference method, and by Saxena[145] using a grillage model for the raft, but there is little doubt that the displacement finite element approach adopted by Cheung and Zienkiewicz[126] is currently the most powerful technique for general raft analysis, chiefly because of the ease with which otherwise awkward boundary conditions can be handled. In principle the stiffness matrix of the continuum is simply added to that of the raft, and the combined matrix inverted to give direct values of total and differential raft settlement. Related values of raft rotation, bending moment and contact pressure can then be readily obtained.

Numerical results for rafts of arbitrary bending stiffness in frictionless contact with a homogeneous isotropic elastic half-space have been given by Cheung and Zienkiewicz[126] and Wang et al.[146] The case of adhesive contact, and also that of loss of contact between the raft and the supporting medium, have been considered by Cheung and Nag.[147] Approximate results for rafts on heterogeneous media have been obtained by Zbirohowski–Koscia and Gunasekera,[71] Wood and Larnach[148] and Hain et al.[149] based on a modified half-space continuum, and a more rigorous treatment of the layered problem has been presented by Wardle and Fraser[150] and Fraser and Wardle.[151]

An example of the results obtained by Fraser and Wardle[151] for uniformly loaded rectangular rafts is shown in Fig. 5. Maximum values of differential settlement $\Delta\rho$ and bending moment per unit width M^* are expressed in terms of the relative raft stiffness K_r, defined as

$$K_r = \frac{4E(1 - \nu_s^2)}{3E_s(1 - \nu^2)} \left(\frac{h}{B}\right)^3 \tag{5}$$

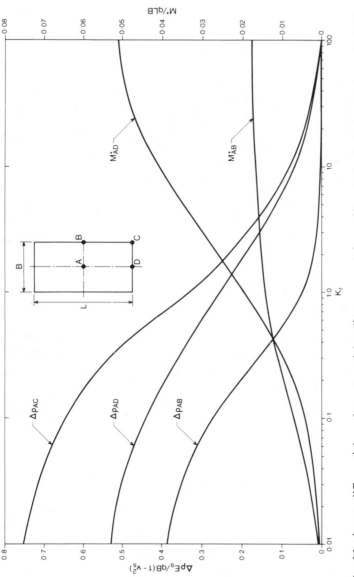

FIG. 5. Maximum differential settlement and bending moment for uniformly loaded rectangular raft ($L/B = 2$) on homogeneous isotropic half-space, after Fraser and Wardle.[151]

where E, ν denote the elastic parameters of a raft of breadth B, length L and thickness h subjected to a uniformly distributed load q, and E_s, ν_s denote the elastic soil parameters. These results relate to rafts in frictionless contact with a homogeneous isotropic half-space, although Fraser and Wardle have given correction factors to be applied in the case of a single elastic layer of finite depth in adhesive contact with a rigid base.

In each of the finite element based methods referred to above, the raft is divided into rectangular elements and the distribution of contact pressure around each node is assumed to be uniform. This general approach has been extended by Hooper[152] to include isoparametric raft elements, which can be used to model rafts of virtually any plan shape. It has also been extended by Svec and Gladwell[153] to include triangular raft elements with a cubic polynomial representation of contact pressure across any given triangular region. Further results for a raft on a homogeneous half-space have been presented by Svec and McNeice,[154] whilst more difficult problems involving loss of contact and thick rafts have been analysed by Svec.[155,156]

It is possible, of course, to analyse rafts using three-dimensional finite elements, but this method is usually prohibitively expensive in terms of data handling and computer costs. Comparative raft analyses carried out by Wardle and Fraser,[150] for example, demonstrated that computer solution times using three-dimensional elements were greater than those based on the raft-on-continuum approach by a factor of about twenty. In almost all practical cases, therefore, this latter approach is to be preferred.

There appear to be no analytical solutions to consolidation problems for rectangular rafts. However, the rate of total raft settlement can be estimated using solutions for uniformly loaded rectangular areas on a poro-elastic medium. In this connection, rigorous solutions have been obtained by Gibson and McNamee[157, 158] for a half-space and by Booker[159] for a layer of finite depth. Approximate solutions for a finite layer have also been presented by Davis and Poulos.[75] Three-dimensional analyses of raft consolidation problems, although theoretically quite feasible using finite element or boundary element methods, hardly constitute a practical approach. Some form of simplified plate-on-continuum method, such as using the diffusion theory in conjunction with a modified half-space continuum, would be adequate for most purposes.

5.5 STRIP AND PAD FOUNDATIONS

An extensive study of beams on Winkler foundations has been made by Hetényi,[1] and the results are pertinent to the behaviour of strip footings in the longitudinal direction. Solutions for infinite beams subjected to arbitrary applied loading are generally straightforward, but beams of finite length are more difficult to handle. The relative merits of various solution methods have been discussed by Fraser,[160] Just *et al.*[161] and Fraser,[162] and alternative approximate procedures have been described by Levinton[163] and Popov.[164] The problem of foundation tilt has been considered by Weissmann,[165] and the effect of local loss of support has been discussed by Lytton and Meyer.[166]

Three-dimensional solutions for strip and pad foundations on the surface of an elastic continuum can be obtained using the methods outlined in the previous section on non-circular rafts. However, if the width of a foundation is small compared with its length, it is usually sufficient to carry out a simpler two-dimensional analysis. This approach can also be used to analyse large raft foundations—particularly those which support a regular grid of widely spaced columns—by dividing the raft into a series of strips and treating each strip as a separate footing.

For the idealised case of an infinitely long strip, the transverse bending moments, contact pressures and settlements can be obtained from a plane–strain analysis. Various analytical solutions for the limiting condition of load applied directly to the surface of semi-infinite or layered elastic continua have been summarised by Poulos and Davis.[26] For the opposite limit of a completely rigid strip indenting a homogeneous isotropic elastic half-plane under symmetrical applied loading, solutions have been obtained by Sadowsky[167] for frictionless contact and by Muskhelishvili[168] for adhesive contact. The effect of the elasticity of the indentor has been dealt with by Khadem and O'Connor[169] and Adams and Bogy[170] for both frictionless and adhesive contact, and the influence of finite friction within the contact area has been investigated in an approximate manner by Conway and Farnham[171] and more rigorously by Spence.[172]

The case of moment loading applied to a rigid strip has been considered by Fröhlich[173] for frictionless contact and by Muskhelishvili[168] for both frictionless and adhesive contact. In addition, detailed results for an orthotropic half-plane subjected to a wide range of stress and displacement boundary conditions have been presented by Gerrard and Harrison[174] and Gerrard and Wardle.[175] Solutions to the problem of a rigid strip indenting a layer of finite thickness have been given by Paria,[176] Smith,[177] Conway et al.[178] and Milovic et al.,[179] the latter including the effect of inclined applied loading.

Where the bending stiffness of the strip is finite, half-plane solutions for the case of frictionless contact have been obtained by Borowicka[65,66,180] and Lee and Phillips[181] for various forms of applied loading, and the unbonded contact problem for a central uniform load has been solved by Gladwell.[182] Analytical results for a uniformly loaded strip in adhesive contact with a half-plane have been given by Lee.[183] As in the case of circular rafts, the finite element method is particularly well-suited to deal with the plane problem of long strip foundations; in this connection, a plane–strain analysis of a rectangular raft has been carried out by Crowser et al.[184]

A useful approximation in assessing the behaviour of strip foundations in the longitudinal direction is to assume that conditions of plane stress apply. This approach was adopted by Prager,[185] Biot,[186] Reissner[187] and Bosson[188] in dealing with beams of infinite length and arbitrary bending stiffness in frictionless contact with a homogeneous isotropic elastic half-plane. Results of three-dimensional analyses have been presented by Pichumani and Triandafilidis,[189] based partly on analytical solutions obtained by Herrmann[190] for continuous beams subject to periodic loading. Biot[186] also included an approximate three-dimensional analysis by assuming either uniform settlement or uniform contact pressure in the transverse direction, and this analysis has been extended by Vesić.[191,192] Similar assumptions were made by Gorbunov–Possadov and Serebrjanyi,[135] Panchanathan and Chandrasekaran,[193] Glassman[194] and Brown[195,196] in obtaining solutions for beams of finite length subjected to uniform and concentrated loading. The corresponding problem with adhesive contact between the footing and supporting medium does not appear to have been solved.

Results for a strip footing of length L, breadth B, subjected to a uniform loading q, are shown in Fig. 6, where the aspect ratio $L/B = 10$. Values of differential settlement $\Delta\rho$ and central bending moment M are expressed in terms of the relative stiffness of the footing K_s, defined as

$$K_s = \frac{16EI(1 - \nu_s^2)}{\pi E_s L^4} \tag{6}$$

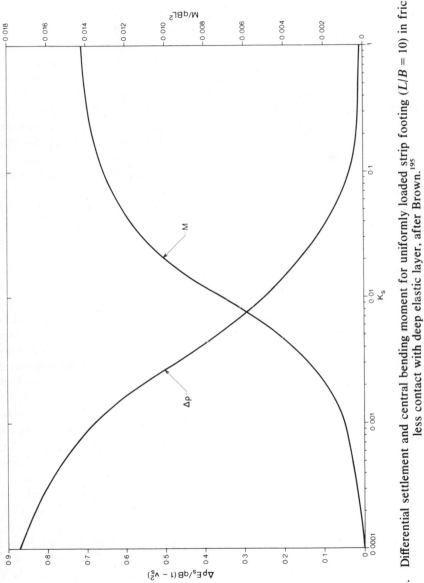

FIG. 6. Differential settlement and central bending moment for uniformly loaded strip footing ($L/B = 10$) in frictionless contact with deep elastic layer, after Brown.[195]

where EI denotes the bending stiffness of the footing cross-section and E_s, ν_s denote the elastic parameters of the homogeneous half-space material. Under uniform applied loading, the maximum bending moment occurs at the centre for all but the most flexible footings.

The time-settlement variation across the transverse section of a long strip footing can be assessed to some extent from the results relating to uniform strip loading of a poro-elastic continuum. A number of analytical solutions to this problem have been obtained by Davis and Poulos[75] using diffusion theory, and by Biot,[197] Biot and Clingan,[198] McNamee and Gibson,[80] Schiffman et al.[199] and Gibson et al.,[81] using the rigorous coupled theory. Problems that include the stiffness effect of the footing are particularly difficult to solve, although results for the time response in the longitudinal direction of an infinitely long beam have been given by Biot and Clingan.[200] In general, consolidation analyses of strip footings are best tackled using the finite element method or some other numerical technique.

5.6 PILED FOUNDATIONS

The interactive analysis of piled foundations is a substantially more difficult task than the corresponding analysis of plain rafts or footings. This naturally stems from the inherent geometrical complexity of a group of piles, which in turn leads to problems in theoretical formulation and computation. Consequently, the design of piled foundations is still more of an art than a science, and is likely to remain so for some considerable time.

In traditional methods of pile group analysis, little or no account is taken of pile–soil interaction; the group is reduced to some simplified form and then dealt with by standard methods of structural analysis. However, the present discussion concerns only the more recent developments in which the piles are assumed to be embedded in an elastic continuum. In this connection, comparative analyses of planar pile groups have been carried out by Poulos[201] using various solution methods.

One type of simplified continuum approach to pile group analysis is to apply the average structural load directly to the soil at some specified level down the pile length. This method has been discussed by Whitaker,[202] but of course can only give limiting values of total and differential settlement. However, it has been extended by Hooper and

Wood[203] in the analysis of a piled raft foundation simply by considering a plain raft positioned at various distances above pile base level.

An alternative simplified approach is to replace the pile group by a solid pier of equivalent length, as suggested by Poulos,[204] or by a solid block of equivalent material, to represent more accurately the compressional stiffness of the pile-reinforced soil. This latter approach can also be used to estimate—to a first approximation—the distribution of load carried by the raft and piles respectively, as outlined by Hooper.[152]

These simplified methods serve essentially as practical design aids, and can be particularly useful during the early design stage of a building foundation. Indeed, in many cases, no further pile group analysis is necessary. Occasionally, the pile group geometry is such that a detailed two-dimensional analysis becomes feasible, and analyses of this type have been carried out by Hooper,[205] Naylor and Hooper[206] and Desai et al.[207] using finite element methods. Although the pile group representation is necessarily approximate, these analyses can readily accomodate such parameters as soil layering and pile cap flexibility.

In this connection, some of the results obtained by Hooper[205] for a piled raft foundation are shown in Fig. 7. The building is a 90-m high block of flats with a 9-m deep basement and a 1·5-m thick raft. The 51 concentrically arranged bored piles are 25 m long with shaft and underream diameters of 0·9 m and 2·4 m respectively (Fig. 7(a)). The gross building weight is 228 MN, which corresponds to an average pressure of 368 kN/m². The soil succession consists of 5 m of fill, sand and gravel, followed by about 60 m of London Clay.

The foundation was analysed using an axisymmetric finite element model (Fig. 7(b)), and the clay modulus was assumed to increase linearly with depth. The nominal bending stiffness of the raft was increased by a factor of ten to allow for the very stiff cross-wall superstructure, and the applied loading was taken to be concentrated more towards the centre of the raft than the edge. Reasonable agreement was obtained between measured and computed raft settlements (Fig. 7(c)); similar correlation was achieved between measured pile loads and raft contact pressures and corresponding computed values. The effect of pile length on computed raft settlements is shown in Fig. 7(d), indicating that the piles are considerably more effective in reducing total settlements than differential settle-

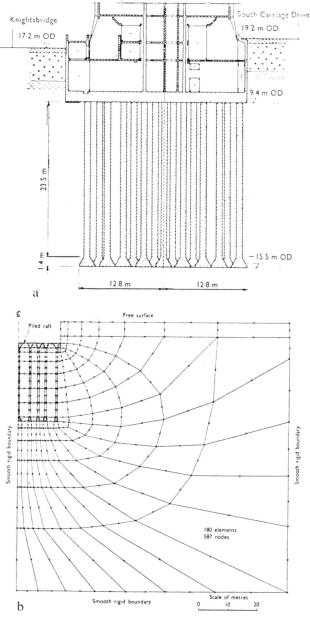

FIG. 7. Approximate analysis of a piled raft foundation (after Hooper[205]). (a) Section. (b) Axisymmetric finite element model.

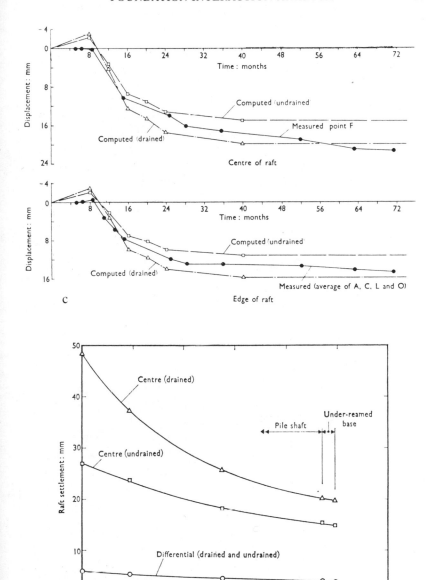

FIG. 7—*cont.* (c) Measured and computed raft settlements. (d) Effect of pile length on computed raft settlements.

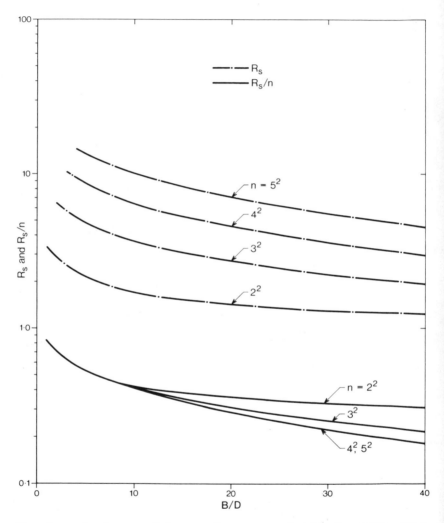

FIG. 8(a).   Settlement of free-standing groups of rigid piles ($L/D = 25$) in incompressible half-space, after Poulos.[204] Completely rigid pile cap.

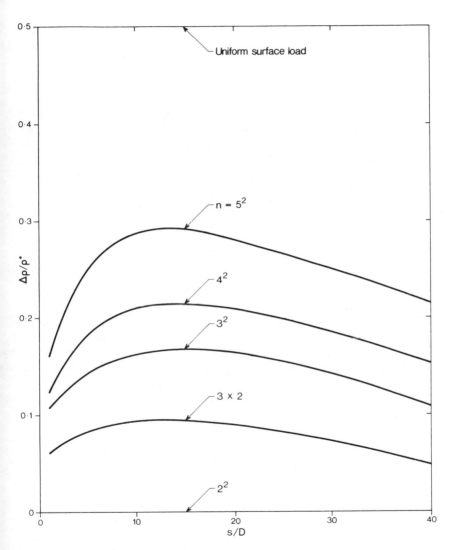

FIG. 8(b).   Completely flexible pile cap.

ments. The results also suggest that the underreamed pile bases are contributing very little to the overall behaviour of the pile group.

Although there appears to be no exact solution to the problem of a single axially loaded pile embedded in a homogeneous isotropic elastic half-space, approximate solutions of the type obtained by Poulos and Davis[208] for rigid piles with and without enlarged bases are sufficiently accurate for most practical purposes. These solutions are based on the integration of expressions given by Mindlin[209] for stresses in the half-space resulting from a concentrated load applied below the plane surface, and imply complete adhesion between pile and soil. This technique also may be applied to piles in layered continua using the Steinbrenner approximation,[7] although its range of applicability is unlikely to be as wide as for plain foundations. It would be of considerable interest, for example, to compare results obtained using this approximation with those based on the equations derived by Chan et al.[210] for the case of a concentrated load applied within a layered half-space. Alternatively, results for a single pile of circular cross-section embedded in a layered medium could be obtained by means of an axisymmetric finite element analysis. In this regard, other methods of accounting for soil heterogeneity have been proposed by Banerjee and Davies[211] and Randolph and Wroth[212].

The corresponding elastic analysis of symmetrical groups of identical piles embedded in a homogeneous isotropic half-space is relatively straightforward as it can be directly based on the settlement interaction between any two of the piles. This so-called interaction factor method has been used by Poulos[204] to analyse free-standing groups of rigid piles, and some of the results are shown in Fig. 8. These relate to axially loaded groups of breadth B embedded in an incompressible medium; the piles are straight-shafted with a pile spacing of s, and $L/D = 25$, where L and D denote the pile length and diameter respectively. Because of the relatively small number of piles in each group, the undrained settlements are a high proportion of the final drained settlements, but this ratio may be considerably reduced for larger groups of piles.

In Fig. 8(a), results are expressed in terms of the group settlement ratio R_s, defined as the ratio of group settlement to the settlement of a single pile carrying the same average load as a pile in the group. They are also plotted in terms of R_s/n, where n denotes the number of piles, in order to indicate the group response for the same total applied load; evidently total settlements are strongly dependent on the breadth of the group, rather than the number of piles within the group. In Fig. 8(b), the maximum differential settlement $\Delta\rho$ is expressed as a ratio of the maximum settlement ρ^*. These results suggest that the effectiveness of piles in reducing differential settlements rapidly decreases as the group size increases, although this decrease may not be so marked in soils whose modulus increases with depth.

The interaction factor method of pile group analysis has been extended by Poulos and Mattes[213,214] to include the effect of pile compressibility, based on the results for a single compressible pile obtained by Mattes and Poulos.[215] The same basic problem has also been solved in a more general way by Butterfield and Banerjee[216] using a boundary integral equation method; this solution has been extended by Banerjee and Driscoll[217] to include raked piles, as well as horizontal and moment loading.

Pile group analysis becomes a good deal more difficult if the pile cap or raft is taken to be in contact with the ground. Approximate solutions for the case of a completely rigid pile cap in frictionless contact with the plane surface of a homogeneous isotropic half-space have been given by Davis and Poulos,[218] based on the results obtained by Poulos[219] for a single rigid pile with a circular pile cap. A more rigorous solution of this problem that also includes the effect of pile compressibility has been obtained by Butterfield and Banerjee,[220] and the case of eccentric loading has been examined in subsequent work by Banerjee.[221]

Results for an axially loaded square group of nine rigid piles are plotted in Fig. 9. Here the group settlement ρ is shown for a total applied load P, together with the apportionment of load between the piles and the base of the pile cap. Clearly the load carried on the pile cap base can be very substantial even at moderate pile spacings, but related piled raft analyses carried out by Hooper[152,205] suggest that this load can often decrease to quite small values as a result of soil consolidation.

Comparatively few analytical solutions have been obtained for the case of finite bending stiffness of the raft, pile cap or footing. Results for piled strip footings have been given by Brown and Wiesner[222] for uniform applied loading and by Wiesner and Brown[223] for a concentrated load positioned at various points along the length of the footing. An approximate solution to the piled raft problem has been proposed by Hongladaromp et al.[224] in which the piles are represented

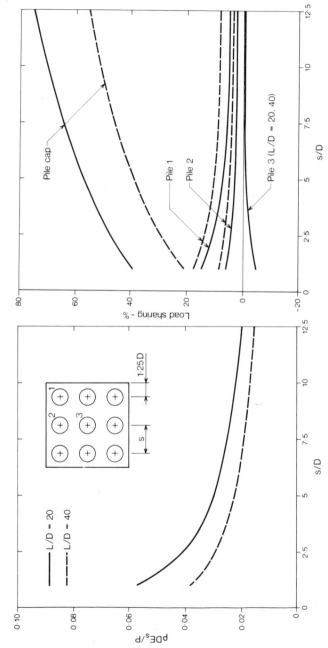

FIG. 9. Settlement and load sharing for rigid piles with rigid pile cap in contact with incompressible half-space, after Butterfield and Banerjee.[220]

by Winkler springs, but this is bound to give rise to unsatisfactory results when the raft is relatively stiff. An even simpler model based on Winkler springs has been used by Hight and Green[225] in an analysis of a piled raft foundation. A more comprehensive approach has been adopted by Hain[226] in which the finite element method outlined by Cheung and Zienkiewicz[126] for plain rafts is combined with the interaction factor method for piles; in this way, the simple soil stiffness matrix can be replaced by the appropriate soil–pile stiffness matrix. A series of comparative calculations carried out by Brown et al.[227] confirms that this is likely to be a reliable method of analysis, although the computational effort required for most practical problems would be considerable.

Results of the type shown in Figs. 8 and 9 are undoubtedly useful in demonstrating many of the basic features of pile group behaviour, but their direct application in practice is rather limited. It should be emphasised that there are two additional assumptions, besides that of linear elasticity, implicit in all the general analytical solutions referred to earlier. First, the reinforcing effect of the piles within the soil mass is neglected. Second, the Steinbrenner approximation is used to deal with soil layering. Clearly some form of fully three-dimensional analysis would be required if these assumptions were to be relaxed. One such analysis has been carried out by Ottaviani,[228] for example, using three-dimensional finite elements to model small piled footings embedded in an elastic layer, but this approach is completely impractical for the general analysis of large pile groups. A more efficient method appears to be that developed by Banerjee[229] and Banerjee and Davies[230] using an integral equation formulation, as only the boundary surfaces need to be specified in detail. This method is discussed more fully in Chapter 9. However, computing costs may still be excessive for the pile group sizes normally encountered in practice. Intuitively, it seems likely that the pile reinforcing effect would be relatively small except at very close pile spacings, but that the Steinbrenner approximation may not be nearly as effective in dealing with pile groups in layered strata as it is with plain foundations.

5.7 NON-LINEAR FOUNDATION ANALYSIS

The previous discussion on soil–structure interaction has been concerned almost entirely with linear elastic analysis, apart from the references to primary consolidation and interfacial slip between structure and soil. In recent years, however, there have been many attempts to deal with non-linear material properties, and it is useful here to summarise some of this work.

Most of the non-linear foundation analyses have concentrated on the soil rather than the structure, although some work relating to plain rafts and ground slabs has been reported. For example, the ultimate load-carrying capacity of concrete pavements has been investigated by Meyerhof[231] using a simplified analytical method, and a theoretical study of an infinite visco-elastic raft founded on a poro-elastic half-space has been presented by Marvin.[232] Solutions for the case of a rigid-perfectly plastic circular raft on Winkler springs and an elastic half-space have been obtained by Augusti[233] and Krajcinovic[234] respectively. Davies[235] has applied the yield line theory to edge-loaded circular concrete rafts, and the flexural cracking of circular concrete rafts on an elastic continuum has been discussed by Hooper.[236] The ultimate strength of edge-loaded rectangular concrete rafts has been assessed by Ranganatham and Hendry,[237] Gangadharan and Reddy[238] and Reddy and Murphree[239] on the basis of yield line theory. In addition to this work on raft foundations, the effect of differential settlements on brick structures has been analysed by Green et al.,[240] based on the limiting tension finite element formulation described by Zienkiewicz et al.[241]

Probably the simplest method of accounting for non-linear soil behaviour is to carry out a piece-wise linear analysis in which the elastic moduli can be modified at each load increment. Various analyses of this type have been performed by Huang,[242,243] Morgenstern and Phukan,[244] Girijavallabhan and Reese,[245] Radhakrishnan and Reese,[246] Desai and Reese,[247] Duncan and Chang,[248] Majid and Craig,[249] Desai[250] and Domaschuk and Valliappan[251] using finite element methods. They all deal with the two-dimensional problem of strip or circular footings, although Ellison et al.[252] and Desai[253] have analysed a single axially loaded pile.

Plasticity theory has been used by Höeg et al.,[254] Zienkiewicz et al.[255] and Valliappan et al.[256] to investigate the non-linear behaviour of strip footings, and numerical results for a flexible circular foundation on a layer of strain-softening clay have been obtained by Höeg.[257] The effect of plastic yield beneath the edge regions of rigid foundations has been considered by Schultze[258] and Biernatowski[259] for strip footings and by Brown[260] for circular rafts. The case of a symmetrically loaded rigid strip footing in frictionless contact with a deep stratum of purely cohesive soil whose undrained shear strength c_u increases linearly with depth has been investigated by Davis and Booker.[261] Distributions of computed contact pressure p at failure are shown in Fig. 10 for different degrees of strength heterogeneity; these contrast sharply with the familiar saddle-shaped curves obtained for perfectly elastic materials—except in the special case of zero surface modulus—and, where the applied load is distributed uniformly, imply bending moments that are hogging rather than sagging.

A quite different approach to the strip footing problem has been followed by Naylor and Zienkiewicz[262] in using a critical state model for the clay. This model allows for progressive yield and apparently can be readily constructed from conventional soil test data. Ohta and Hata[263] have proceeded along similar lines, although their theoretical soil model also incorporates the consolidation equations.

Many other forms of non-linear soil analysis have been developed. The theoretical aspects of secondary compression and creep have been discussed by Freudenthal and Spillers,[264] Garlanger,[265] Mesri,[266] Komamura and Huang[267] and others, while Gibson et al.,[268] Selvadurai[269] and Carter et al.[270] have considered various large-deformation problems. Visco-elastic models have been used by Soydemir and Schmid,[271] Christian and Watt,[272] Ramaswamy and Vaidyanathan,[273] Booker and Small[274] and Brown[275] to study the creep response of footings and by Booker and Poulos[276] and Ottaviani and Cappellari[277] to investigate the long-term behaviour of piles. A theory of elasto-plastic consolidation has been described by Small et al.,[278] and the formulation of a visco-plastic soil model, together with its application to the strip footing problem, has been presented by Zienkiewicz et al.[279] and Saran et al.[280]

In principle, therefore, it appears that virtually any type of non-linear problem can now be analysed. Nevertheless, the difficulties in obtaining the necessary soil parameters are very considerable, and severely limit the practical application of the more sophisticated methods of analysis. In this connection, the results for circular and strip foundations reported by Huang,[242,243] Morgenstern and Phukan,[244] Höeg et al.[254] and Naylor and Zienkiewicz[262] based on different soil models all indicate that vertical stresses are insensitive to the degree of non-linearity, so that satisfactory results will be obtained for many foundation problems using the simpler methods of linear elastic analysis.

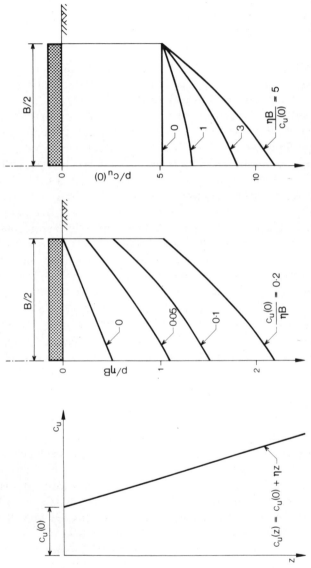

Fig. 10. Contact pressure distribution at failure for rigid strip footing, after Davis and Booker.[261]

5.8   INFLUENCE OF SUPERSTRUCTURE STIFFNESS

The discussion thus far has concentrated on the analysis of plain and piled foundations subject to various forms of applied loading, with little attention given to the effect of superstructure stiffness on the results of any such analysis. Yet this effect can often be significant, and in some cases dominates the foundation response. There is little doubt, however, that the most difficult aspect of interactive analysis is one of achieving an adequate representation of the superstructure, as only in the very simplest cases can the complete structure be modelled.

There are occasions, of course, where the superstructure stiffness is so low that it can be neglected. This is usually the case for steel storage tanks; see, for example, Burland et al.[281] A similar feature was encountered by Hooper[152] in the foundation analysis of a low-rise concrete-framed office building with open-plan floors and a large plan area.

If the basic structure consists essentially of a series of plane walls or frames on isolated footings, with comparatively few interconnecting members, then it may be realistic to carry out a two-dimensional interactive analysis. In this connection, finite element results have been obtained by Smith et al.[282] for a plane unperforated shear wall founded on a piled strip footing, using a Winkler spring modelling of the piles. Analytical solutions for plane coupled shear walls supported on a Winkler foundation have been given by Coull,[283,284] Tso,[285] Tso and Chan[286] and Arvidsson,[287] based on the standard method of representing the wall openings and connecting beams by an equivalent continuous medium. A rather more satisfactory approach to this problem might be to represent the wall by finite elements and to model the soil as an elastic continuum, either by finite elements or by separately computing the appropriate soil stiffness matrix.

Results obtained by Chandrasekaran and Khedkar[288] from a linear elastic analysis based on the latter method are shown diagrammatically in Fig. 11. They relate to a symmetrical six-storey shear wall of uniform thickness, subjected to a typical distribution of vertical loading and wind loading, and founded on isolated footings in contact with a homogeneous half-space. Also shown are the results of an independent analysis based on completely restrained footings. In this example, the effect of soil–structure interaction is to increase lateral wall movements and maximum beam moments by factors of 2·6 and 1·4 respectively.

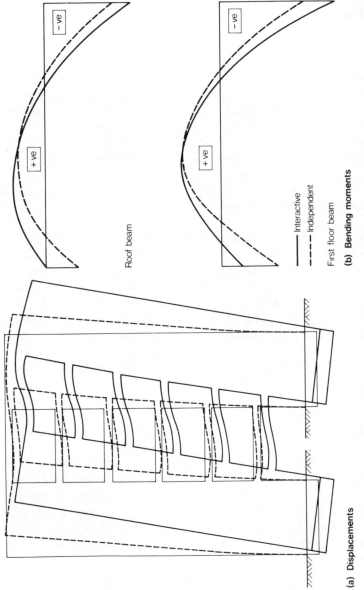

Roof beam

First floor beam

—— Interactive
---- Independent

(b) Bending moments

(a) Displacements

FIG. 11.   Comparative results of shear wall analysis, after Chandrasekaran and Khedkar.[288]

There have been many attempts to analyse the effect of column settlements on open plane frames. Meyerhof[289] proposed an approximate method based on empirical load–settlement curves for footings of various sizes. A more rigorous iterative method was proposed by Chamecki[290] for any skeletal frame on isolated footings, and a computer program based on this method has been described by Larnach.[291] Related methods of analysis have also been reported by Yokoo and Yamagata,[292–294] Sommer[295] and Litton and Buston,[296] and several other aspects of plane frame analysis have been discussed by Grasshof,[297] Sved and Kwok,[298] Getzler,[299] Heil[300] and Chamecki.[301] Further analyses of plane frames supported either on Winkler springs or an elastic continuum have been described by Lee and Harrison,[302] Lee and Brown,[303] Seetharamulu and Kumar,[304] King and Chandrasekaran,[305] Poulos,[306] Brown,[307] Wood and Larnach,[308] Miyahara and Ergatoudis[309] and Ungureanu et al.[310] Almost all these analyses deal with comparatively small open frames; for larger frames, especially those with infill panels or external cladding, it would usually be necessary to analyse some form of equivalent structure, along the lines proposed by Meyerhof.[311]

Various methods of analysing open space frames supported on isolated footings have been described by Morris,[312] De Jong and Morgenstern,[313] Majid and Cunnell,[314] Buragohain et al.,[315] Jain et al.,[316] and Wood et al.[317] Where the frame is supported by a raft foundation, solution methods have been discussed by Haddadin,[318] King and Chandrasekaran,[319,320] Hain and Lee,[321] Fraser and Wardle[322] and Wardle and Fraser.[323] Once again, the emphasis in all these studies is on small open frames; even then, the problem size in terms of computer storage and solution times can be very considerable.

Results obtained by Wardle and Fraser,[323] from an elastic analysis of a rafted space frame, demonstrate several noteworthy features of soil–structure interaction, particularly with respect to the effect of superstructure stiffness. Figure 12 summarises the structural details of the raft and the open seven-storey superstructure, and the soil is taken to be an elastic layer of depth equal to the raft width. Alternative ways of modelling the superstructure are also shown. In case 1, the superstructure stiffness is taken as zero and the column loads calculated on the basis of rigid column supports are applied directly to the raft. Case 2 relates to an interactive analysis of the complete superstructure, and in case 3 the superstructure is condensed to a single-storey frame using the method described by Meyerhof.[311] In

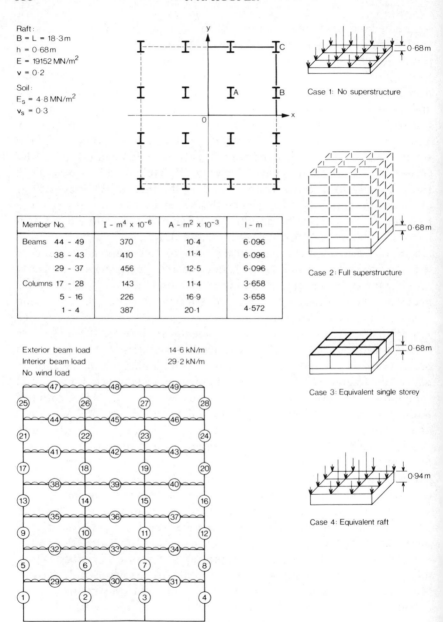

FIG. 12. Results of interactive space frame analysis, after Wardle and Fraser.[323]

TABLE 1

RESULTS OF INTERACTIVE SPACE FRAME ANALYSIS, AFTER WARDLE AND FRASER[323]

Case no.	h (m)	K_r	Raft settlement (mm)			Column load (kN)			Raft bending moment (kNm/m)	
			ρ^*	$\Delta\rho_{ox}$	$\Delta\rho_{oc}$	A	B	C	Centre	Column A
1	0·68	0·26	132	14·6	30·9	2490	1245	622	244	396
			(140)	(23·0)	(47·7)				(421)	(397)
2	0·68	0·26	124	7·6	13·7	2150	1278	896	119	250
3	0·68	0·26	124	8·0	14·8	2185	1278	860	126	258
4	0·94	0·68	124	7·1	15·5	2490	1245	622	332	481
			(128)	(11·2)	(23·8)				(554)	(520)

Figures in brackets are for uniformly loaded raft.

case 4, the superstructure is fully condensed to give a thicker equivalent raft, which has the effect of increasing the relative raft stiffness K_r, defined by eqn. (5), from 0·26 to 0·68.

The results of the four analyses are tabulated in Table 1. In each case the maximum total settlements ρ^* are similar, but the differential settlements $\Delta\rho$ of the 0·68 m-thick raft with no superstructure (case 1) are approximately double the values obtained when superstructure stiffness is taken into account. This leads to corresponding reductions in raft bending moment, but only at the expense of substantially increased loads in the corner columns and, by implication, increased bending moments in the beams and columns of the corner bays. In practice, however, the extent of this load transfer might be considerably reduced due to the settlements that inevitably occur during construction and to creep deformations in the structural members.

With regard to core-column structures of at least moderate height supported on a raft, the bending moments and shear forces are usually large enough to call for a comparatively thick raft. In these circumstances, it is likely that the raft itself will be dominant in controlling differential settlements, especially across the short spans, with little additional contribution from the superstructure. Moreover, this contribution may decrease quite rapidly with increasing building height as a result of column shortening. However, where the core and columns are supported on separate foundations, the influence of superstructure stiffness generally becomes more significant. In such cases, some form of interactive frame analysis may be appropriate.

Any rigorous foundation analysis of cross-wall structures with essentially orthogonal primary walls seems virtually intractable, but a useful approximate method is to reduce the wall structure to an equivalent raft and solve for the latter. The effect of the computed differential settlements on the wall structure can then be assessed, and the procedure repeated if the resulting redistribution of structural load is significant. An alternative approach is to specify zero rotations in the plane of the walls at foundation level; this is particularly useful in giving an upper bound to the effect of superstructure stiffness. In this connection, the results of interactive analyses of cross-wall structures, supported either on plain raft or piled raft foundations, have been reported by Klepikov et al.,[324] Hooper[205] and Hooper and Wood.[15,203]

Results obtained by Hooper and Wood[15] from an elastic analysis based on the equivalent raft concept are shown in Fig. 13. The building is a 22-storey residential block of cross-wall construction founded on a 0·76 m-thick raft of asymmetric plan shape. The soil succession is gravel (5 m), London clay (25 m), Woolwich and Reading Beds (15 m), Thanet Sands (12 m) and Chalk, and the analysis was based on a plain raft foundation at the surface of a modified half-space continuum. The plotted raft settlement profiles relate to the total building weight of 112 MN, which was taken to be a uniformly distributed load of 246 kN/m².

In Fig. 13(c), the computed curves of total raft settlement corresponding to undrained and fully drained soil conditions form an envelope to the major part of the measured time-settlement curve. It is also clear from Fig. 13(d) that the computed differential settlements are completely dominated by the assumed degree of superstructure stiffness. The very pronounced dishing of the settlement profile for the case of completely flexible loading (68 mm between nodes 17 and 21, drained analysis) is only slightly modified by the presence of the 0.76 m-thick unrestrained raft; this is consistent with the low value of relative raft stiffness. Computed differential settlements are considerably reduced for the case where rotational restraints are applied to the 0·76 m-thick raft, and are further reduced to very low values (5 mm between nodes 17 and 21, drained analysis), similar to those actually measured, for the case of an unrestrained 4·6 m-thick raft. The derivation of this equivalent raft thickness was based on the combined bending stiffness of the first two storeys, assuming complete shear connections between floors. The small values of computed foundation tilt stem from the asymmetric plan shape of the uniformly loaded raft, together with the high rigidity of the superstructure.

These results indicate the extent to which differential settlements can be influenced by superstructure stiffness; for this particular building measured differential settlements were approximately one-tenth of those computed for a raft foundation with no contribution from the superstructure. It should be emphasised, however, that the associated reductions in differential raft settlements and bending moments will be accompanied by increased wall stresses. Such increases may be very considerable in the outer regions of the walls at foundation level, and hence should be allowed for in design.

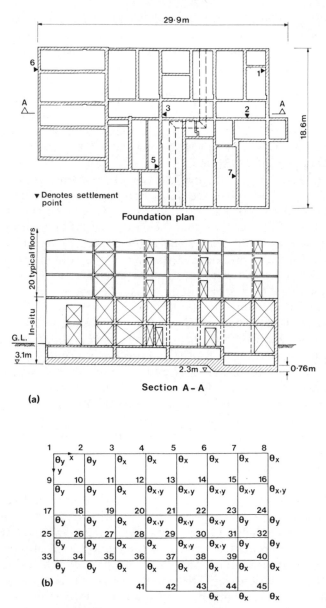

FIG. 13.   Foundation settlements of a cross-wall structure, after Hooper and Wood.[15] (a) Details of structure. (b) Raft mesh and nodal restraints ($\theta_{x,y}$ denotes zero rotation about x,y axes).

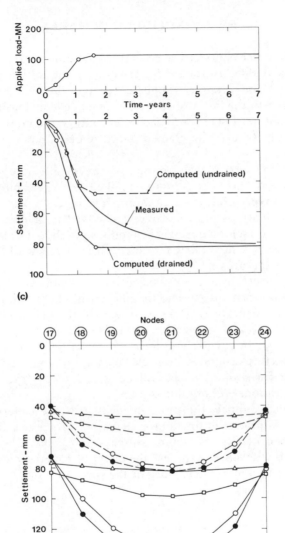

FIG. 13—*cont.* (c) Total raft settlements. (d) Computed settlement profiles; key: ● no raft, ○ 0·76 m raft (unrestrained), □ 0·76 m raft (restrained), △ 4·6 m raft (unrestrained), ----- undrained, ——— drained.

5.9 CONCLUDING REMARKS

Perhaps the most beneficial aspect of foundation interaction analysis is that of providing a rational framework for the design of plain raft foundations. The relative simplicity of the raft-on-continuum approach suggests that its use in design practice could well become a matter of routine. This in turn should encourage the more widespread use of plain rafts, often in place of more complex and costly types of foundation.

Although the analysis and design of single piles is comparatively straightforward, this is far from being the case where pile groups are concerned. In principle it is now possible to carry out fully three-dimensional analyses of pile groups embedded in layered soil media, but the cost of such analyses is generally so high as to preclude their application to practical design problems associated with the majority of building structures.

These rigorous methods of analysis are probably of greatest value when used to carry out parametric and sensitivity studies of different pile group configurations. This also applies to the various sophisticated analytical and numerical methods relating to plain or piled foundation systems with non-linear material properties. For most building structures founded on piles, some approximate and relatively inexpensive form of analysis will be adequate for the purpose of achieving a rational foundation design.

In a similar manner, the direct modelling of the superstructure is completely impractical in almost all cases. However, a simplified representation of superstructure stiffness can usually be made that will enable the designer to assess the effect of foundation movements on the integrity and serviceability of the superstructure.

REFERENCES

1. HETÉNYI, M. (1946). *Beams on Elastic Foundations*, University of Michigan Press, Ann Arbor.
2. TIMOSHENKO, S. P. and WOINOWSKY-KRIEGER, S. (1959). *Theory of Plates and Shells*, McGraw-Hill, New York.
3. KERR, A. D. (1964). Elastic and viscoelastic foundation models, *J. Appl. Mech., Trans. ASME*, **31**, 491–498.

4. GIBSON, R. E. (1967). Some results concerning displacements and stresses in a non-homogeneous elastic half-space, *Géotechnique*, **17**(1), 58–67.
5. GIBSON, R. E. and KALSI, G. S. (1974). The surface settlement of a linearly inhomogeneous cross-anisotropic elastic half-space, *Z. Angew. Math. Phys.*, **25**, 843–847.
6. GIBSON, R. E. and SILLS, G. C. (1975). Settlement of a strip load on a non-homogeneous orthotropic incompressible elastic half-space, *Quart. J. Mech. Appl. Math.*, **28**(2), 233–243.
7. STEINBRENNER, W. (1934). Tafeln zur Setzungsberechnung, *Die Strasse*, **1**, 121–124.
8. TERZAGHI, K. (1943). *Theoretical Soil Mechanics*, Wiley, New York.
9. DAVIS, E. H. and POULOS, H. G. (1968). The use of elastic theory for settlement prediction under three-dimensional conditions, *Géotechnique*, **18**(1), 67–91.
10. HOOPER, J. A. (1975). Elastic settlement of a circular raft in adhesive contact with a transversely isotropic medium, *Géotechnique*, **25**(4), 691–711.
11. PEUTZ, M. G. F., VAN KEMPEN, H. P. M. and JONES, A. (1968). Layered systems under normal surface loads, *Highway Res. Rec. No. 228*, 34–45.
12. GERRARD, C. M. and HARRISON, W. J. (1971). The analysis of a loaded half space comprised of anisotropic layers, *Tech. Paper No. 10*, Div. Appl. Geomech., CSIRO, Australia.
13. POULOS, H. G. (1967). Stresses and displacements in an elastic layer underlain by a rough rigid base, *Géotechnique*, **17**(4), 378–410.
14. UESHITA, K. and MEYERHOF, G. G. (1968). Surface displacement of an elastic layer under uniformly distributed loads, *Highway Res. Rec. No. 228*, 1–10.
15. HOOPER, J. A. and WOOD, L. A. (1976). Foundation analysis of a cross-wall structure, *Proc. Int. Conf. Performance Blg. Struct.*, Glasgow, 1976, **1**, 229–248.
16. GALLETLY, G. D. (1959). Optimum design of thin circular plates on an elastic foundation, *Proc. Instn. Mech. Engrs.*, **173**(27), 687–696.
17. FREDERICK, D. (1959). On some problems in bending of thick circular plates on an elastic foundation, *J. Appl. Mech., Trans. ASME*, **78**, 195–200.
18. SINHA, S. N. (1963). Large deflections of plates on elastic foundations, *J. Engng. Mech. Div., Proc. ASCE*, **89**(EM1), 1–24.
19. BOLTON, R. (1972). Stresses in circular plates on elastic foundations, *J. Engng. Mech. Div., Proc. ASCE*, **98**(EM3), 629–640.
20. LEONARDS, G. A. and HARR, M. E. (1959). Analysis of concrete slabs on ground, *J. Soil Mech. Fdns. Div., Proc. ASCE*, **85**(SM3), 35–58.
21. VOLTERRA, E. (1952). Bending of a circular beam resting on an elastic foundation, *J. Appl. Mech., Trans. ASME*, **19**, 1–4.
22. REISMANN, H. (1954). Bending of circular and ring-shaped plates on an elastic foundation, *J. Appl. Mech., Trans. ASME*, **21**, 1–7.
23. VLASOV, V. Z. and LEONT'EV, N. N. (1966). *Beams, Plates and Shells on Elastic Foundations*, Israel Prog. Sci. Translations, Jerusalem.

24. RADES, M. (1971). Dynamic response of a rigid circular plate on a Kerr-type foundation model, *Int. J. Engng. Sci.*, **9**, 1061–1073.

25. GIROUD, J. P. (1972). *Tables pour le Calcul des Fondations*, Dunod, Paris, **1**.

26. POULOS, H. G. and DAVIS, E. H. (1974). *Elastic Solutions for Soil and Rock Mechanics*, Wiley, New York.

27. SNEDDON, I. N. (1946). Boussinesq's problem for a flat-ended cylinder, *Proc. Camb. Phil. Soc.*, **42**(1), 29–39.

28. SCHIFFMAN, R. L. and AGGARWALA, B. D. (1961). Stresses and displacements produced in a semi-infinite elastic solid by a rigid elliptical footing, *Proc. 5th Int. Conf. SMFE*, Paris, 1961, **1**, 795–801.

29. CHAN, H. C. and CHEUNG, Y. K. (1974). Contact pressure of rigid footings on elastic foundations, *Civ. Engng.*, 51–59.

30. MOSSAKOVSKI, V. I. (1954). The fundamental general problem of the theory of elasticity for a half-space with a circular curve determining boundary conditions, *Prikl. Mat. Mekh.*, **18**, 187–196 (in Russian).

31. KEER, L. M. (1967). Mixed boundary-value problems for an elastic half-space, *Proc. Camb. Phil. Soc.*, **63**, 1379–1386.

32. SPENCE, D. A. (1968). Self similar solutions to adhesive contact problems with incremental loading, *Proc. Roy. Soc.*, **A305**, 55–80.

33. POPOVA, O. V. (1972). Stress and displacement distributions in a homogeneous half-space below a circular foundation, *Soil Mech. Fdn. Engng.*, **9**(2), 86–89.

34. EGOROV, K. E., BARVASHOV, V. A. and FEDOROVSKY, V. G. (1973). Some applications of the elasticity theory to design of foundations, *Proc. 8th Int. Conf. SMFE*, Moscow, 1973, **1**(3), 69–74.

35. SPENCE, D. A. (1975). The Hertz contact problem with finite friction, *J. Elasticity*, **5**(3), 297–319.

36. KHADEM, R. and O'CONNOR, J. J. (1969). Axial compression of an elastic circular cylinder in contact with two identical elastic half-spaces, *Int. J. Engng. Sci.*, **7**, 785–800.

37. ZLATIN, A. N. and UFLIAND, I. S. (1976). Axisymmetric contact problem on the impression of an elastic cylinder into an elastic layer, *J. Appl. Math. Mech.*, **40**(1), 67–72.

38. GERRARD, C. M. and HARRISON, W. J. (1970). Circular loads applied to a cross anisotropic half-space, *Tech. Paper No. 8*, Div. Appl. Geomech., CSIRO, Australia.

39. GERRARD, C. M. and WARDLE, L. J. (1973). Solutions for point loads and generalized circular loads applied to a cross-anisotropic half space, *Tech. Paper No. 13*, Div. Appl. Geomech., CSIRO, Australia.

40. CARRIER, W. D. and CHRISTIAN, J. T. (1973). Rigid circular plate resting on a non-homogeneous elastic half-space, *Géotechnique*, **23**(1), 67–84.

41. EGOROV, K. E. and NICHIPOROVICH, A. A. (1961). Research on the deflexion of foundations, *Proc. 5th Int. Conf. SMFE*, Paris, 1961, **1**, 861–866.

42. ENGLAND, A. H. (1962). A punch problem for a transversely isotropic layer, *Proc. Camb. Phil. Soc.*, **58**(3), 539–547.

43. EGOROV, K. E. and SEREBRJANYI, R. V. (1963). Determination of stresses in rigid circular foundation, *Proc. 2nd Asian Reg. Conf. SMFE*, Japan, 1963, **1**, 246–250.

44. POULOS, H. G. (1968). The behaviour of a rigid circular plate resting on a finite elastic layer, *Civ. Engng. Trans. Instn. Engrs.*, Australia, **CE10**(2), 213–219.

45. MILOVIC, D. M. (1972). Stresses and displacements in an anisotropic layer due to a rigid circular foundation, *Géotechnique*, **22**(1), 169–174.

46. DHALIWAL, R. S. (1970). Punch problem for an elastic layer overlying an elastic foundation, *Int. J. Engng. Sci.*, **8**(4), 273–288.

47. SHELEST, L. A. (1975). Distribution of stresses and displacements in a foundation bed of finite thickness beneath a circular rigid foundation, *Soil Mech. Fdn. Engng.*, **12**(6), 404–407.

48. CHEN, W. T. and ENGEL, P. A. (1972). Impact and contact stress analysis in multilayer media, *Int. J. Solids Struct.*, **8**(11), 1257–1281.

49. BUTTERFIELD, R. and BANERJEE, P. K. (1971). A rigid disc embedded in an elastic half-space, *Geotech. Engng.*, **2**(1), 35–52.

50. BANERJEE, P. K. (1971). Foundations within a finite elastic layer, *Civ. Engng. Publ. Wks. Rev.*, **66**, 1197–1202.

51. EGOROV, K. E. (1965). Calculation of bed for foundation with ring footing, *Proc. 6th Int. Conf. SMFE*, Montreal, 1965, **2**, 41–45.

52. VALOV, G. M. (1968). Infinite elastic layer and half-space under the action of a ring-shaped die, *J. Appl. Math. Mech.*, **32**(5), 917–930.

53. SHIBUYA, T., KOIZUMI, T. and NAKAHARA, I. (1974). An elastic contact problem for a half-space indented by a flat annular rigid stamp, *Int. J. Engng. Sci.*, **12**(9), 759–771.

54. DHALIWAL, R. S. and SINGH, B. M. (1977). Annular punch on an elastic layer overlying an elastic foundation, *Int. J. Engng. Sci.*, **15**(4), 263–270.

55. BROWN, P. T. (1969). Numerical analyses of uniformly loaded circular rafts on elastic layers of finite depth, *Géotechnique*, **19**(2), 301–306.

56. BROWN, P. T. (1969). Numerical analyses of uniformly loaded circular rafts on deep elastic foundations, *Géotechnique*, **19**(3), 399–404.

57. BROWN, P. T. (1974). Influence of soil inhomogeneity on raft behaviour, *Soils Fdns.*, **14**(1), 61–70.

58. HOOPER, J. A. (1974). Analysis of a circular raft in adhesive contact with a thick elastic layer, *Géotechnique*, **24**(4), 561–580.

59. ZIENKIEWICZ, O. C. (1971). *The Finite Element Method in Engineering Science*, McGraw-Hill, London.

60. HOOPER, J. A. (1976). Parabolic adhesive loading of a flexible raft foundation, *Géotechnique*, **26**(3), 511–525.

61. BURAGOHAIN, D. N. and SHAH, V. L. (1977). Curved interface elements for interaction problems, *Proc. Int. Symp. Soil Struct. Interaction*, Roorkee, 1977, 197–201.

62. VARMA, B. S., KHADILKAR, B. S. and CHANDRASEKARAN, V. S. (1977). Circular footings on transversely isotropic soils, *Proc. Int. Symp. Soil Struct. Interaction*, Roorkee, 1977, 265–270.

63. WILSON, E. L. (1965). Structural analysis of axisymmetric solids, *J. Am. Inst. Aero. Astro.*, **3**(12), 2269–2274.

64. BOROWICKA, H. (1936). Influence of rigidity of a circular foundation slab on the distribution of pressures over the contact surface, *Proc. 1st Int. Conf. SMFE*, 1936, Cambridge, Mass., **2**, 144–149.

65. BOROWICKA, H. (1939). Druckverteilung unter elastischen Platten, *Ingenieur-Archiv.*, **10**(2), 113–125.

66. BOROWICKA, H. (1943). Über ausmittig belastete, starre Platten auf elastisch-isotropem Untergrund, *Ingenieur-Archiv.*, **14**(1), 1–8.

67. GLADWELL, G. M. L. and IYER, K. R. P. (1974). Unbonded contact between a circular plate and an elastic half-space, *J. Elasticity*, **4**(2), 115–130.

68. PICKETT, G. and McCORMICK, F. J. (1951). Circular and rectangular plates under lateral load and supported by an elastic solid foundation, *Proc. 1st US Nat. Congr. Appl. Mech.*, ASME, Chicago, 1951, 331–338.

69. SMITH, I. M. (1970). A finite element approach to elastic soil–structure interaction, *Canad. Geotech. J.*, **7**(2), 95–105.

70. BOSWELL, L. F. and SCOTT, C. R. (1975). A flexible circular plate on a heterogeneous elastic half-space: influence coefficients for contact stress and settlement, *Géotechnique*, **25**(3), 604–610.

71. ZBIROHOWSKI-KOSCIA, K. F. and GUNASEKERA, D. A. (1970). Foundation settlement and ground reaction calculations using a digital computer, *Civ. Engng. Publ. Wks. Rev.*, **65**, 152–157.

72. CHAKRAVORTY, A. K. and GHOSH, A. (1975). Finite difference solution for circular plates on elastic foundations, *Int. J. Num. Meth. Engng.*, **9**(1), 73–84.

73. BANERJEE, A. and JANKOV, Z. D. (1975). Circular mats under arbitrary loading, *J. Struct. Div.*, *Proc. ASCE*, **101**(ST10), 2133–2145.

74. GIBSON, R. E. and LUMB, P. (1953). Numerical solution of some problems in the consolidation of clay, *Proc. Instn. Civ. Engrs.*, **2**(1), 182–198.

75. DAVIS, E. H. and POULOS, H. G. (1972). Rate of settlement under two- and three-dimensional conditions, *Géotechnique*, **22**(1), 95–114.

76. DE JONG, G. J. (1957). Application of stress functions to consolidation problems, *Proc. 4th. Int. Conf. SMFE*, London, 1957, **1**, 320–323.

77. MANDEL, J. (1957). Consolidation des couches d'argiles, *Proc. 4th Int. Conf. SMFE*, London, 1957, **1**, 360–367.

78. MANDEL, J. (1961). Tassements produits par la consolidation d'une couche d'argile de grande épaisseur, *Proc. 5th Int. Conf. SMFE*, Paris, 1961, **1**, 733–736.

79. PARIA, G. (1957). Axisymmetric consolidation for a porous elastic material containing a fluid, *J. Maths. Phys.*, **36**, 338–346.

80. McNAMEE, J. and GIBSON, R. E. (1960). Plane strain and axially symmetric problems of the consolidation of a semi-infinite clay stratum, *Quart. J. Mech. Appl. Math.*, **13**(2), 210–227.

81. GIBSON, R. E., SCHIFFMAN, R. L. and PU, S. L. (1970). Plane strain and axially symmetric consolidation of a clay layer on a smooth impervious base, *Quart. J. Mech. Appl. Math.*, **23**(4), 505–520.

82. BIOT, M. A. (1941). General theory of three-dimensional consolidation, *J. Appl. Phys.*, **12**(2), 155–164.

83. AGBEZUGE, L. K. and DERESIEWICZ, H. (1975). The consolidation settlement of a circular footing, *Israel J. Tech.*, **13**, 264–269.
84. CHIARELLA, C. and BOOKER, J. R. (1975). The time-settlement behaviour of a rigid die resting on a deep clay layer, *Quart. J. Mech. Appl. Math.*, **28**(3), 317–328.
85. SANDHU, R. S. and WILSON, E. L. (1969). Finite element analysis of seepage in elastic media, *J. Engng. Mech. Div., Proc. ASCE*, **95**(EM3), 641–652.
86. CHRISTIAN, J. T. and BOEHMER, J. W. (1970). Plane strain consolidation by finite elements, *J. Soil Mech. Fdns. Div., Proc. ASCE*, **96**(SM4), 1435–1457.
87. HWANG, C. T., MORGENSTERN, N. R. and MURRAY, D. W. (1971). On solutions of plane strain consolidation problems by finite element methods, *Canad. Geotech. J.*, **8**(1), 109–118.
88. HWANG, C. T., MORGENSTERN, N. R. and MURRAY, D. W. (1972). Application of the finite element method to consolidation problems, *Proc. Symp. Appl. FEM Geotech. Engng.*, Vicksburg, Mississippi, 1972, **2**, 739–765.
89. YOKOO, Y., YAMAGATA, K. and NAGAOKA, H. (1971). Finite element method applied to Biot's consolidation theory, *Soils Fdns.*, **11**(1), 29–46.
90. YOKOO, Y., YAMAGATA, K. and NAGAOKA, H. (1971). Variational principles for consolidation, *Soils Fdns.*, **11**(4), 25–35.
91. YOKOO, Y., YAMAGATA, K. and NAGAOKA, H. (1971). Finite element analysis of consolidation following undrained deformation, *Soils Fdns.*, **11**(4), 37–58.
92. BOOKER, J. R. (1973). A numerical method for the solution of Biot's consolidation theory, *Quart. J. Mech. Appl. Math.*, **26**(4), 457–470.
93. GHABOUSSI, J. and WILSON, E. L. (1973). Flow of compressible fluid in porous elastic media, *Int. J. Num. Meth. Engng.*, **5**, 419–442.
94. VALLIAPPAN, S. and LEE, I. K. (1975). Consolidation of non-homogeneous anisotropic layered soil media, *Proc. 2nd Australia–NZ Conf. Geomech.*, Brisbane, 1975, 67–71.
95. SMITH, I. M. and HOBBS, R. (1976). Biot analysis of consolidation beneath embankments, *Géotechnique*, **26**(1), 149–171.
96. WESTERGAARD, H. M. (1925). Computation of stresses in concrete roads, *Proc. Highway Res. Bd.*, **5**(1), 90–112.
97. WESTERGAARD, H. M. (1926). Stresses in concrete pavements computed by theoretical analysis, *Public Roads*, **7**(2), 25–35.
98. WESTERGAARD, H. M. (1933). Analytical tools for judging results of structural tests of concrete pavements, *Public Roads*, **14**(10), 185–188.
99. WESTERGAARD, H. M. (1939). Stresses in concrete runways of airports, *Proc. Highway Res. Bd.*, **19**, 197–202.
100. WESTERGAARD, H. M. (1948). New formulas for stresses in concrete pavements of airfields, *Trans. ASCE*, **113**, 425–439.
101. WYMAN, M. (1950). Deflections of an infinite plate, *Canad. J. Res.*, **28**(Sec. A), 293–302.

200 J. A. HOOPER

102. NAGHDI, P. M. and ROWLEY, J. C. (1953). On the bending of axially symmetric plates on elastic foundations, *Proc. 1st Midwestern Conf. Solid Mech.*, Univ. of Illinois, 1953, 119–123.
103. LIVESLEY, R. K. (1953). Some notes on the mathematical theory of a loaded elastic plate resting on an elastic foundation, *Quart. J. Mech. Appl. Math.*, **6**(1), 32–44.
104. CAMPBELL, J. E. and HEAPS, H. S. (1955). Transmission of stress through a thick slab supported by a yielding foundation, *Canad. J. Tech.*, **33**(5), 324–334.
105. REISSNER, E. (1957). Stresses in elastic plates over flexible subgrades, *Trans. ASCE*, **122**, 627–653.
106. BROWN, C. B., LAURENT, J. M. and TILTON, J. R. (1977). Beam-plate system on Winkler foundation, *J. Engng. Mech. Div., Proc. ASCE*, **103**(EM4), 589–600.
107. HOGG, A. H. A. (1938). Equilibrium of a thin plate, symmetrically loaded, resting on an elastic foundation of infinite depth, *Phil. Mag.*, 7th Ser., **25**, 576–582.
108. HOLL, D. L. (1938). Thin plates on elastic foundations, *Proc. 5th Int. Congr. Appl. Mech.*, Cambridge, Mass, 1938, 71–74.
109. PICKETT, G. and RAY, G. K. (1951). Influence charts for concrete pavements, *Trans. ASCE*, **116**, 49–73.
110. RICHART, F. E. and ZIA, P. (1962). Effect of local loss of support on foundation design, *J. Soil. Mech. Fdns. Div., Proc. ASCE*, **88**(SM1), 1–27.
111. WEITSMAN, Y. (1969). On the unbonded contact between plates and an elastic half space, *J. Appl. Mech., Trans. ASME*, **36**, 198–202.
112. PU, S. L. and HUSSAIN, M. A. (1970). Note on the unbonded contact between plates and an elastic half space, *J. Appl. Mech., Trans. ASME*, **37**, 859–861.
113. KEER, L. M., DUNDURS, J. and TSAI, K. C. (1972). Problems involving a receding contact between a layer and a half space, *J. Appl. Mech., Trans. ASME*, **39**, 1115–1120.
114. PARKES, E. W. (1956). A comparison of the contact pressures beneath rough and smooth rafts on an elastic medium, *Géotechnique*, **6**(4), 183–189.
115. HOGG, A. H. A. (1944). Equilibrium of a thin slab on an elastic foundation of finite depth, *Phil. Mag.*, 7th Ser., **35**, 265–276.
116. VINT, J. and ELGOOD, W. N. (1935). The deformation of a bloom plate resting on an elastic base when a load is transmitted to the plate by means of a stanchion, *Phil. Mag.*, 7th Ser., **19**, 1–21.
117. MURPHY, G. (1937). Stresses and deflections in loaded rectangular plates on elastic foundations, *Iowa Engng. Exp. Stn. Bull. No. 135.*
118. FLETCHER, H. J. and THORNE, C. J. (1952). Thin rectangular plates on elastic foundation, *J. Appl. Mech., Trans. ASME*, **19**, 361–368.
119. MACKEY, S. and CHUNG, T. K. (1969). Considerations in the analysis of raft foundations, *Civ. Engng. Publ. Wks. Rev.*, **64**, 877–885.
120. EWELL, W. W. and OKUBO, S. (1951). Deflections in slabs on elastic foundations, *Proc. Highway Res. Bd.*, **30**, 125–133.

121. HUDSON, W. R. and MATLOCK, H. (1966). Analysis of discontinuous orthotropic pavement slabs subjected to combined loads, *Highway Res. Rec. No. 131*, 1–48.
122. HUDSON, W. R. and STELZER, C. F. (1968). A direct computer solution for slabs on foundation, *J. Am. Concr. Inst.*, **65**(3), 188–200.
123. PANAK, J. J., FOWLER, D. W. and MATLOCK, H. (1972). Slab foundation subjected to complex loads, *J. Am. Concr. Inst.*, **69**(10), 630–637.
124. PANAK, J. J. and RAUHUT, J. B. (1975). Behaviour and design of industrial slabs on grade, *J. Am. Concr. Inst.*, **72**(5), 219–224.
125. ALLEN, D. N. DE G. and SEVERN, R. T. (1960–63). The stresses in foundation rafts, *Proc. Instn. Civ. Engrs.*, **15**, 35–48 (Pt. 1); **20**, 293–304 (Pt. 2); **25**, 257–266 (Pt. 3).
126. CHEUNG, Y. K. and ZIENKIEWICZ, O. C. (1965). Plates and tanks on elastic foundations—an application of finite element method, *Int. J. Solids Struct.*, **1**, 451–461.
127. SEVERN, R. T. (1966). The solution of foundation mat problems by finite element methods, *Struct. Engr.*, **44**(6), 223–228.
128. WINTER, E. (1974). Calculated and measured settlements of a mat foundation in Arlington, Va., USA, *Proc. Conf. Settlement Struct.*, Cambridge, 1974, 451–459.
129. WEGMULLER, A. W., WOODWARD, J. H. and BAYLISS, J. R. (1976). Design of large paper machine mat foundation, *J. Struct. Div., Proc. ASCE*, **102**(ST1), 231–250.
130. HUANG, Y. H. and WANG, S. T. (1973). Finite-element analysis of concrete slabs and its implications for rigid pavement design, *Highway Res. Rec. No. 446*, 55–69.
131. FREDERICK, D. (1957). Thick rectangular plates on an elastic foundation, *Trans. ASCE*, **122**, 1069–1085.
132. YANG, T. Y. (1970). Flexible plate finite element on elastic foundation, *J. Struct. Div., Proc. ASCE*, **96**(ST10), 2083–2101.
133. YANG, T. Y. (1972). A finite element analysis of plates on a two-parameter foundation model, *Computers Structures*, **2**, 593–614.
134. JONES, R. and XENOPHONTOS, J. (1977). The Vlasov foundation model, *Int. J. Mech. Sci.*, **19**(6), 317–323.
135. GORBUNOV-POSSADOV, M. I. and SEREBRJANYI, R. V. (1961). Design of structures on elastic foundations, *Proc. 5th Int. Conf. SMFE*, Paris, 1961, **1**, 643–648.
136. ABSI, E. (1970). Étude de problèmes particuliers, *Annls. L'Inst. Tech. Bâtim. Trav. Publ.*, No. 265, 173–188.
137. BROWN, P. T. (1972). Analysis of rafts on clay, Ph.D. Thesis, University of Sydney.
138. ABSI, E., GARNIER, J. and GIROUD, J. -P. (1976). Contraintes dues à une fondation superficielle rectangulaire et rigide, *Annls. L'Inst. Tech. Bâtim. Trav. Publ.*, No. 338, 25–72.
139. CONWAY, H. D. and FARNHAM, K. A. (1968). The relationship between load and penetration for a rigid, flat-ended punch of arbitrary cross-section, *Int. J. Engng. Sci.*, **6**, 489–496.

140. SOVINC, I. (1969). Displacements and inclinations of rigid footings on a limited elastic layer of uniform thickness, *Proc. 7th Int. Conf. SMFE, Mexico*, 1969, **1**, 385–389.
141. MILOVIC, D. M. and TOURNIER, J. P. (1973). Stresses and displacements due to rigid rectangular foundation on a layer of finite thickness, *Soils Fdns.*, **13**(4), 29–43.
142. PICKETT, G., JANES, W. C., RAVILLE, M. E. and MCCORMICK, F. J. (1951). Deflections, moments and reactive pressures for concrete pavements, *Kansas State Coll. Engng. Exp. Stn. Bull. No. 65*.
143. JANES, R. L. (1962). Digital computer solution for pavement deflections, *J. Portland Cem. Assn.*, **4**(3), 30–36.
144. VALANTAGUL, C., HONGLADAROMP, T. and LEE, S. -L. (1971). Separation of contact surface in bending of slabs on grade, *Proc. 4th Asian Reg. Conf. SMFE*, Bangkok, 1971, 87–94.
145. SAXENA, S. K. (1973). Pavement slabs resting on elastic foundation, *Highway Res. Rec. No. 466*, 163–178.
146. WANG, S. K., SARGIOUS, M. and CHEUNG, Y. K. (1972). Advanced analysis of rigid pavements, *J. Transp. Engng., Proc. ASCE*, **98**(TE1), 37–44.
147. CHEUNG, Y. K. and NAG, D. K. (1968). Plates and beams on elastic foundations—linear and non-linear behaviour, *Géotechnique*, **18**(2), 250–260.
148. WOOD, L. A. and LARNACH, W. J. (1974). The effects of soil–structure interaction on raft foundations, *Proc. Conf. Settlement Struct.*, Cambridge, 1974, 460–470.
149. HAIN, S. J., VALLIAPPAN, S. and LEE, I. K. (1976). Analysis of rafts on non-homogeneous non-linear soil, *Proc. 2nd Int. Conf. FEM Engng.*, Univ. of Adelaide, 1976, 28.1–28.15.
150. WARDLE, L. J. and FRASER, R. A. (1974). Finite element analysis of a plate on a layered cross-anisotropic foundation, *Proc. 1st Int. Conf. FEM Engng.*, Univ. of NSW, 1974, 565–578.
151. FRASER, R. A. and WARDLE, L. J. (1976). Numerical analysis of rectangular rafts on layered foundations, *Géotechnique*, **26**(4), 613–630.
152. HOOPER, J. A. (1977). Unpublished work.
153. SVEC, O. J. and GLADWELL, G. M. L. (1973). A triangular plate bending element for contact problems, *Int. J. Solids Struct.*, **9**(3), 435–446.
154. SVEC, O. J. and MCNEICE, G. M. (1972). Finite element analysis of finite sized plates bonded to an elastic half space, *J. Comp. Meth. Appl. Mech. Engng.*, **1**(3), 265–277.
155. SVEC, O. J. (1974). The unbonded contact problem of a plate on the elastic half space, *J. Comp. Meth. Appl. Mech. Engng.*, **3**, 105–113.
156. SVEC, O. J. (1976). Thick plates on elastic foundations by finite elements, *J. Engng. Mech. Div., Proc. ASCE*, **102**(EM3), 461–477.
157. GIBSON, R. E. and MCNAMEE, J. (1957). The consolidation settlement of a load uniformly distributed over a rectangular area, *Proc. 4th Int. Conf. SMFE*, London, 1957, **1**, 297–299.

158. GIBSON, R. E. and MCNAMEE, J. (1963). A three-dimensional problem of the consolidation of a semi-infinite clay stratum, *Quart. J. Mech. Appl. Math.*, **16**(2), 115–127.
159. BOOKER, J. R. (1974). The consolidation of a finite layer subject to surface loading, *Int. J. Solids Struct.*, **10**(9), 1053–1065.
160. FRASER, D. J. (1969). Beams on elastic foundations. A computer-orientated solution for beams with free ends, *Civ. Engng. Trans. Instn. Engrs. Australia*, **CE11**(1), 25–30.
161. JUST, D. J., STARZEWSKI, K. and RONAN, P. B. (1971). Finite element method of analysis of structures resting on elastic foundations, *Proc. Symp. Interaction Struct. Fdn.*, Birmingham, 1971, 108–117.
162. FRASER, R. A. (1976). Outline of solutions available for the design of raft foundations, *Tech. Paper No. 25*, Div. Appl. Geomech., CSIRO, Australia.
163. LEVINTON, Z. (1949). Elastic foundations analysed by the method of redundant reactions, *Trans. ASCE*, **114**, 40–52.
164. POPOV, E. P. (1951). Successive approximations for beams on an elastic foundation, *Trans. ASCE*, **116**, 1083–1095.
165. WEISSMANN, G. F. (1972). Tilting foundations, *J. Soil Mech. Fdns. Div.*, *Proc. ASCE*, **98**(SM1), 59–78.
166. LYTTON, R. L. and MEYER, K. T. (1971). Stiffened mats on expansive clay, *J. Soil Mech. Fdns. Div.*, *Proc. ASCE*, **97**(SM7), 999–1019.
167. SADOWSKY, M. (1928). Zweidimensionale Probleme der Elastizität-stheorie, *Z. Angew. Math. Mech.*, **8**(2), 107–121.
168. MUSKHELISHVILI, N. I. (1963). *Some Basic Problems of the Mathematical Theory of Elasticity*, 4th edition, Noordhoff, Groningen.
169. KHADEM, R. and O'CONNOR, J. J. (1969). Adhesive or frictionless compression of an elastic rectangle between two identical elastic half-spaces, *Int. J. Engng. Sci.*, **7**, 153–168.
170. ADAMS, G. G. and BOGY, D. B. (1976). The plane solution for the elastic contact problem of a semi-infinite strip and half plane, *J. Appl. Mech., Trans. ASME*, **43**(4), 603–607.
171. CONWAY, H. D. and FARNHAM, K. A. (1967). The contact stress problem for indented strips and slabs under conditions of partial slipping, *Int. J. Engng. Sci.*, **5**, 145–154.
172. SPENCE, D. A. (1973). An eigenvalue problem for elastic contact with finite friction, *Proc. Camb. Phil. Soc.*, **73**, 249–268.
173. FRÖHLICH, O. K. (1953). On the settling of buildings combined with deviation from their originally vertical position, *Proc. 3rd Int. Conf. SMFE*, Zurich, 1953, **1**, 362–365.
174. GERRARD, C. M. and HARRISON, W. J. (1970). Stresses and displacements in a loaded orthorhombic half space, *Tech. Paper No. 9*, Div. Appl. Geomech., CSIRO, Australia.
175. GERRARD, C. M. and WARDLE, L. J. (1973). Solutions for line loads and generalized strip loads applied to an orthorhombic half space, *Tech. Paper No. 14*, Div. Appl. Geomech., CSIRO, Australia.
176. PARIA, G. (1960). The non-symmetric punch problem in layered media by the Wiener-Hopf method, *Quart. J. Math.*, **11**, 116–123.

177. SMITH, S. F. (1964). On a flat punch indenting an elastic layer in plane strain, *Quart. J. Math.*, **15**, 223–237.
178. CONWAY, H. D., VOGEL, S. M., FARNHAM, K. A. and SO, S. (1966). Normal and shearing contact stresses in indented strips and slabs, *Int. J. Engng. Sci.*, **4**, 343–359.
179. MILOVIC, D. M., TOUZOT, G. and TOURNIER, J. P. (1970). Stresses and displacements in an elastic layer due to inclined and eccentric load over a rigid strip, *Géotechnique*, **20**(3), 231–252.
180. BOROWICKA, H. (1936). The distribution of pressure under a uniformly loaded elastic strip resting on elastic-isotropic ground, *Final Rep. 2nd Congr. Int. Assn. Bridge Struct. Engng.*, Berlin, 1936, 840–845.
181. LEE, I. K. and PHILLIPS, J. T. (1969). An analysis of flexible strips on a linear elastic foundation, *Civ. Engng. Trans. Instn. Engrs. Australia*, **CE11**(1), 1–8.
182. GLADWELL, G. M. L. (1976). On some unbonded contact problems in plane elasticity theory, *J. Appl. Mech., Trans. ASME*, **43**, 263–267.
183. LEE, I. K. (1963). Elastic settlements of footings with a rough interface, *Proc. 4th Australia–NZ Conf. SMFE*, 225–232.
184. CROWSER, J. C., SCHUSTER, R. L. and SACK, R. L. (1974). Settlement and contact pressure distribution of a mat-supported silo group on an elastic subgrade, *Proc. Conf. Settlement Struct.*, Cambridge, 1974, 344–352.
185. PRAGER, W. (1927). Zur Theorie elastisch gelagerter Konstruktionen, *Z. Angew. Math. Mech.*, **7**(5), 354–360.
186. BIOT, M. A. (1937). Bending of an infinite beam on an elastic foundation, *J. Appl. Mech., Trans. ASME*, **4**(1), A1–A7.
187. REISSNER, M. E. (1937). On the theory of beams resting on a yielding foundation, *Proc. Nat. Acad. Sci.*, **23**, 328–333.
188. BOSSON, G. (1939). The flexure of an infinite elastic strip on an elastic foundation, *Phil. Mag.*, 7th Ser., **27**(180), 37–50.
189. PICHUMANI, R. and TRIANDAFILIDIS, G. E. (1971). Interaction between rigid pavement slab and its foundation—a finite element analysis, *Proc. Symp. Interaction Struct. Fdn.*, Birmingham, 1971, 146–159.
190. HERRMANN, L. R. (1964). A three-dimensional elasticity solution for continuous beams, *J. Franklin Inst.*, **278**(2), 75–83.
191. VESIĆ, A. B. (1961). Bending of beams resting on isotropic elastic solid, *J. Engng. Mech. Div., Proc. ASCE*, **87**(EM2), 35–53.
192. VESIĆ, A. B. (1961). Beams on elastic subgrade and the Winkler's hypothesis, *Proc. 5th Int. Conf. SMFE*, Paris, 1961, **1**, 845–850.
193. PANCHANATHAN, S. and CHANDRASEKARAN, V. S. (1963). The effect of depth to rigid layer on the distribution of contact pressure beneath rigid and flexible foundations, *Proc. 2nd Asian Reg. Conf. SMFE*, Japan, 1963, **1**, 192–195.
194. GLASSMAN, A. (1972). Behaviour of crossed beams on elastic foundations, *J. Soil Mech. Fdns. Div., Proc. ASCE*, **98**(SM1), 1–11.
195. BROWN, P. T. (1972). Longitudinal bending of uniformly loaded strip footings on deep elastic foundations, *Australian Geomech. J.*, **G2**(1), 28–31.

196. BROWN, P. T. (1975). Strip footing with concentrated loads on deep elastic foundations, *Geotech. Engng.*, **6**(1), 1–13.
197. BIOT, M. A. (1941). Consolidation settlement under a rectangular load distribution, *J. Appl. Phys.*, **12**, 426–430.
198. BIOT, M. A. and CLINGAN, F. M. (1941). Consolidation settlement of a soil with an impervious top surface, *J. Appl. Phys.*, **12**, 578–581.
199. SCHIFFMAN, R. L., CHEN, A. T. -F. and JORDAN, J. C. (1969). An analysis of consolidation theories, *J. Soil Mech. Fdns. Div., Proc. ASCE*, **95**(SM1), 285–312.
200. BIOT, M. A. and CLINGAN, F. M. (1942). Bending settlement of a slab resting on a consolidating foundation, *J. Appl. Phys.*, **13**, 35–40.
201. POULOS, H. G. (1974). Analysis of pile groups subjected to vertical and horizontal loads, *Australian Geomech. J.*, **G4**(1), 26–32.
202. WHITAKER, T. (1976). *The Design of Piled Foundations*, 2nd edition, Pergamon Press, Oxford.
203. HOOPER, J. A. and WOOD, L. A. (1977). Comparative behaviour of raft and piled foundations, *Proc. 9th Int. Conf. SMFE*, Tokyo, 1977, **1**, 545–550.
204. POULOS, H. G. (1968). Analysis of the settlement of pile groups, *Géotechnique*, **18**(4), 449–471.
205. HOOPER, J. A. (1973). Observations on the behaviour of a piled-raft foundation on London Clay, *Proc. Instn. Civ. Engrs.*, **55**(2), 855–877.
206. NAYLOR, D. J. and HOOPER, J. A. (1974). An effective stress finite element analysis to predict the short and long term behaviour of a piled-raft foundation on London Clay, *Proc. Conf. Settlement Struct.*, Cambridge, 1974, 394–402.
207. DESAI, C. S., JOHNSON, L. D. and HARGETT, C. M. (1974). Analysis of pile-supported gravity lock, *J. Geotech. Engng. Div., Proc. ASCE*, **100**(GT9), 1009–1029.
208. POULOS, H. G. and DAVIS, E. H. (1968). The settlement behaviour of single axially loaded incompressible piles and piers, *Géotechnique*, **18**(3), 351–371.
209. MINDLIN, R. D. (1936). Force at a point in the interior of a semi-infinite solid, *Physics*, **7**, 195–202.
210. CHAN, K. S., KARASUDHI, P. and LEE, S. L. (1974). Force at a point in the interior of a layered elastic half space, *Int. J. Solids Struct.*, **10**(11), 1179–1199.
211. BANERJEE, P. K. and DAVIES, T. G. (1977). The behaviour of axially and laterally loaded single piles embedded in nonhomogeneous soils, *Rep. Univ. Coll. Cardiff.*
212. RANDOLPH, M. F. and WROTH, C. P. (1977). A fundamental approach to predicting the deformation of vertically loaded piles, *Rep. TR38*, Camb. Univ.
213. POULOS, H. G. and MATTES, N. S. (1971). Settlement and load distribution analysis of pile groups, *Australian Geomech. J.*, **G1**(1), 18–28.
214. POULOS, H. G. and MATTES, N. S. (1971). Displacements in a soil mass due to pile groups, *Australian Geomech. J.*, **G1**(1), 29–35.

215. MATTES, N. S. and POULOS, H. G. (1969). Settlement of single compressible pile, *J. Soil Mech. Fdns. Div.*, *Proc. ASCE*, **95**(SM1), 189–207.
216. BUTTERFIELD, R. and BANERJEE, P. K. (1971). The elastic analysis of compressible piles and pile groups, *Géotechnique*, **21**(1), 43–60.
217. BANERJEE, P. K. and DRISCOLL, R. M. (1976). Three-dimensional analysis of raked pile groups, *Proc. Instn. Civ. Engrs.*, **61**(2), 653–671.
218. DAVIS, E. H. and POULOS, H. G. (1972). The analysis of pile raft systems, *Australian Geomech. J.*, **G2**(1), 21–27.
219. POULOS, H. G. (1968). The influence of a rigid pile cap on the settlement behaviour of an axially-loaded pile, *Civ. Engng. Trans. Instn. Engrs. Australia*, **CE10**(2), 206–208.
220. BUTTERFIELD, R. and BANERJEE, P. K. (1971). The problem of pile group–pile cap interaction, *Géotechnique*, **21**(2), 135–142.
221. BANERJEE, P. K. (1975). Effects of the pile cap on the load displacement behaviour of pile groups when subjected to eccentric loading, *Proc. 2nd Australia–NZ Conf. Geomech*, Brisbane, 1975, 179–184.
222. BROWN, P. T. and WIESNER, T. J. (1975). The behaviour of uniformly loaded piled strip footings, *Soils Fdns*, **15**(4), 13–21.
223. WIESNER, T. J. and BROWN, P. T. (1976). Behaviour of piled strip footings subject to concentrated loads, *Australian Geomech. J.*, **G6**(1), 1–5.
224. HONGLADAROMP, T., CHEN, N-J. and LEE, S-L. (1973). Load distribution in rectangular footings on piles, *Geotech. Engng.*, **4**(2), 77–90.
225. HIGHT, D. W. and GREEN, P. A. (1976). The performance of a piled-raft foundation for a tall building in London, *Proc. 6th Euro. Conf. SMFE*, Vienna, 1976, **1.2**, 467–472.
226. HAIN, S. J. (1975). Analysis of rafts and raft-pile foundations, *Proc. Symp. Soil Mech.*, Univ. of NSW, 1975, 213–253.
227. BROWN, P. T., POULOS, H. G. and WIESNER, T. J. (1975). Piled raft foundation design, *Proc. Symp. Raft Fdns.*, Perth, Australia, 1975, 13–21.
228. OTTAVIANI, M. (1975). Three-dimensional finite element analysis of vertically loaded pile groups, *Géotechnique*, **25**(2), 159–174.
229. BANERJEE, P. K. (1976). Analysis of vertical pile groups embedded in non-homogeneous soil, *Proc. 6th Euro. Conf. SMFE*, Vienna, 1976, **1.2**, 345–350.
230. BANERJEE, P. K. and DAVIES, T. G. (1977). Analysis of pile groups embedded in Gibson soil, *Proc. 9th Int. Conf. SMFE*, Tokyo, 1977, **1**, 381–386.
231. MEYERHOF, G. G. (1962). Load-carrying capacity of concrete pavements, *J. Soil Mech. Fdns. Div.*, *Proc. ASCE*, **88**(SM3), 89–116.
232. MARVIN, E. L. (1972). Viscoelastic plate on poroelastic foundation, *J. Engng. Mech. Div.*, *Proc. ASCE*, **98**(EM4), 911–928.
233. AUGUSTI, G. (1970). Mode approximations for rigid–plastic structures supported by an elastic medium, *Int. J. Solids Struct.*, **6**(6), 809–827.
234. KRAJCINOVIC, D. (1976). Rigid–plastic circular plates on elastic foundation, *J. Engng. Mech. Div.*, *Proc. ASCE*, **102**(EM2), 213–224.
235. DAVIES, J. D. (1962). Yield line theory applied to edge loaded circular foundation slabs, *Civ. Engng. Publ. Wks. Rev.*, **57**, 1285–1286.

236. HOOPER, J. A. (1976). The effect of flexural cracking on differential raft settlements, *Proc. Instn. Civ. Engrs.*, **61**(2), 567–574.
237. RANGANATHAM, B. V. and HENDRY, A. W. (1963). The ultimate flexural strength of reinforced concrete rafts, *Mag. Concr. Res.*, **15**(45), 159–170.
238. GANGADHARAN, A. C. and REDDY, D. V. (1964). Ultimate-load analysis of an edge-loaded rectangular slab resting on soil, *Concr. Constr. Engng.*, **59**(12), 445–448.
239. REDDY, D. V. and MURPHREE, E. L. (1968). Ultimate-load analysis of edge-loaded foundation slabs, *Struct. Engr.*, **46**(1), 13–16.
240. GREEN, D. G., MACLEOD, I. A. and STARK, W. G. (1976). Observation and analysis of brick structures on soft clay, *Proc. Int. Conf. Performance Blg. Struct.*, Glasgow, 1976, **1**, 321–336.
241. ZIENKIEWICZ, O. C., VALLIAPPAN, S. and KING, I. P. (1968). Stress analysis of rock as a 'no tension' material, *Géotechnique*, **18**(1), 56–66.
242. HUANG, Y. H. (1968). Stresses and displacements in nonlinear soil media, *J. Soil Mech. Fdns. Div., Proc. ASCE*, **94**(SM1), 1–19.
243. HUANG, Y. H. (1969). Finite element analysis of nonlinear soil media, *Proc. Symp. Appl. FEM Civ. Engng.*, Nashville, Tennessee, 663–690.
244. MORGENSTERN, N. R. and PHUKAN, A. L. T. (1968). Stresses and displacements in a homogeneous non-linear foundation, *Proc. Int. Symp. Rock Mech.*, Madrid, 1968, 313–320.
245. GIRIJAVALLABHAN, C. V. and REESE, L. C. (1968). Finite element method for problems in soil mechanics, *J. Soil Mech. Fdns. Div., Proc. ASCE*, **94**(SM2), 473–496.
246. RADHAKRISHNAN, N. and REESE, L. C. (1969). Behaviour of strip footings on layered cohesive soils, *Proc. Symp. Appl. FEM Civ. Engng.*, Nashville, Tennessee, (1969). 691–728.
247. DESAI, C. S. and REESE, L. C. (1970). Analysis of circular footings on layered soils, *J. Soil Mech. Fdns. Div., Proc. ASCE*, **96**(SM4), 1289–1310.
248. DUNCAN, J. M. and CHANG, C. -Y. (1970). Nonlinear analysis of stress and strain in soils, *J. Soil Mech. Fdns. Div., Proc. ASCE*, **96**(SM5), 1629–1653.
249. MAJID, K. I. and CRAIG, J. S. (1971). An incremental finite element analysis of structural interaction with soil of non-linear properties, *Proc. Symp. Interaction Struct. Fdn.*, Birmingham, 1971, 131–145.
250. DESAI, C. S. (1971). Nonlinear analyses using spline functions, *J. Soil Mech. Fdns. Div., Proc. ASCE*, **97**(SM10), 1461–1480.
251. DOMASCHUK, L. and VALLIAPPAN, P. (1975). Nonlinear settlement analysis by finite element, *J. Geotech. Engng. Div., Proc. ASCE*, **101**(GT7), 601–614.
252. ELLISON, R. D., D'APPOLONIA, E. and THIERS, G. R. (1971). Load-deformation mechanism for bored piles, *J. Soil Mech. Fdns. Div., Proc. ASCE*, **97**(SM4), 661–678.
253. DESAI, C. S. (1974). Numerical design-analysis for piles in sands, *J. Geotech. Engng. Div., Proc. ASCE*, **100**(GT6), 613–635.
254. HÖEG, K., CHRISTIAN, J. T. and WHITMAN, R. V. (1968). Settlement of strip load on elastic-plastic soil, *J. Soil Mech. Fdns. Div., Proc. ASCE*, **94**(SM2), 431–445.

255. ZIENKIEWICZ, O. C., VALLIAPPAN, S. and KING, I. P. (1969). Elasto-plastic solutions of engineering problems. Initial stress, finite element approach, *Int. J. Num. Meth. Engng.*, **1**(1), 75–100.
256. VALLIAPPAN, S., BOONLAULOHR, P. and LEE, I. K. (1976). Non-linear analysis for anisotropic materials, *Int. J. Num. Meth. Engng.*, **10**(3), 597–606.
257. HÖEG, K. (1972). Finite element analysis of strain-softening clay, *J. Soil Mech. Fdns. Div.*, *Proc. ASCE*, **98**(SM1), 43–58.
258. SCHULTZE, E. (1961). Distribution of stress beneath a rigid foundation, *Proc. 5th Int. Conf. SMFE*, Paris, 1961, **1**, 807–813.
259. BIERNATOWSKI, K. (1973). The state of stress and displacement in the contact surface between a rigid foundation and subsoil, *Proc. 8th Int. Conf. SMFE*, Moscow, 1973, **1**(3), 15–17.
260. BROWN, P. T. (1968). The effect of local bearing failure on behaviour of rigid circular rafts, *Civ. Engng. Trans. Instn. Engrs. Australia*, 1968, **CE10**(2), 190–192.
261. DAVIS, E. H. and BOOKER, J. R. (1973). The effect of increasing strength with depth on the bearing capacity of clays, *Géotechnique*, **23**(4), 551–563.
262. NAYLOR, D. J. and ZIENKIEWICZ, O. C. (1971). Settlement analysis of a strip footing using a critical state soil model in conjunction with finite elements, *Proc. Symp. Interaction Struct. Fdn.*, Birmingham, 1971, 93–107.
263. OHTA, H. and HATA, S. (1973). Immediate and consolidation defor-mations of clay, *Proc. 8th Int. Conf. SMFE*, Moscow, 1973, **1**(3), 193–196.
264. FREUDENTHAL, A. M. and SPILLERS, W. R. (1962). Solutions for the infinite layer and the half-space for quasi-static consolidating elastic and viscoelastic media, *J. Appl. Phys.*, **33**(9), 2661–2668.
265. GARLANGER, J. E. (1972). The consolidation of soils exhibiting creep under constant effective stress, *Géotechnique*, **22**(1), 71–78.
266. MESRI, G. (1973). Coefficient of secondary compression, *J. Soil Mech. Fdns. Div.*, *Proc. ASCE*, **99**(SM1), 123–137.
267. KOMAMURA, F. and HUANG, R. J. (1974). New rheological model for soil behaviour, *J. Geotech. Engng. Div.*, *Proc. ASCE*, **100**(GT7), 807–824.
268. GIBSON, R. E., ENGLAND, G. L. and HUSSEY, M. J. L. (1967). The theory of one-dimensional consolidation of saturated clays, *Géotech-nique*, **17**(3), 261–273.
269. SELVADURAI, A. P. S. (1977). Axisymmetric flexure of an infinite plate resting on a finitely deformed incompressible elastic half space, *Int. J. Solids Struct.*, **13**(4), 357–365.
270. CARTER, J. P., SMALL, J. C. and BOOKER, J. R. (1977). A theory of finite elastic consolidation, *Int. J. Solids Struct.*, **13**(5), 467–478.
271. SOYDEMIR, C. and SCHMID, W. E. (1970). Deformation and stability of viscoelastic soil media, *J. Soil Mech. Fdns. Div.*, *Proc. ASCE*, **96**(SM6), 2081–2098.
272. CHRISTIAN, J. T. and WATT, B. J. (1972). Undrained visco-elastic analysis of soil deformations, *Proc. Symp. Appl. FEM Geotech. Engng.*, Vicksburg, Mississippi, 1972, **2**, 533–577.

273. RAMASWAMY, S. V. and VAIDYANATHAN, R. (1977). Settlement of footings on compacted clays, *Proc. Int. Symp. Soil Struct. Interaction*, Roorkee, 1977, 251–257.
274. BOOKER, J. R. and SMALL, J. C. (1977). Finite element analysis of primary and secondary consolidation, *Int. J. Solids Struct.*, **13**(2), 137–149.
275. BROWN, P. T. (1977). Structure–foundation interaction and soil creep, *Proc. 9th Int. Conf. SMFE*, Tokyo, 1977, **1**, 439–442.
276. BOOKER, J. R. and POULOS, H. G. (1976). Analysis of creep settlement of pile foundations, *J. Geotech. Engng. Div., Proc. ASCE*, **102**(GT1), 1–14.
277. OTTAVIANI, M. and CAPPELLARI, G. (1976). Time-behaviour of axially loaded bored piles in a cohesive soil, *Proc. 6th Euro. Conf. SMFE*, Vienna, 1976, **1**(2), 529–532.
278. SMALL, J. C., BOOKER, J. R. and DAVIS, E. H. (1976). Elasto-plastic consolidation of soil, *Int. J. Solids Struct.*, **12**(6), 431–448.
279. ZIENKIEWICZ, O. C., HUMPHESON, C. and LEWIS, R. W. (1975). Associated and non-associated visco-plasticity and plasticity in soil mechanics, *Géotechnique*, **25**(4), 671–689.
280. SARAN, S., PANDE, G. N. and ZIENKIEWICZ, O. C. (1977). Shallow foundation problems, *Proc. Int. Symp. Soil Struct. Interaction*, Roorkee, 1977, 223–230.
281. BURLAND, J. B., SILLS, G. C. and GIBSON, R. E. (1973). A field and theoretical study of the influence of non-homogeneity on settlement, *Proc. 8th Int. Conf. SMFE*, Moscow, 1973, **1**(3), 39–46.
282. SMITH, R. G., THORBURN, S. and TINCH, W. R. (1970). The influence of pile stiffness on the foundation stresses of a multi-storey shear wall, *Build. Sci.*, **5**(1), 21–30.
283. COULL, A. (1971). Interaction of coupled shear walls with elastic foundations, *J. Am. Concr. Inst.*, **68**(6), 456–461.
284. COULL, A. (1971). Coupled shear walls subjected to differential settlement, *Build. Sci.*, **6**(4), 209–212.
285. TSO, W. K. (1972). Stresses in coupled shear walls induced by foundation deformation, *Build. Sci.*, **7**(3), 197–203.
286. TSO, W. K. and CHAN, P. C. K. (1972). Flexible foundation effect on coupled shear walls, *J. Am. Concr. Inst.*, **69**(11), 678–683.
287. ARVIDSSON, K. (1976). Elastically founded shear walls with two rows of openings, *J. Am. Concr. Inst.*, **73**(3), 151–154.
288. CHANDRASEKARAN, V. S. and KHEDKAR, S. P. (1977). Shear wall foundation interaction, *Proc. Int. Symp. Soil Struct. Interaction*, Roorkee, 1977, 123–127.
289. MEYERHOF, G. G. (1947). The settlement analysis of building frames, *Struct. Engr.*, **25**(9), 369–409.
290. CHAMECKI, S. (1956). Structural rigidity in calculating settlements, *J. Soil Mech. Fdns. Div., Proc. ASCE*, **82**(SM1), 1–19.
291. LARNACH, W. J. (1970). Computation of settlements in building frames, *Civ. Engng. Publ. Wks. Rev.*, **65**, 1040–1043.
292. YOKOO, Y. and YAMAGATA, K. (1954). On the differential settlement of structure due to the consolidation of clay stratum, *Proc. 4th Jap. Nat. Congr. Appl. Mech.*, 1954, 111–114.

293. YOKOO, Y. and YAMAGATA, K. (1956). On the calculation formulae of differential settlement of structures, *Proc. 6th Jap. Nat. Congr. Appl. Mech.*, 1956, 161–164.

294. YOKOO, Y. and YAMAGATA, K. (1956). A theoretical study on differential settlements of structures, *Bull. No. 14*, Disaster Prev. Res. Inst., Kyoto University.

295. SOMMER, H. (1965). A method for the calculation of settlements, contact pressures and bending moments in a foundation including the influence of the flexural rigidity of the superstructure, *Proc. 6th Int. Conf. SMFE*, Montreal, 1965, **2**, 197–201.

296. LITTON, E. and BUSTON, J. M. (1968). The effect of differential settlement on a large, rigid, steel-framed, multi-storey building, *Struct. Engr.*, **46**(11), 353–356.

297. GRASSHOF, H. (1957). Influence on flexural rigidity of superstructure on the distribution of contact pressure and bending moments of an elastic combined footing, *Proc. 4th Int. Conf. SMFE*, London, 1957, **1**, 300–306.

298. SVED, G. and KWOK, H. L. (1963). The effect of non-linear foundation settlement on the distribution of bending moments in a building frame, *Proc. 4th Australia–NZ Conf. SMFE*, Adelaide, 1963, 18–22.

299. GETZLER, Z. (1968). Influence of structural rigidity ratio on foundation design, *Concrete*, **2**(6), 234–237.

300. HEIL, H. (1969). Studies on the structural rigidity of reinforced concrete building frames on clay, *Proc. 7th Int. Conf. SMFE*, Mexico, 1969, **2**, 115–121.

301. CHAMECKI, S. (1969). Calcul des tassements progressifs des fondations, et tenant compte de l'interaction des structures et du sol, *Annls. L'Inst. Tech. Bâtim. Trav. Publ.*, No. 261, 1319–1334.

302. LEE, I. K. and HARRISON, H. B. (1970). Structure and foundation interaction theory, *J. Struct. Div., Proc. ASCE*, **96**(ST2), 177–197.

303. LEE, I. K. and BROWN, P. T. (1972). Structure–foundation interaction analysis, *J. Struct. Div., Proc. ASCE*, **98**(ST11), 2413–2431.

304. SEETHARAMULU, K. and KUMAR, A. (1973). Interaction of foundation beam and soil with frames, *Proc. 8th Int. Conf. SMFE*, Moscow, 1973, **1**(3), 231–234.

305. KING, G. J. W. and CHANDRASEKARAN, V. S. (1974). An assessment of the effects of interaction between a structure and its foundation, *Proc. Conf. Settlement Struct.*, Cambridge, 1974, 368–383.

306. POULOS, H. G. (1975). Settlement analysis of structural foundation systems, *Proc. 4th SE Asian Conf. Soil Engng.*, Kuala Lumpur, 1975, 4.54–4.62.

307. BROWN, P. T. (1975). The significance of structure–foundation interaction, *Proc. 2nd Australia–NZ Conf. Geomech.*, Brisbane, 1975, 79–82.

308. WOOD, L. A. and LARNACH, W. J. (1975). The interactive behaviour of a soil–structure system and its effect on settlements, *Proc. Symp. Geotech. Struct.*, Univ. of NSW, 1975.

309. MIYAHARA, F. and ERGATOUDIS, J. G. (1976). Matrix analysis of structure–foundation interaction, *J Struct. Div., Proc. ASCE*, **102**(ST1), 251–265.

310. UNGUREANU, N., CIONGRADI, and STRAT, L. (1977). Framed structure–foundation beams–soil interaction, *Proc. Int. Symp. Soil Struct. Interaction*, Roorkee, 1977, 101–108.
311. MEYERHOF, G. G. (1953). Some recent foundation research and its application to design, *Struct. Engr.*, **31**, 151–167.
312. MORRIS, D. (1966). Interaction of continuous frames and soil media, *J. Struct. Div., Proc. ASCE*, **92**(ST5), 13–44.
313. DE JONG, J. and MORGENSTERN, N. R. (1971). The influence of structural rigidity on the foundation loads of the CN tower, Edmonton. *Canad. Geotech. J.*, **8**(4), 527–537.
314. MAJID, K. I. and CUNNELL, M. D. (1976). A theoretical and experimental investigation into soil–structure interaction, *Géotechnique*, **26**(2), 331–350.
315. BURAGOHAIN, D. N., RAGHAVAN, N. and CHANDRASEKARAN, V. S. (1977). Interaction of frames with pile foundations, *Proc. Int. Symp. Soil Struct. Interaction*, Roorkee, 1977, 109–115.
316. JAIN, O. P., TRIKHA, D. N. and JAIN, S. C. (1977). Differential foundation settlement of high-rise buildings, *Proc. Int. Symp. Soil Struct. Interaction*, Roorkee, 1977, 237–243.
317. WOOD, L. A., LARNACH, W. J. and WOODMAN, N. J. (1977). Observed and computed settlements of two buildings, *Proc. Int. Symp. Soil Struct. Interaction*, Roorkee, 1977, 129–136.
318. HADDADIN, M. J. (1971). Mats and combined footings—analysis by the finite element method, *J. Am. Concr. Inst.*, **68**(12), 945–949.
319. KING, G. J. W. and CHANDRASEKARAN, V. S. (1974). Interactive analysis of a rafted multi-storey space frame resting on an inhomogeneous clay stratum, *Proc. 1st Int. Conf. FEM Engng.*, Univ. NSW, 1974, 493–509.
320. KING, G. J. W. and CHANDRASEKARAN, V. S. (1977). Interactive analysis using a simplified soil model, *Proc. Int. Symp. Soil Struct. Interaction*, Roorkee, 1977, 93–99.
321. HAIN, S. J. and LEE, I. K. (1974). Rational analysis of raft foundation, *J. Geotech. Engng. Div., Proc. ASCE*, **100**(GT7), 843–860.
322. FRASER, R. A. and WARDLE, L. J. (1975). A rational analysis of shallow foundations considering soil–structure interaction, *Australian Geomech. J.*, **G5**(1), 20–25.
323. WARDLE, L. J. and FRASER, R. A. (1975). Methods for raft foundation design including soil–structure interaction, *Proc. Symp. Raft Fdns.*, Perth, Australia, 1975, 1–11.
324. KLEPIKOV, S. N., BOBRITSKY, G. M., RIVKIN, S. A. and MALIKOVA, T. A. (1973). Analysis of the foundation slabs and upper structure interaction, *Proc. 8th Int. Conf. SMFE*, Moscow, **1**(3), 127–132.

Chapter 6

DIAPHRAGM WALLS

T. H. HANNA

Department of Civil and Structural Engineering, The University of Sheffield, UK

SUMMARY

Bentonite suspensions, first used as drilling muds in the oil industry, have, over the last 25 years, been used increasingly to support the sides of trenches for the construction of diaphragm walls. This chapter describes the excavating systems available, and the methods used to assess the stability of the fluid-filled trench. The construction, analysis and design of the walls are discussed, and several examples of their application are described.

6.1 INTRODUCTION

In the oil well drilling industry the use of bentonite suspensions has allowed deep borings to be supported when unlined and has enabled the drill cuttings to be brought to the ground surface. In the civil engineering industry, this well established process was first applied to the support of holes for bored piles. Later it was applied to the support of trenches, when it was found possible to use mechanical grabs for excavation. The bentonite fluid is used to support the sides of the trench and, after excavation, fresh concrete is placed in the trench to replace the fluid.

The main stages in the construction of a wall panel are illustrated in Fig. 1. Firstly, excavation is carried out adjacent to a wall panel with the aid of a grab, and the trench so formed is filled with a bentonite suspension. The excavated materials are brought to the surface and

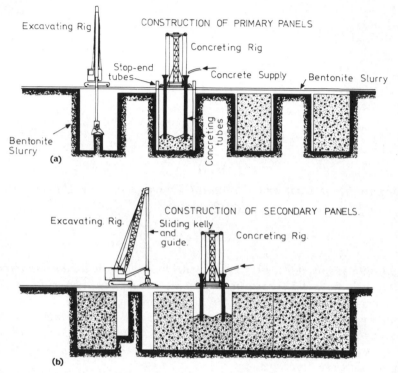

FIG. 1. Sequence of construction of diaphragm walls. (Soletanche Co (UK) Ltd). (a) Construction of primary panels. (b) Construction of secondary panels.

when the excavation is complete a stop-end is inserted. The reinforcement cage, pre-assembled, is lowered into the excavation and finally concrete is poured through tremie pipes filling the trench by gravity action and self-compaction, and displacing the bentonite suspension which is withdrawn for treatment and further use. This process is repeated until the wall is completed in a series of panels from 2 m to about 10 m in length.

This method of wall construction has had great success since its first use about 25 years ago.[1] Several factors have led to this successful development:

(a) The availability of bentonite, produced from montmorillonitic clays and refined and converted to sodium montmorillonite.

(b)   The need for walls for deep basements, underpasses, and tunnels.
(c)   The development of trenching equipment.
(d)   The development of plant for processing of bentonites.
(e)   The economics of the method.
(f)   The development of the science of bentonite support of trenches.

The art of diaphragm wall construction has kept ahead of the science, and will continue to do so. For this reason, a number of technical problems exist to which only enlightened opinion is available. These include bond between steel and concrete, arching in short trenches, mix design and specification and site controls, as well as methods of wall analysis. However, much literature on the general subject is available starting with the *Symposium on Grouts and Drilling Muds* (1963), the *Conference on Diaphragm Walls and Anchorages* (1974) and the *Seminar on Diaphragm Walls and Anchorages* (1976) as well as in numerous technical journals.

When dealing with the design and construction of diaphragm walls it must be appreciated that consideration has to be given to a wide range of interrelated topics such as site investigation, mud disposal, concrete mix and placement and the behaviour of the complete wall/soil/excavation system. In this brief review only the more important features can be touched on and the reader is referred to a number of other sources for further information.[2-8]

6.2   EXCAVATING SYSTEMS

All excavating systems rely on mechanical devices which enable the progressive excavation of a trench in such a way that the stabilising fluid is introduced simultaneously as trench excavation proceeds. The supporting fluid is generally a bentonite suspension which can form a membrane of low permeability at the interface between the soil and the suspension.

The primary requirement is minimum disturbance of the ground in the vicinity of the trench. In addition, the rate of trench cutting must be slow enough to permit the formation of the 'filter cake' membrane at the soil/bentonite interface. Three types of excavation method are used: *percussive, rotary* and *traditional.* The first two are used in rock to loosen and break down the ground into small pieces which are mixed up with the bentonite at the excavation face. This suspension is then brought to the ground surface by either direct or reverse circulation methods. The chippings are removed, and the bentonite cleaned where necessary and returned to the excavation. *Traditional* excavating tools such as grabs, buckets, shovels, etc. are brought to ground level for discharge and may be supported by either a rope or a kelly bar. The lifting and lowering rates of the drill tools must be controlled to minimise turbulence in the bentonite which may give rise to localised collapse of the trench sides in certain ground conditions, particularly in sands.

During trench cutting operations there may be deviations from the vertical and horizontal as well as local deviations from the average trench face. Many practical factors control such deviations, some of which can be assessed in advance of construction works, and much useful guidance is given by Fuchsberger.[9]

An important purpose of the stabilising fluid is to form an impervious membrane at the trench wall and a bentonite suspension with or without an additive is normally used. It is prepared by adding a measured quantity of powdered bentonite to clean water, mixing it and allowing it to hydrate. The time required for hydration depends on the method of mixing and useful information is given in a number of references.[10,11,16] In use, the bentonite suspension changes colour and properties, becoming more viscous and dense because of physical and chemical contamination with soil and cement particles. Such changes occur on every site, and control of the bentonite suspension is necessary. This usually entails the removal of the heavy soil particles, and a number of tests have been developed to enable field checks to be made on the bentonite. These include tests for viscosity, strength, density, pH, fluid loss and sand content. The range of bentonite properties suggested by the Federation of Piling Specialists appears to give good results.[5]

Experience has shown that practically any ground can be stabilised by the correct choice of bentonite suspension. In coarse soil, such as sand or gravel, the bentonite suspension must not penetrate too far into the trench walls, yet it must stabilise the individual soil grains. Such suspensions require a high viscosity and high gel strength. Occasionally inert fillers are added to the suspension to control excessive loss of bentonite into the trench walls.

It will be appreciated that one of the main differences between an open and a bentonite-filled trench is the hydrostatic pressure of the suspension. Use is made of this head when dealing with high ground water tables. In order that a filter cake shall form on the trench side it is essential that the pressure head of bentonite suspension is greater than that of the ground water. To allow for changes in the level of the bentonite suspension during trenching, the bentonite head is usually kept 1 to 1·5 m above ground water level. Artesian and sub-artesian pressures are usually reduced by pumping, whilst allowances may have to be made for steady seepage conditions and tidal variations, as well as for dilution of the bentonite by sea water.[4,9]

6.3 STABILITY OF FLUID-FILLED TRENCHES

Much research work over the years has shown that the stability of the trench is controlled by a number of interrelated factors as follows:

(a) The hydrostatic pressure exerted by the slurry.
(b) The passive resistance of the bentonite slurry.
(c) The resistance to deformation of the filter cake on the sides of the trench, which tends to act as a vertical membrane.
(d) Electro-osmotic forces, because the bentonite suspension is electro-negative relative to the soil.
(e) The penetration of the bentonite suspension into the sides of the trench, especially in sands and gravels, to form a filter cake. This tends to provide an impermeable membrane, so that the pressure of the bentonite suspension is transmitted directly to the soil skeleton.

In an attempt to quantify the stability of a fluid-filled trench Nash and Jones[12] developed the following theory based upon the equilibrium between the hydrostatic force of the bentonite suspension and the force required to restrain a wedge of soil from sliding. If it is assumed that the face of the excavation trench has a water-tight membrane against it then the bentonite suspension will apply a thrust against the trench side (Fig. 2). For a trench in clay it may be assumed that no effective stress changes occur because the cutting is open for a few days only (i.e. $\phi = 0$). For this condition the factor of safety, F, can be found from equilibrium of the wedge ABC and is given by:

$$F = \frac{4c_u}{H(\gamma - \gamma_b)} \tag{1}$$

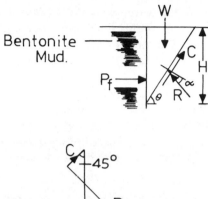

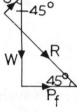

FIG. 2.   Stability of a slurry-filled trench: forces acting on wedge. (After Nash and Jones.[12])

where H is the depth of the trench, γ the bulk density of the soil, γ_b the density of the bentonite suspension in the trench and c_u the undrained shear strength of the clay. This analysis ignores the presence of tension cracks in clay near the surface.

The preceding analysis has been extended to cohesionless soils by Morgenstern and Amir-Tahmasseb,[13] allowing for a variable position of the ground water table (Fig. 3). They show that the density of the bentonite suspension necessary to stop the wedge of soil from sliding at the angle is given by:

$$n^2 \frac{\gamma_b}{\gamma_w} = \frac{\dfrac{\gamma}{\gamma_w} \cot \theta [\sin \theta - \cos \theta \tan \phi'] + m^2 \operatorname{cosec} \theta \tan \phi'}{\cos \theta + \sin \theta \tan \phi'} \tag{2}$$

where $\theta \simeq 45 + \phi'/2$, ϕ' is the effective angle of internal friction of the soil and γ_w the density of water.

The stability of a trench filled with bentonite suspension is dependent to some extent on factors not included in the above equations. The density of the bentonite suspension may be increased by the addition of heavy minerals, although there are limits to the maximum specific gravity which can be used. The suspension tends to gel and

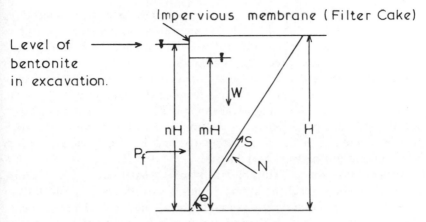

FIG. 3.   Stability of a sliding wedge on sand.[13]

gain some shear strength but the agitation caused by the operation of the excavating machinery will tend to eliminate this potential strength. It is well known that arching can develop and this aids stability. It should be appreciated that arching can be both in the vertical and horizontal directions but its contribution is difficult to quantify and rely on. The use of trench guide walls to support the top 1–1·5 m of ground tends to close tension cracks in the ground. Where the ground is permeable it is known that the bentonite can penetrate and cause a light cementing action which contributes to trench stability. Because of the many unknown factors which cannot be fully quantified it is prudent to have factors of safety between 1·5 and 3 or greater against instability of the trench.

6.4   CONSTRUCTION OF WALLS

There are two main wall classifications—cast *in situ* and precast. The general requirements for concrete in such walls do not differ significantly from concrete in more conventional structures. However, in addition to purely structural considerations, quality control of the concrete becomes very important and controls the final quality of the wall. In addition to strength, workability is the most important parameter, and most of the problems associated with cast *in situ* walls can be eliminated by use of a very workable mix which is not sensitive to segregation. As mentioned earlier, the concrete is poured into the trench through the bentonite suspension via tremie pipes and displaces the bentonite suspension from the bottom of the excavation upwards. Mechanical compaction is not possible. For this reason the concrete must have a consistency such that it will flow under gravity and completely displace all of the bentonite suspension. Experience shows that in general the specific gravity of the bentonite should not exceed 1·3. Also, the setting time of the mix must be such that placement can occur allowing for the inevitable delays which arise on and during transport to site.

Suggestions for a suitable mix are based on many years' experience. In general the slump should be on the range 150–200 mm and the water/cement ratio should not exceed 0·6. The aggregate should comprise about 40% sand with well rounded particles less than 20 mm maximum size. The cement content is normally 400 kg/m^3 or greater. Plasticisers are used, and these can include air entraining agents. Retarders can be used to prevent premature stiffening of cements or to delay the setting time under difficult placing conditions.

The concrete is placed through tremie pipes which are raised in stages as the concrete level rises. Such pipes must be flush-jointed and are usually 0·15 to 0·3 m diameter. For the concrete to fill a panel it must flow laterally, and to ensure a uniform flow the horizontal distances of flow should not exceed about 2·5 m. Thus two or more tremie pipes may be required for long panels. The actual volume of concrete placed will usually exceed the theoretical volume of the trench because in most ground conditions some overbreak occurs, especially in sands and gravels.

Steel reinforcement cages are assembled on site and special details can be incorporated in the cage to accommodate keys or anchor holes. It will be appreciated that, prior to concrete placement, the steel is covered with bentonite. It is therefore important to have the minimum number of transverse bars present since they tend to trap bentonite around the intersections.

The dimensions of a panel are determined by a number of factors as follows:

(a) Short panels are more stable than long ones, and are necessary in soft ground.
(b) It is essential that the concrete for one complete panel can be placed before setting occurs.

(c)   Economic considerations dictate that the number of stop-ends should be small, that the reinforcement cage should be easily handled on site, that anchor arrangements should be realistic, and that a whole number of panels should be completed in a working day.

With all walls a joint exists at the end of each panel and over the years many joint systems have been used. In general each panel acts as an independent unit except where shear connectors are used. The standard joint comprises the semi-circular end of the previous panel formed by removing the stop-end. The concrete of the next panel is poured against this to form the joint. Details of special joints are given by Xanthakos.[2] It is important that adequate attention is given to detailing the stop-ends, and Slivinski and Fleming[4] recommend that the stop-ends should always be considered to be outside the panel length (Fig. 4). Assuming a steel cover of 90 mm, Fig. 4 shows the relationships between wall thickness, panel length, cage length and mean concreted length. Slivinski and Fleming also stress the importance of cage orientation, with no possibility of wrong identification of the earth and excavation faces.

With all walls it is necessary to have a system of temporary guide walls at the ground surface. They serve a dual purpose in guiding the alignment of the trench and in supporting the top of the excavation where the ground is usually weak. Guide walls are spaced about 50 mm further apart than the width of the trenching grab. They must resist the soil pressure as well as the forces from the adjacent tracks on which the equipment travels. Usually walls have a minimum depth of 1 m. In difficult ground it may be necessary to lower the ground water table temporarily or to raise the guide walls above ground level to provide the excess hydrostatic head in the trench that is necessary for stability.

Concrete used for diaphragm walls can be considered to be impermeable. The permeability of a panel depends on cracks and other defects such as segregation. With good site practices both of these can be limited to a few special cases only. The joints between panels are not completely impermeable but the penetration of the bentonite into the soil usually keeps the flow of water very small. Most leaks tend to develop near corners, where differential movements between panels are greatest. Differential movements between adjacent panels are difficult to quantify, and depend on factors such as the dimensions

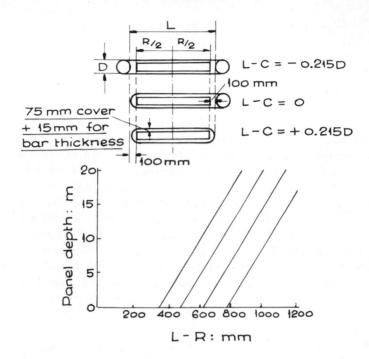

FIG. 4.   Panel dimensions; L = panel length, R = standard cage length, C = mean concreted length, D = panel width.[4]

of the panel, the support system, the excavation method, and the plan layout of the wall. Normal practice is to allow the damp patches to appear and then to inject grout into the soil behind the wall to seal off the water flow.[9]

    The quality of the finished wall depends on many factors, in particular the ground conditions. The generally accepted tolerance is 1 in 80 or better in verticality. Where the ground is soft it is possible for concrete to bypass the stop-ends, causing bulges below the guide walls. In addition irregularities in the face of the wall, caused by loose pockets, will require trimming. The usual tolerance is 100 mm for protrusions on the finished wall.

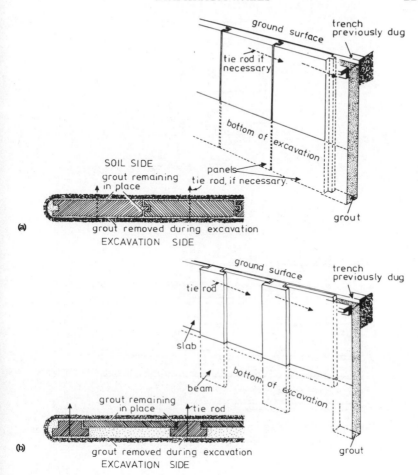

FIG. 5.   Prefabricated wall (a) with identical panels; (b) with beam and slab panels.

Connections between the wall and floors are achieved by leaving recesses formed by insert boxes attached to the reinforcement cages.

A number of prefabricated diaphragm systems[14,15] have been in use since about 1970 and they give a number of distinct advantages over the cast *in situ* wall:

(a)   Better quality control in precasting allows the use of thinner diaphragms.

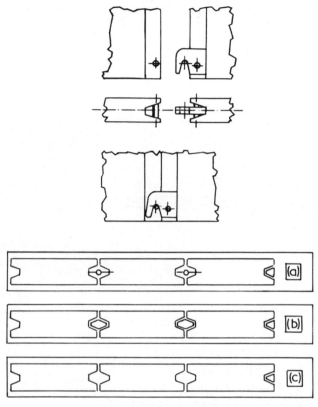

FIG. 6. Method of securing adjacent precast panels; (a) waterstop joint; (b)
reinforced concrete key; (c) sealing grout.[14]

(b) Water-tightness of the joints can be improved.
(c) Walls can be constructed to much finer tolerances, and the
     shape can be tailored to form part of the final structure.

With such systems the precast units are lowered into the bentonite
suspension and the slurry displaced by a grout. Usually the grout is
made up from the bentonite suspension with the addition of cement.
This grout provides a thin lining behind the precast wall and adds to
the overall water-tightness of the system.

Several precast systems are available. The simplest comprises
identical precast panels with a tongued and grooved joint as shown in
Fig. 5. In the Prefasif system (Fig. 6) a special locking hook at the

Area 0.7 - 5.0 m²   Area 2.5 - 3.0 m²

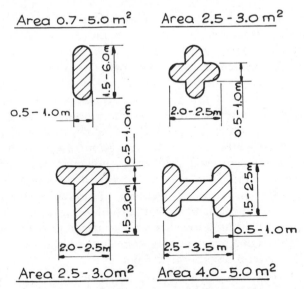

Area 2.5 - 3.0 m²   Area 4.0 - 5.0 m²

FIG. 7. Typical load bearing unit element cross sections.[39]

bottom of each panel gives alignment, while a special grouted joint with waterstop between panels (if required) controls permeability.

At present the sizes of panels are limited by lifting capability on site.

Where diaphragm wall units are used for load bearing purposes, all of the construction precautions associated with walls have to be followed. The use of special grabs permits the construction of a wide range of load bearing units, some of which are illustrated in Fig. 7.

Post-tensioned diaphragm walls have been used to a limited extent, particularly in Italy and Switzerland, and useful information is given by Gysi.[16] Whilst such construction methods give rise to considerable economy it is important to appreciate that it is necessary to position the cable correctly in the wall as severe stressing of the wall will otherwise occur.

6.5 WALL ANALYSIS AND DESIGN

The formation of an excavation causes a three-dimensional change in effective stress in the ground, and this change is usually time-dependent. Associated with this stress change, there are movements of the

surrounding ground in the form of a vertical settlement behind the
wall, a lateral displacement of the wall and a vertical heave of the
base of the excavation (Fig. 8). These movements are controlled by
the wall support system in association with empirical data from actual
field observations.

With a wall of the diaphragm type, in addition to changes in the
stress state in the ground, the construction method and sequence are
important. Many different methods of temporary and permanent
support are used which may or may not permit free movement of the
wall. It is the mode of wall movement, however, which determines
the magnitude of the earth pressure and its distribution on the wall;
hence, at the design stage it is essential to appreciate the method of
construction to be followed. For example, a cantilever wall is
frequently used with underpasses, and no permanent bracing is used
at the top, although a permanent bottom slab is usually provided. This
slab provides lateral constraint and prevents lateral movement. With
deeper walls, bracing, in the form of struts or anchors, is required at
one or several levels to control lateral movement and make economical use of the wall material. In recent years ground anchors have
started to replace the conventional internal strut. The primary reason
is that whilst the anchor is more expensive than the strut, the
excavation is left free from bracing and more economical muck-
shifting and construction methods can be used.

A survey of failures of braced excavations will reveal that there are
several failure modes that must be checked at the design stage. These
include excessive movement of the wall and the surrounding ground,

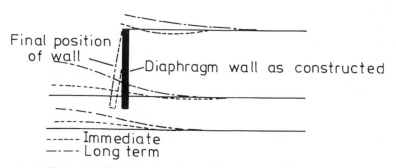

FIG. 8.   Diagrammatic representation of the pattern of short- and long-term
vertical movements that occur around a deep excavation in clay where the
top of the wall is prevented from moving.[41]

yield of the strut or anchor support system, and instability of the base of the excavation particularly in soft clays or in water-bearing sands and gravels. It should be appreciated, therefore, that as much attention should be given to the selection and design of the dewatering system as to the structural analysis of the wall and its support system.

It is now recognised that, in general, the triangular earth pressure distribution according to Coulomb does not develop behind a firmly braced wall because of the limited wall movements possible, and because of the construction sequence followed. In the case of the cantilever type of wall, the pressure state which develops is very similar to that given by the Rankine or Coulomb theories. The problem arises in giving a value to the passive pressure mobilised in front of the wall[17] and considerable care is therefore necessary in determining the appropriate earth pressure values. Walls supported near the top by one row of raker struts tend to move and induce a strain state which generates a Coulomb pressure distribution. However, when several layers of struts or ground anchors are used the ground deformations are limited and there are divided opinions on the earth pressure distribution most appropriate for design purposes.[18,19] The important point to bear in mind is that the pressure distribution can be expected to vary widely with different field installation methods. At present there is no authoritative documentary evidence on this subject and it is felt that the following tentative methods are safe. For multi-strutted walls the empirical envelopes due to Peck[20] should be used and they are, in general, supported both by field and laboratory data.[19,21] The earth pressure diagram (Fig. 9) is an

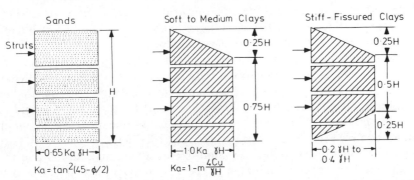

FIG. 9. Apparent earth pressure diagrams suggested by Terzaghi and Peck[50] for computing strut loads in braced excavations.

apparent or artificial diagram and is used to calculate the individual strut loads as indicated in the figure. In sand strata the maximum intensity of pressure is taken to be $0.65\ K_a\gamma H$ while in clays the value depends on clay consistency. In soft to firm clays a value of $K_a\gamma H$ is used while in stiff to hard clays values between 0.2 and $0.4\ \gamma H$ are in use. Some designers tend to use an earth pressure coefficient greater than the active value, K_a, and a value between active and at-rest states has found support. There are many other approximate methods of strutted wall analysis such as the method of James and Jack,[22] in which the wall is considered as a beam resting on springs. With all these methods it has been found impossible to predict either strut loads or ground movements with any degree of accuracy for most field cases.

At present the design of multi-anchored walls tends to follow the above procedures for strutted walls in which it is assumed that the horizontal component of the anchor forces equals the net earth and water pressure load on the wall. Quite clearly there are differences between the two methods of support. First, the force system in the wall differs because the vertical component of the anchor force is carried by the wall. Great care must therefore be taken to ensure that the wall has sufficient bearing capacity to sustain these forces as well as any structural loadings that it may be required to carry. At the present time it is possible to check bearing capacity by use of the methods developed by Meyerhof[23] for the bearing capacity of footings under inclined and eccentric loads, but it must be appreciated that such checks are very approximate. Secondly, ground anchors are relatively flexible members and during construction they are pre-stressed to loads determined theoretically in an attempt to control movement. Because of the limitations of the design methods referred to above, these theoretical load values are approximate, and it must be expected that during and subsequent to the completion of construction, changes in load in the individual anchors will occur. Associated with a load change is a change in the length of the anchor tendon. Consequently the wall must move or the anchor must yield or a combination of both may occur. In reality the problem is much more complex and Fig. 10 illustrates the probable changes which take place. A third point to bear in mind is that while the anchors and the wall may behave as anticipated at the design stage, it is possible that the anchors may be located in soil which is tending to creep towards the excavation. In such cases very large movements are possible yet

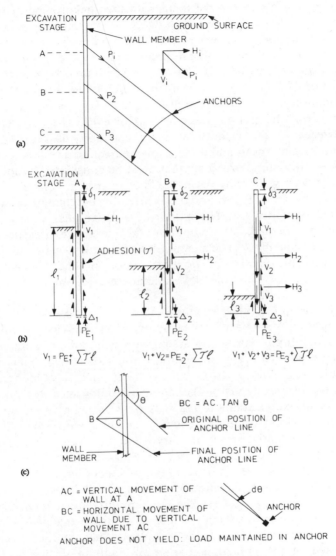

FIG. 10. Movements and load changes in a multi-anchored wall; (a) the wall system; (b) vertical movements and load changes during excavation; (c) rotation of the anchor. *Note*: $H_1 + H_2 + H_3 =$ the area under the pressure diagram (Fig. 9). Values of P_1, P_2 and P_3 may change as excavation continues. Test loading of an anchor will temporarily modify the load distribution.

it is not possible to control them with the anchor support technique.

The stability of the base of a deep excavation can be checked by standard methods[24] and where the diaphragm wall extends well below the base of the excavation problems seldom arise. Piping conditions must also be assessed if the water head is sufficient to produce velocities of flow that can result in a 'quick' condition. Much useful information is given by Tomlinson.[25] Various methods are available for lowering the ground water table. However, where the water table is lowered in the vicinity of an excavation, distress to adjacent structures may result and a range of methods of control is available.[26]

There are several problems that must be considered by the designer of diaphragm walls. Surcharge loads arise from a number of sources, particularly from the foundations of adjacent buildings within the retained ground mass. The extent of such surcharges varies widely and many methods of computing the surcharge pressures against walls are available.[17] The position of the ground water table and its variation with time are particularly important since diaphragm walls must resist not only the soil loading referred to earlier, but also hydrostatic loading.

From the preceding paragraphs it will be clear that the computed earth pressures are based primarily on empirical evidence, and checks on the stability of struts and anchor ties is based on assumed loading diagrams derived from observation and experience. Thus it is impossible to predict performance with any degree of confidence. In addition, few methods make adequate allowance for the overall stability of the excavation and it is possible with anchored walls to have the complete support system lying within a mass of soil which is part of a potential zone of movement or even slip. Under such circumstances the limit methods of analysis are far from adequate and over the past decade or so major advances have been made using analytical modelling techniques. With such modelling methods there are three basic approaches. The conventional one involves the analysis of each design problem on the assumption that it is completely isolated from the rest of the problem. Such methods have been discussed in the preceding paragraphs and in them the wall was considered in isolation and loaded by assumed earth and water pressures. It should be noted that it was not possible to examine the influences of the wall on the adjacent ground nor was it possible to predict movements of the structures and adjacent ground. Such a method of analysis remains attractive because it is simple, but it is inadequate because it does not determine the factor of safety of the support system, and it is not possible to compare different designs with a view to reducing ground and structural movements. An improvement on the conventional method is to allow for the interaction between the soil and the structure using a bed of springs.[27] Many difficulties are found, particularly in determining appropriate spring constants for the soil, in simulating construction details, and in the inability to predict soil movement.

Many of the above problems may be overcome in part by the use of the finite element method. Details of the finite element method are given in Chapters 1, 2, and 3, and the following is a brief summary of its application to the diaphragm wall problem. The first step in the analysis comprises the idealisation of the problem by constructing a finite element mesh which is used to represent the ground and the structure. Here it is important to note that the number of elements must be sufficient to allow for expected stress concentrations and structure flexibility as well as simulation of the boundary conditions such as three-dimensional effects.[28] The interface between the soil and the structure must also be idealised to allow for relative displacements and a number of mathematical models have been proposed.[28,29] The choice of constitutive models of behaviour for both the structural material and the ground is still somewhat uncertain and a range of models is in use. Much development work is in progress, particularly in the development of plastic models,[28] non-linear elastic models[30] and creep models,[28] as well as in making adequate allowances for joints and discontinuities in rock masses.[31]

It is well known that the sequence of construction influences earth pressure and the behaviour of earth retaining structures.[32] In particular, the effects of excavation, ground water lowering, strutting or anchoring and the placement of surcharge loads, are controlling and, by dividing the construction sequence into several stages, a finite element analysis may be used for each to predict the conditions applicable to the start of the next stage. Useful data are provided on excavation simulation,[33] strutting[34] and anchoring.[21] In simulation of the excavation process it is usual to alter the material properties and stiffness of the element to approximate its new role rather than to change the element mesh. The changing of the role of an element is achieved by recalculating the stiffness of the element before each load increment, using material constants specified for that increment. For example, in an excavation situation, when the element is excavated, it is assigned a near zero stiffness. Useful guidance on this subject is given in a number of publications.[21,33,34]

Whilst it is to be expected that all methods of analysis will be used in future, it is clear that the finite element method has particular advantages for evaluating the performance of different support systems, solving problems where no previous case records exist and for predicting zones of ground movement in the vicinity of deep excavations as well as providing a method of construction control. The diaphragm wall construction process provides many challenges to the designer and constructor. For example, there are uncertainties as to how one should allow for the following effects:

    (a)   the wall stiffness,
    (b)   the flexibility of the wall system and its influence on ground and wall movements,
    (c)   prestress level in tie-backs or in struts,
    (d)   the distribution of total and effective stresses on the wall, and
    (e)   the stress analysis of the wall member.

Diaphragm walls are relatively stiff members but numerical studies suggest that large increases in stiffness are needed to cause reductions in wall and ground movements. Perhaps it is well to remember that with diaphragm walls there are no gaps behind the wall as may occur with sheet piling or timber lagging. Such walls are also relatively impervious and little water will flow through the wall to cause consolidation effects in the retained soil mass. It is the view of the author that factors such as these may be equally or more important in the control of wall movement than the stiffness of the wall. The overall movements are therefore controlled not only by the wall but also by the stiffness of the support elements (ground anchors, struts, and the connections between the individual panels of the wall). Consequently it is the overall stiffness which is the determining factor, and a recent finite element study has shown how an increase in the stiffness of the ground anchors is as effective as increasing the wall stiffness in limiting movements.

The prestressing of struts or ground anchor supports automatically provides a 'stiff' system and tends to control movement. However, more data are still needed on the subject of earth pressure distribution. The required level of prestress is difficult to quantify at present, and the general approach followed by most authorities is to design according to the trapezoidal envelope suggested by Peck[20] for braced excavations, but using an earth pressure coefficient of about $(K_0 + K_a)/2$.[19,35] It will be noted that the recommendation is not in agreement with some practices,[18] but where a triangular pressure distribution is assumed, the forces in the lower struts or anchors are too large. With the trapezoidal distribution, higher loads are applied near the top of the wall. These forces are very effective during the early stages of the excavation in controlling lateral movements. Increasing the level of prestress may produce harmful effects such as causing overloading of the wall vertically and for these reasons the optimum design envelope for determining the prestress load determination is trapezoidal.[19,36]

As mentioned in earlier paragraphs the actual earth pressure distribution on a diaphragm wall is a function of a number of interdependent factors. Perhaps the most important factor is prestress level and, in particular, whether the total load transmitted to the wall by the support system is greater than or equal to the active thrust. As a guide to design it is usual to assume that the earth pressure distribution is the same as the hypothetical envelope used to determine the prestressing forces. Generally the system stiffness is such that active pressures cannot develop but some reduction from the at-rest pressure state occurs due to yield during the excavation processes. In addition to the earth pressure loads, water pressures may be significant particularly in the coarse grained soils and in such cases the earth pressures are computed using effective unit weights and the full hydrostatic pressures are added. In some geological situations seepage can take place beneath the wall. This causes a decrease in the water pressures but an increase in the soil pressures. The net effect is a small decrease in load on the wall.

The structural analysis of the wall may be performed in a number of ways depending on the sensitivity of the structure. The simplest approach is to assume hinges to form at each point of support.[37] In such an assumption the maximum moment occurs near mid-span. A more realistic approach is to treat the wall as a continuous beam with assumed loading as previously discussed, and to analyse the system using numerical techniques.[27] The simplest and most usual method is to represent the soil by a series of springs, and by this means allowance is made for the interaction between the wall, the soil and the support system. Where the project justifies it, the finite element technique can be used.[28-30,34,36] An area of analysis still in its infancy is the prediction of the vertical bearing capacity of diaphragm walls. At present simple checks are made to ensure adequate wall penetration beneath the base of the excavation using well-proven bearing capacity theory for eccentric and inclined load of surface and shallow footings.[23] Recent work has shown that the finite element idealisation may also incorporate such a component but that it is very expensive with respect to computer time.[38]

The prediction of ground movements near to an excavation is difficult and estimates are made based on experience. Normally with diaphragm walls the movements are small for most excavation situations, but where deep excavations are to be formed, or the ground is of a firm consistency, accurate methods of movement prediction are necessary, especially where adjacent structures and services are particularly sensitive to movement. The finite element method of analysis is the only technique that can be used, and much useful work has already been done in the analysis of excavations supported by diaphragm walls. Not only do such analyses provide a displacement field in the surrounding soil but they also permit a more rational approach to the general problem of determining the necessary penetration of the wall beneath the base of the excavation.

The use of diaphragm wall units as load bearing foundations presents very different design problems. Loads may be horizontal, vertical or inclined and many of the design principles in use for large diameter bored piles are applicable. Applied load is carried in both side friction and in end bearing, and at the design stage it is important to consider a number of factors. For example, if an excavation is formed on one side of the diaphragm, stress changes will occur in the ground, and in time a new stress system will develop. For such a situation it appears unwise to make use of experience gained from the traditional large diameter bored pile. The following comments refer to the load bearing units as envisaged by Kienberger[39] (Fig. 11). An estimate of ultimate load capacity in clays may be based on the semi-empirical design methods in use for bored piles which are as follows. The ultimate load carried by the shaft, Q_s, is

$$Q_s = A \cdot \alpha \cdot c_u \qquad (3)$$

where A is the area of panel in contact with the soil, α a factor relating c_u to the ultimate shaft friction and c_u the average undrained shear strength of the clay along the embedded length of the unit. The base load, Q_b, is

$$Q_b = A_b \cdot c_{ub} N_c \qquad (4)$$

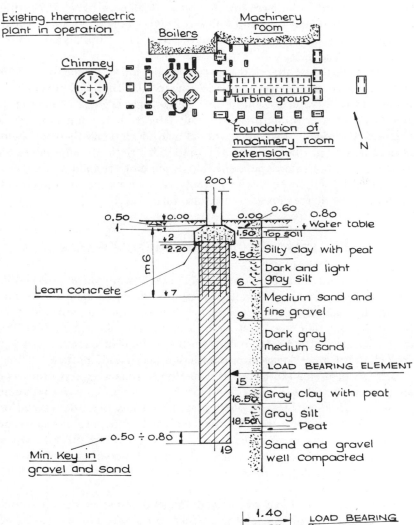

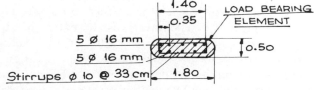

FIG. 11. Detail of diaphragm panels used as load bearing units for power
station foundation.[2]

where A_b is the plan area of the base of the unit, N_c a bearing capacity factor equal to about 9·0 and c_{ub} the 'bulk' undrained shear strength of the clay at foundation base level. The ultimate load Q_u is the sum of Q_s and Q_b.

At working loads, the load transfer to the soil is primarily in skin friction which depends on the properties of the contact between the soil and the concrete. Research[40] has shown that the use of bentonite does not lead to a reduction in the available skin friction mobilised.

In coarse grained soils, conventional bearing capacity theory should be used assuming that end bearing and side friction components of resistance are not interdependent. There are more refined methods of analysis which might be used for analysis, such as the finite element technique, but at present such uses are rare.

6.6   APPLICATIONS OF DIAPHRAGM WALLS

There are numerous applications of diaphragm walls on record and in the following paragraphs some examples are given. Here it is important to understand the practical details of construction and how they may influence the final structure. It may be argued that the detailed design is more dependent on the method of execution of the work than for most forms of construction because diaphragm walling is carried out under the cover of a bentonite suspension yet the wall has to be ready for use immediately it is uncovered. It is to the credit of the constructors of such walls that the art has been developed to such a stage that the uncertainties which do exist do not prevent diaphragm walling techniques from being used competitively. Much research and development must take place, however, before the design and construction of diaphragm walls can be tackled in a fully quantitative way.

Diaphragm walls have been used for some time in building construction. Excavation causes the adjacent ground to move and one of the most important considerations is the extent of such movements and the way in which they affect nearby services and the foundations of adjacent structures. Usually such walls function as permanent perimeter walls and act as cut-offs in water-bearing strata, as well as carrying vertical loads. The wall must, therefore, control lateral seepage and reduce the quantity of water seeping through the bottom of the excavation. This may be achieved in water-bearing ground by extending the wall through the pervious strata and keying it into an impervious stratum below. In the great majority of cases it will be impossible to get a perfectly water-tight wall keyed into an imperme-able stratum, and in such cases consideration has to be given to dewatering.

Various bracing systems are in use depending on the shape and depth of the excavation. In general the use of a bracing system controls but does not eliminate wall movements. Unbraced or cantilever walls are occasionally used, especially where allowable wall movements are large. Normally some form of bracing is required and Fig. 12 illustrates a range of bracing combinations. Strutting

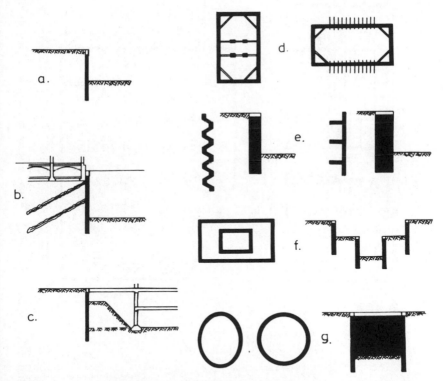

FIG. 12.   Bracing combinations used for deep excavations;[2] (a) cantilever wall; (b) tie back support wall; (c) temporary beam support; (d) internal and corner bracing; (e) counterfort or key-patterned walls. (f) cantilever walls to different levels; (g) circular and elliptical walls.

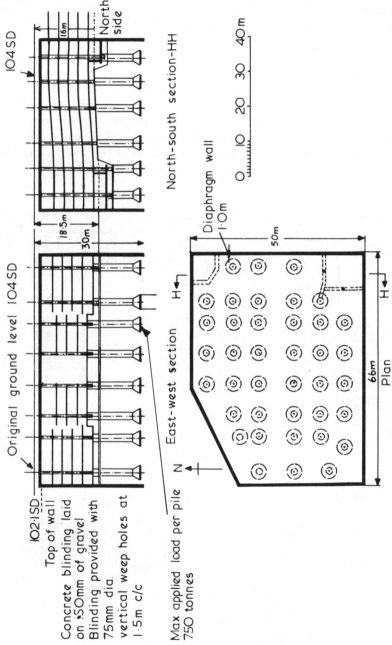

FIG. 13(a).   Underground car park at House of Commons, London; layout of walls and bracing support piles.[42]

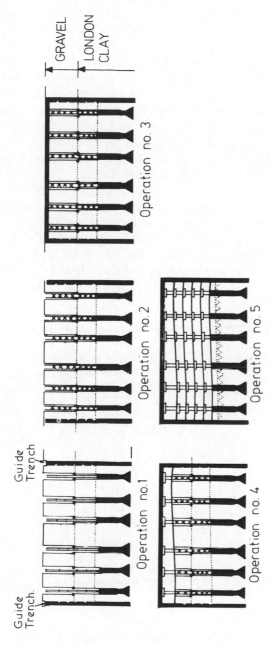

FIG. 13(b). Underground car park at House of Commons, London; construction procedures.[42] *Operation No. 1*: Form guide trenches and construct perimeter reinforced concrete diaphragm wall. Install 2·3 m diam. steel cylinder casings through the gravel and obtain a seal into the clay. Auger and install smaller diameter steel cylinders in the clay down to level 78·50 SD. Auger shafts below this level and form bells. Place cylinder reinforcement and concrete the bells and shafts. Erect structural steel columns. *Operation No. 2*: Concrete the shafts from base plate level of steel columns to the level of the soffit of the lowest suspended floor. Backfill the shafts with granular materials. *Operation No. 3*: Excavate ground surface to soffit of roof slab. Construct roof slab on ground. *Operation No. 4*: Excavate to next floor level. Cut away steel cylinders and remove gravel filling. Construct floor slab on prepared ground surface. *Operation No. 5*: Continue this procedure on successive floors downwards.

across the excavation restricts working space but these methods are widely used. Much useful information is given in the Institution of Structural Engineers report, *Design and Construction of Deep Basements*[41] as well as in other publications.[34] Corner bracing may be used in association with two or three rows of horizontal struts for small excavations. For deep excavations, counterforts can be incorporated in the wall and these increase the depth before bracing is required as well as increasing the bending resistance of the wall. In addition there are special walls such as an excavation within an excavation, or a circular shaped excavation. The above mentioned bracing systems may be combined as required to suit the site conditions.

The following example shows the versatility of the diaphragm wall in the construction of an underground car park in central London. The retaining wall was built using slurry bentonite methods. This wall was strutted by constructing the floors successively from the top downwards, each floor being cast on the ground which had been levelled. The underlying soil was then mined out. Details of the

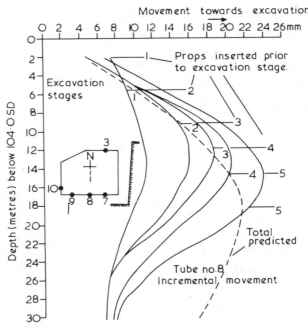

FIG. 14(a).   Observed ground movements of diaphragm wall.[42]

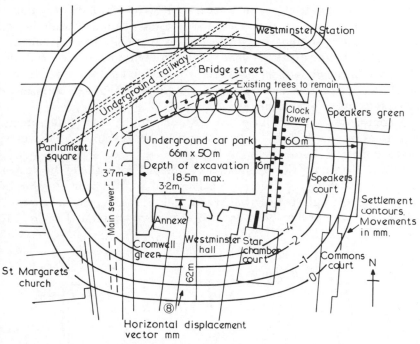

FIG. 14(b).   Radius of influence of ground movements.[42]

structural layout of the garage building are given in Fig. 13(a) and the
sequence of construction is detailed in Fig. 13(b). It should be
appreciated that the adjacent buildings and services were of great
importance and consequently the movements caused during construction, and their capacity to cause damage, had to be assessed with
care.[42] The structure was very carefully monitored during construction
and checks were made using numerical modelling methods. Very
good agreement between measured wall movements and predicted
values was found (Fig. 14(a)). It was also found that the zone of
ground movement around the excavation extended beyond the edge
of the excavation to a distance equal to several times the depth of the
excavation (Fig. 14(b)).

In built-up areas, where space is at a premium, the diaphragm wall
has been found useful in subway systems and underground railways.
The design of a modern subway requires consideration of a large
number of factors which include nearness of adjacent structures and

services, disruption of surface access and ground conditions. For a subway under a street, it is usual to build the diaphragm wall during off-peak hours so that traffic disruption is small. Excavation and strutting then take place between the walls, the tunnel top and bottom slabs are installed, and services are provided. Where there are heavy surcharge loads from nearby structures, keyed walls may be used to provide a large moment of resistance.

Underpasses are an ideal use for diaphragm walls, and many examples are on record. Underpasses may be covered or uncovered and a typical section is given in Fig. 15.

One of the first major applications of diaphragm walls in the design of quay walls was at Liverpool.[43] The wall was 2 km in length. An arch type of wall facing was used (Fig. 16). A cellular type of diaphragm wall system was used for an ore terminal at Redcar.[43] It

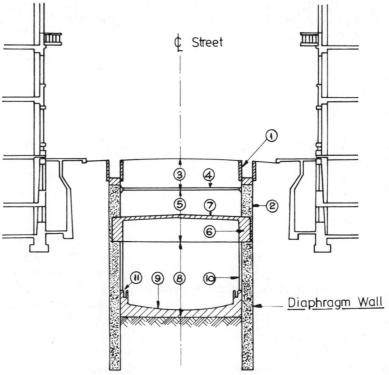

FIG. 15.   Construction stages for subway tunnel.[2]

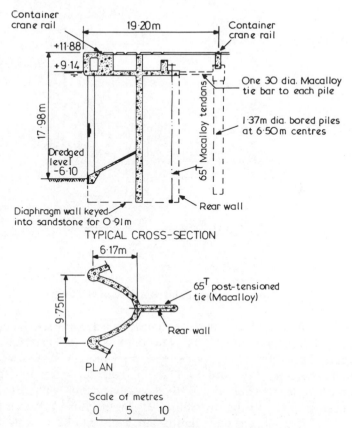

FIG. 16. Royal Seaforth Dock: east quay diaphragm wall.[49]

comprised 21 cells, each 30 m × 15 m in plan (Fig. 17). Special precautions were taken to ensure that tearing of the panels would not occur.

The use of diaphragm walls as load bearing elements is reported by Kienberger,[39] and the arrangement of the foundations for UNO City, Vienna, demonstrates the technique. Y-shaped structures of 56–116 m in height are grouped around a pavilion with load concentrations at the middle and ends of the Y. The arrangement of a typical diaphragm load bearing unit is given in Fig. 18. To prove the design, load tests were carried out on three elements.

The diaphragm wall technique has been used for circular-shaped

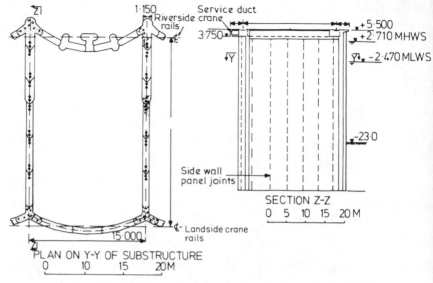

FIG. 17. Redcar ore terminal: diaphragm wall details.[49]

structures and one of the early examples is a car park in central London.[44] Forty two panels 0·814 m thick and 19·8 m deep were constructed from curved tracks. A temporary bracing system was not required. The final structure is 47 m in diameter internally, giving seven parking levels (Fig. 19). With a relatively thin wall, consideration has to be given to the possibility of ring buckling.[41] Recently the use of unreinforced, circular diaphragm walls has been considered,

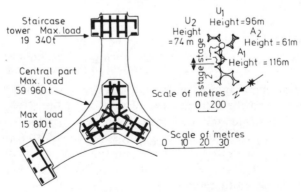

FIG. 18. UNO City, Vienna: foundation plan of load bearing units.[39]

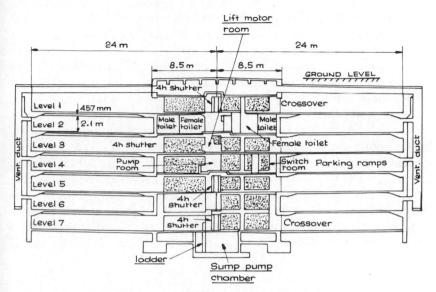

FIG. 19. Bloomsbury Square underground parking garage, London.[2]

and tests using centrifuge methods have been made by Rigden and Rowe[45] for a 0·8 m-thick wall designed as a thin shell. These tests showed that such a wall would have been stable for an unsupported height of 18 m. Several factors were considered, such as the influence of non-circular construction, non-uniform loading and fixity below dredge level. By use of a flexible bentonite clay fill below excavation level, the fixity restraint was removed and the factor of safety against cracking under vertical bending increased considerably. Unfortunately no field comparisons are available, but the work has shown the value of such testing methods in arriving at economic designs.

In all construction work, the overall cost is usually the controlling factor, and for a particular site several factors control feasibility of a particular solution. Much useful data are given by Puller[46] on the economics of basement construction.

6.7 OTHER FACTORS

In addition to the various factors mentioned in earlier sections, there are several topics of great significance which deserve mention. The diaphragm wall specification is very important. There are three major documents worthy of study. First, the Federation of Piling Specialists (FPS) produced a document on cast *in situ* diaphragm walling in 1973.[47] In this specification design, materials, construction and records to be taken are covered in detail. Secondly, in 1974 a paper based on the FPS document was prepared.[5] In it some of the differences of views held are discussed but the paper covers four main areas, design, materials, workmanship and performance requirements. A review of current practice for controlling the preparation and placing of bentonite slurry and concrete diaphragm walls is given in a more recent review by Hodgson.[6] Topics such as bentonite types, mixing, recycling and test methods, the use of additives, and concrete requirements are discussed in detail. Thirdly, The Institution of Structural Engineers' report, *Design and Construction of Deep Basements*[41] covers many aspects of diaphragm wall use in deep excavations and draws attention to safety, legal and contractual issues.

REFERENCES

1. VEDER, C. (1974). Closing Address, *Diaphragm Walls and Anchorages Conference*, Inst. Civ. Engrs., London, 1974, 221–222.
2. XANTHAKOS, P. (1974). Underground construction in fluid trenches, *National Educational Seminar*, University of Illinois at Chicago Circle, April 1974, 291 pp.
3. KAPP, M. S. (1969). The application of cast *in situ* diaphragm walls in the United States, *Proc. 7th Int. Conf. Soil Mech. Found. Eng.*, 1969, Specialty Session 14, 92–97.
4. SLIVINSKI, Z. and FLEMING, W. G. K. (1974). Practical considerations affecting the construction of diaphragm walls, *Diaphragm Walls and Anchorages Conference*, Inst. Civ. Engrs., London, 1974, 1–10.
5. FLEMING, W. G. K., FUCHSBERGER, M., KIPPS, O. and SLIVINSKI, Z. (1974). Diaphragm wall specification, *Diaphragm Walls and Anchorages Conference*, Inst. Civ. Engrs., London, 1974, 207–212.
6. HODGSON, F. T. (1976). Design and control of bentonite/clay suspensions and concrete in diaphragm wall construction, *Seminar on Diaphragm Walls and Anchorages*, Inst. Civ. Engrs., London, 1976, 31–38.
7. LITTLE, A. L. (1976). Slurry trench practice for diaphragm walls and cutoffs, *Seminar on Diaphragm Walls and Anchorages*, Inst. Civ. Engrs., London, 1976, 45–51.
8. NASH, J. K. T. L. (1974). Stability of trenches filled with fluids, and diaphragm wall construction techniques, *Proc. ASCE*, **100**(CO4), 533–542.
9. FUCHSBERGER, M. (1976). Some practical aspects of diaphragm wall construction, *Diaphragm Walls and Anchorages Conference*, Inst. Civ. Engrs., London, 1976, 75–79.

10. HUTCHINSON, M. T., DAW, G. P., SHOTTON, P. G. and JAMES, A. N. (1974). The properties of bentonite slurries used in diaphragm walling and their control, *Diaphragm Walls and Anchorages Conference*, Inst. Civ. Engrs., London, 1974, 33–39.
11. JEFFERIES, S. A. (1972). The composition and use of slurries in civil engineering practice, Ph.D. Thesis, University of London.
12. NASH, J. K. T. L. and JONES, A. (1963). The support of trenches using fluid mud, *Symposium on Grouts and Drilling Muds in Engineering Practice*, Inst. Civ. Engrs., London, 1963, 177–180.
13. MORGENSTERN, N. R. and AMIR-TAHMASSEB, I. (1965). The stability of a slurry trench in cohesionless soil, *Géotechnique*, 15(4), 387–395.
14. COLAS DES FRANCS, E. (1974). Prefasif prefabricated diaphragm walls, *Diaphragm Walls and Anchorages Conference*, Inst. Civ. Engrs., London, 1974, 81–87.
15. LEONARD, M. S. M. (1974). Precast diaphragm walls used for the A13 motorway, Paris, *Diaphragm Walls and Anchorages Conference*, Inst. Civ. Engrs., London, 1974, 89–93.
16. GYSI, H. J. (1974). Discussion, *Diaphragm Walls and Anchorages Conference*, Inst. Civ. Engrs., London, 1974, 104–105.
17. FLEMING, W. G. K. (1976). General analysis of available methods for designing diaphragm walls and anchored walls, *Seminar on Diaphragm Walls and Anchorages*, Inst. Civ. Engrs., London, 1976, 1–8.
18. LITTLEJOHN, G. S. and MACFARLANE, I. M. (1974). A case history study for multi-tied diaphragm walls, *Diaphragm Walls and Anchorages Conference*, Inst. Civ. Engrs., London, 1974, 113–121.
19. HANNA, T. H. and MATALLANA, G. M. (1970). The behaviour of tied-back retaining walls, *Canadian Geotech. J.*, 7(5), 372–396.
20. PECK, R. B. (1969). Deep excavations and tunnelling in soft ground, State-of-the-Art Report, *Proc. 7th Int. Conf. Soil Mech. Found. Eng.*, Mexico City, 1969, 225–230.
21. CLOUGH, G. W., WEBER, P. R. and LAMONT, J. (1972). Design and observation of a tied-back wall, *Performance of Earth and Earth-Supported Structures*, Purdue University, 1972, 1(2), 1367–1389.
22. JAMES, E. L. and JACK, B. J. (1974). A design study of diaphragm walls, *Diaphragm Walls and Anchorages Conference*, Inst. Civ. Engrs., London, 1974, 41–49.
23. MEYERHOF, G. G. (1953). The bearing capacity of foundations under eccentric and inclined loads, *Proc. 2nd Int. Conf. Soil Mech. Found. Eng.*, Zurich, 1953, 1, 440–445.
24. MEYERHOF, G. G. (1972). Stability of slurry trench cuts in soft clay, *Performance of Earth and Earth-Supported Structures*, Purdue University, 1972, 1(2), 1451–1466.
25. TOMLINSON, M. J. (1970). Lateral support of deep excavations, *Proc. Conf. on Ground Engineering*, Inst. Civ. Engrs., London, 1970, 55–64.
26. ZEEVAERT, L. (1972). *Foundation Engineering for Difficult Subsoil Conditions*, Van Nostrand Reinhold Co, New York.
27. HALIBURTON, T. A. (1968). Numerical analysis of flexible retaining structures, *Proc. ASCE*, 94(SM3), 1233–1251.

28. ZIENKIEWICZ, O. C. (1971). *The Finite Element Method in Engineering Science*, 2nd Edition, McGraw-Hill, New York.
29. CLOUGH, G. W. and DUNCAN, J. M. (1971). Finite element analysis of retaining wall behaviour, *Proc. ASCE*, **97**(SM12), 1657–1673.
30. DESAI, C. S. (1971). Non-linear analyses using spline functions, *Proc. ASCE*, **97**(SM10), 1461–1480.
31. GOODMAN, R. E., TAYLOR, R. L. and BREKKE, T. (1968). A model for the mechanics of jointed rock, *Proc. ASCE*, **94**(SM3), 637–659.
32. TERZAGHI, K. (1934). Large retaining wall tests, *Engineering News Record*, **III**, February, 136–140.
33. CHANG, C. Y. (1969). Finite element analyses of soil movements caused by deep excavation and dewatering, Ph.D. Thesis, University of California, Berkeley.
34. WONG, I. H. (1971). Analysis of braced excavations, Ph.D. Thesis, Massachusetts Institute of Technology.
35. TSUI, Y. (1974). A fundamental study of tied-back wall behaviour, Ph.D. Thesis, Duke University, North Carolina.
36. CLOUGH, G. W. (1972). Application of the finite element method to earth–structure interaction, *State-of-the-Art Report, Applications of the Finite Element Method in Geotechnical Engineering*, Vicksburg, Mississippi, **3**, 1057–1116.
37. The Institution of Structural Engineers (1951). Earth retaining structures, *Civil Engineering Code of Practice, No. 2.*
38. ARBER, N. R. (1976). A study of the load behaviour of a basement-type retaining wall embedded in sand, Ph.D. Thesis, University of Sheffield.
39. KIENBERGER, H. (1974). Diaphragm walls as load bearing foundations, *Diaphragm Walls and Anchorages Conference*, Inst. Civ. Engrs., London, 19–22.
40. CIRIA. The use and influence of bentonite on bored pile construction, to be published.
41. The Institution of Structural Engineers, (1975). *Design and Construction of Deep Basements*, Inst. Struct. Engrs., London.
42. BURLAND, J. B. and HANCOCK, R. J. R. (1977). Underground car park at the House of Commons, London: Geotechnical Aspects, *Structural Engineer*, **55**(2), 87–100.
43. AGAR, M. and IRWIN-CHILDS, F. (1970). Seaforth Docks, Liverpool: planning and design, *Proc. I.C.E.*, **54**(1), 255–274.
44. Anon (1972). Bloomsbury underground car park, *Ground Engineering*, **5**(2).
45. RIGDEN, W. J. and ROWE, P. W. (1974). Model performance of an unreinforced diaphragm wall, *Diaphragm Walls and Anchorages Conference*, Inst. Civ. Engrs., London, 1974, 63–68.
46. PULLER, M. J. (1974). Economics of basement construction, *Diaphragm Walls and Anchorages Conference*, Inst. Civ. Engrs., London, 1974, 171–180.
47. ANON (1973). Specification for cast in place concrete diaphragm walling, *Ground Engineering*, **6**(4), 31–34.

48. LITTLE, A. L. (1976). Slurry trench practice for diaphragm walls and cut-offs, *Seminar on Diaphragm Walls and Anchorages*, Inst. Civ. Engrs., London, 1976, 45–51.
49. FISHER, F. A. (1974). Diaphragm wall projects at Seaforth, Redcar, Bristol and Harrow, *Diaphragm Walls and Anchorages Conference*, Inst. Civ. Engrs., London, 1974, 11–18.
50. TERZAGHI, K. and PECK, R. B. (1967). *Soil Mechanics in Engineering Practice*, 2nd Edition, John Wiley & Sons, New York.

Chapter 7

GROUND ANCHORS

T. H. HANNA

Department of Civil and Structural Engineering, The University of Sheffield, UK

SUMMARY

Ground anchors have been used extensively for stabilisation of rock cuttings and underground openings for more than 40 years, although their use in soils and soft rocks is more recent. Nevertheless, design is relatively unsophisticated, and construction practice is well ahead of methods of analysis. This chapter describes the form and construction of ground anchors in common use, and the methods used to assess their bearing capacity. Procedures for testing anchors are described and methods for checking the overall stability of the structure are discussed.

7.1 INTRODUCTION

The use of anchors is a construction process by which forces are introduced into the adjacent ground in such a manner that a compressive stress exists between the ground and the structure. In this manner the structure is anchored to the ground and where the design is properly executed the stability of the structure is achieved economically. Thus the self weight of the ground is used to replace the dead weight of the structure. There are many examples where ground anchoring techniques are attractive. They include resistance to uplift from water pressure, for example dry dock floors, the overturning of tall structures from wind loading, such as masts on towers, the load testing of piles and plates, the preloading of soil to cause consolidation, and the stabilisation of steep slopes.

Many factors control the use of ground anchors and the following are important.

1. Is the ground competent to carry the anchor forces?
2. Is it possible to use anchors, bearing in mind the presence of adjacent services, buildings and the susceptibility to damage from construction work?
3. How does anchor forming affect the adjacent facilities? For example, will movements of the ground occur and will damage result?
4. Is it possible to obtain permission to install anchors beneath adjacent property?
5. Is there adequate access to enable the anchors to be constructed?

Anchors have been in use for a long time and one of the earliest patents is that of Mitchell for screw piles which was presented in 1833.[1] Since that time numerous anchoring devices have been proposed and used with varying degrees of success. Ground anchors as used today can be traced back to the pioneering efforts of Drouhin,[2] who successfully used the recently introduced prestressing techniques to heighten a concrete dam in Algeria. Since that time they have been used extensively in the mining industry for the stabilisation of rock cuttings and underground openings and much published work has been presented during the past 40 years,[3] although their use in sands, clays, gravels and soft rocks started less than 25 years ago.[4] Today the use of ground anchors has become widespread in the support both of temporary and of permanent structures, such as retaining walls, and also in resisting tensile forces in many situations, as mentioned previously.

Ground anchor design is still relatively unsophisticated and relies on rule-of-thumb methods to a considerable extent. Consequently construction practice is well ahead of methods of analysis. For this reason little is known about the mechanism of anchor behaviour and, in consequence, designs have to be conservative. Also, the lack of proper understanding of anchoring principles has led, in a number of isolated cases, to failures with consequent damage to adjacent structures. With this recent activity in ground anchor use many systems have been developed and have been shown to be suitable for a range of applications. Some particular anchors have practical uses within a restricted range of ground and loading conditions only. It should be appreciated, therefore, that an incorrect choice of anchor can lead to trouble at a later stage and there may be very serious implications. When faced with an anchor design problem the engineer has to quantify his design with confidence and determine the load carrying capacity of the anchors when forming part of a structure. Analytical methods of prediction are desirable and important developments have taken place primarily through the use of computer modelling techniques.[5] Despite these recent advances there are many major construction operations which cannot be quantified analytically and consequently the field testing of anchors and the monitoring of the behaviour of anchored structures are essential.

The primary purpose of this chapter is to assist the users of ground anchors to understand the main factors which control selection, design and use of a ground anchor scheme. To this end a review is given of the major anchor systems available, the methods of construction employed, and the principles of design and performance evaluation used. Brief information is given on the present state of knowledge and the reader is drawn to other sources of information that are relevant. Recently a number of codes of practice and manuals on ground anchor use have been published and for those contemplating ground anchor use these should be referred to.[6–10]

7.2 DETAILS OF A GROUND ANCHOR

A ground anchor essentially is a structural member which transmits a tensile force to the adjacent ground. The shear strength of the ground *en masse* is used to resist this tensile force. In general an anchor comprises a high strength steel tendon installed at the required inclination to resist the applied load efficiently. The force in the anchor is that necessary for equilibrium between the anchor, the structure and the ground mass, such that the movement of the structure and the ground mass are kept to relatively small values. Because the ground environment may be corrosive, the tendon is usually surrounded by a cement grout and the tensile force is transmitted to the surrounding ground via this grout annulus. Load is mobilised over the fixed anchor length, whilst over the free anchor length the tendon is debonded with respect to the surrounding ground mass, as shown in Fig. 1.

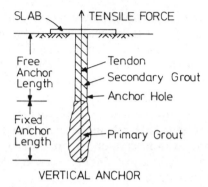

FIG. 1.   Detail of a ground anchor used in a vertical direction.

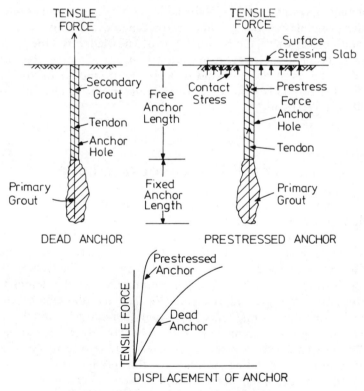

FIG. 2.   A comparison of the principles of operation of dead and prestressed anchors.

A dead anchor is under no load but when the ground moves relative to the anchor member, load is mobilised in it, as shown in Fig. 2. With such an anchor, relatively large movements are required to mobilise its full load-carrying capacity. These movements can be reduced to a significant extent by prestressing methods where the anchor is initially tensioned to the structure or to a ground surface slab or components. This serves a dual purpose in that the anchor system is also subjected to a structural load test. The level of this prestress force is usually taken as a percentage of the design working load.[11]

In use anchors may be called upon to serve for short periods of time, normally a few days to several months, or in some cases many years and these uses are referred to as short- and long-term respectively.[4]

It is important to understand and appreciate the mechanism of load mobilisation within the anchorage zone. There are two common modes. First it is normal, in the majority of anchors, to develop load from the top of the anchorage zone, thus putting the surrounding grout in the anchorage zone in tension. Associated with this tensioning, cracking of the grout may occur with the potential for corrosion attack. In some sophisticated anchor systems the tensile force is transferred to the bottom end of the anchorage zone and these anchors are referred to as compression anchors, as shown in Fig. 3.

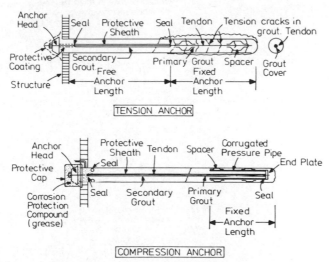

FIG. 3.   General details of tension and compression anchors.

Anchors of this type are seldom used for temporary work and have found much favour throughout the world, particularly in western Europe.[12]

There are three main classifications of anchors as follows: (I) a cylinder filled with grout; (II) a cylinder enlarged by grout injected under a controlled pressure or by the forcing of pea-sized gravel into the sides of the anchor hole; (III) a cylinder enlarged at one or more positions along its length to enable a larger load to be developed. Schematic diagrams for the three anchor systems are shown in Fig. 4.

The first type of anchor is usually used in rock. After the anchor hole has been drilled to the required size and depth it is good practice to check the permeability of the fixed anchor zone by water testing. In this test the fixed zone is sealed by means of a packer located at the top end of the fixed anchorage zone and water is pumped in under controlled pressure. The flow of water is recorded. This test quantifies

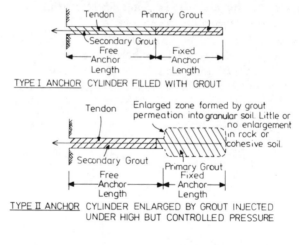

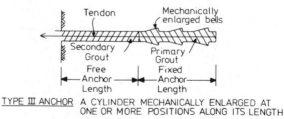

FIG. 4.   Classification of main anchor types in use in the UK.

the structure of the rock mass and provides a measure of the quantities of grout required. If the test fails the hole is grouted under low pressure and after a period of rest the hole is re-drilled and re-tested for water tightness.[13] The anchor tendon may then be placed and its form depends on whether it is for temporary or permanent use. For temporary use it is usual to coat the free anchorage length with a grease-impregnated tape. This gives short-term corrosion protection, yet permits the tendon to stretch during stressing. For permanent anchors the free anchorage length is protected within a polythene sheath surrounding the tendon or else each member forming the tendon is grease-coated and tightly covered with a polypropylene sheath. Within the fixed anchor zone the tendon is stripped, degreased and the tendon arranged to suit the particular anchoring system being employed. Grouting of the tendon is then carried out by tremie methods and may be before or after placement of the tendon.

The second type of anchor may be used in both cohesive and cohesionless soils, and a borehole between 75 and 125 mm in diameter is drilled, water or compressed air being used to bring the spoil to the

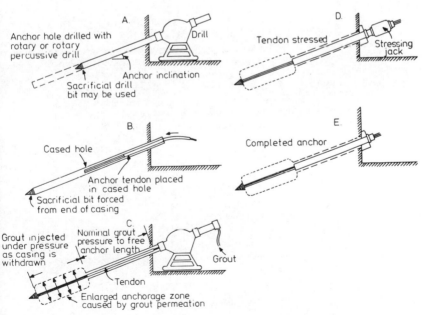

FIG. 5.   Construction sequences—Type II anchors.

surface. The grouting and the method of tendon placement in the hole depends on the anchor system being used. In most systems the grout pipe is attached to the tendon and the grout is pressurised as the borehole casing is withdrawn. By the correct choice of grouting pressure the diameter of the grouted zone may be increased several-fold from that of the original hole and its success depends to a large extent on the permeability of the soil to grout penetration. It is important that the pressure grouting operations are controlled and

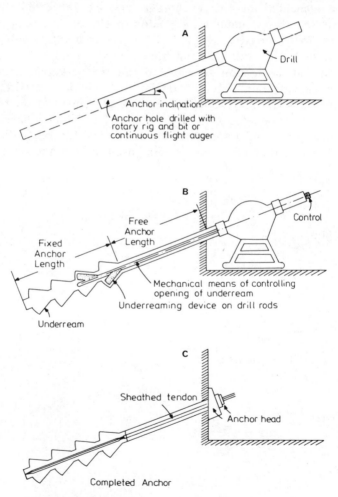

FIG. 6.   Construction sequences—Type III anchors.

there are several systems available, the most reliable being the tube à manchette system, which is a development of a grouting technique. A typical construction sequence is shown in Fig. 5 for this anchor system.

The third anchor system relies on the development of the strength of the clay at the anchor–clay interface. The most common form of anchor comprises the drilling of a cylindrical shaft, followed by the enlargement of the shaft at predetermined positions using a special underreaming device. A typical construction sequence is shown in Fig. 6.

Other methods of anchor formation have been employed and perhaps the most successful has been through the use of the tube à manchette system of grout control using several stages of grouting. This type of anchor in Germany[12] is preferred to the multi-under-reamed anchor which is more commonly used in the UK.[4]

7.3 BEARING CAPACITY OF AN ANCHOR

Static formulae have been arrived at over the years in using semi-empirical techniques and the development of the expressions closely parallels developments that took place many years ago for bored piles. An anchor may fail in several modes and each must be carefully checked at the design stage. These are: (a) within the ground mass; (b) at the ground–grout interface; (c) at the grout–tendon interface; and (d) at the attachment of the tendon to the structure. In addition to providing a safe working load per anchor, it is essential to ensure that the overall stability of the anchored structure is guaranteed. It is this latter requirement that usually determines the overall length of the anchor, the former requirement being controlled by the length of the fixed anchor zone.

Referring to Fig. 7, which shows idealised anchors for the three general cases referred to earlier, the following semi-empirical expressions have found much favour and can be used as a guide to the load carrying capacity of an anchor. In sands and gravels the ultimate carrying capacity of an anchor may be given as

$$q_u = L_m \tan \phi'$$

L being the fixed anchor length in metres, ϕ' being the angle of internal friction of the soil and m an empirical constant which

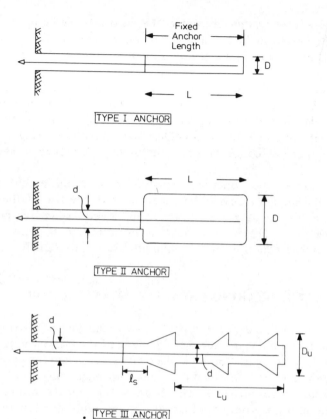

FIG. 7. Anchor dimensions.

depends on the fineness of the soil. For coarse sands and gravels values of 40–60 tonnes/m are recommended and for fine to medium sands 15 tonnes/m.

In stiff clays the type (I) anchor system may be treated as a long cylinder in adhesion and the pullout load

$$q_u = 0 \cdot 3 c_u \pi D \times L$$

where D is the diameter of the fixed anchor, L is the fixed anchor length and c_u is the average undrained shear strength of the clay in the fixed anchorage zone.

When the walls of the fixed anchor cylindrical zone are roughened by the penetration of gravel-sized stones, the shaft adhesion factor rises from about 0·3 to about 0·6 and the diameter of the fixed anchor increases by up to 50%. Thus the pullout load

$$q_u = 0·6c_u\pi D \times L + \frac{\pi}{4}(D^2 - d^2)c_u N_c$$

N_c is a bearing capacity factor $= 9·0$.

The incorporation of multi-underreamed anchorages increases the ultimate bearing capacity as follows

$$q_u = \pi \times D_u L_u c_u f_u + \frac{\pi}{4}(D_u^2 - d^2)c_u n_u + 0·3c_u\pi \, dl_s$$

where the various parameters are as shown in Fig. 7.

For anchors placed in rock strata, in general the bearing capacity is not a problem, provided appropriate methods of anchor hole drilling, tendon placement and subsequent grouting are used. In soft rocks empirical formulae are in use and an excellent review of the general subject of anchors in rock is given by Littlejohn and Bruce;[13] this document, along with the review text by Hobst and Zajic[3] should be consulted. When anchors are closely spaced together the zones of stress in the ground will interact and overlap and it is for this reason that designers tend to work to minimum spacings between anchors. This can be achieved by locating the fixed anchorage zones at different depths or at different inclinations in order to overcome the effects of closeness.[4] In some designs reduction curves based at anchor spacing are recommended and useful guidance is given in the French Code of Practice,[8] in which the allowable load of an anchor may be reduced depending on the anchor spacing.

In addition to checking on the ultimate working load of an isolated anchor there are a number of additional design considerations. These include loss of load and creep, corrosion protection, specification of the tendon and, in some cases, effects of repeated loading. Creep is the movement of the anchor top with time under sustained constant load, whilst load loss is the change of tendon load with time, the tendon being held under constant strain. Much useful data are given by Antill[14] and Bannister[15] on stress relaxation in tendons, whilst Littlejohn and Bruce[13] summarise observations on relaxation effects. Creep of anchors may occur in both cohesive and cohesionless ground conditions and in some situations it may be useful to determine the creep coefficient. Details are given by Ostermayer[12] and also in the German Code of Practice.[9]

In considering the corrosion protection system necessary for an anchor three main factors must be considered. These are the working life of the anchor; the environment in which the anchor is to be used; and the possibility of damage to the corrosion protection system during site operations when the anchor is being formed. As mentioned in an earlier section, anchors are usually divided into temporary and permanent categories and the protection provided reflects this. For temporary anchors it is usual to provide some protection over the free anchor length and normally the tendon is grease-coated and covered with a plastic-type tape. In some cases the surrounding annulus is infilled with a weak filler, such as cement grout. In cases where there is a very severe corrosive environment even temporary anchors will require additional protection and in such a situation the compression-type anchor is more suitable. For permanent anchors one has to allow for change in environment during the lifetime of the anchor system and in general a double protection system is used. Over the fixed anchor length the grout cover is normally employed, whilst the free anchor length is protected by a grease packed plastic sheath placed over the tendon under factory conditions. The annulus between the sheath and the borehole is normally grout-filled after the anchor has been stressed. Much useful information is given by Ostermayer,[12] Portier[16] and Bureau Securitas.[8]

Bar and strand are the more commonly used tendon materials in the UK and the steel characteristics should conform to the relevant British Standards.[4] With bar tendons either single bars or up to four bars may be used, although the latter method requires a large anchor hole and heavy lifting tackle to place the tendon in the hole. Seven-strand wire is common and tendons may consist of between four and as many as 20 strands. Where prestressing wire is used, 10 to 100 wires may be assembled on a single tendon. Working stresses in the tendon steel are based, in part, on experience and generally, for temporary anchors, the factor of safety against failure is at least 1·6. For permanent anchors a safety factor of at least 2 is usually employed. This also permits a larger test overload to be applied during the stressing operation of up to 1·5 times the design working stress. There are numerous details that must be considered at the design stage and these have been reviewed by Hanna.[4]

In general repeated loading is not a serious factor for the majority of prestressed anchor foundations, although there may be situations where it can be significant and each of these should be treated on its merits. Such structures include those subjected to wind and wave action as well as tidal fluctuations. Some guidance is given by Hobst and Zajic.[3]

7.4 GROUND ANCHOR CONSTRUCTION

As with all ground engineering problems it is essential to quantify the site conditions and identify the relevant soil parameters. This requires an adequate site investigation which should provide the following information: (a) the soil and rock succession; (b) the compactness and grading of sandy strata, the strength of cohesive strata and the quality of rock strata, in particular their susceptibility to softening; (c) the presence of joints and discontinuities and fissures in the ground; (d) the bulk and local permeability of clay-type soil and the water-tightness of rocks; (e) the variability of the ground conditions across the site; and (f) the cementing of the sand and silt strata, if any. To provide this information in the detail necessary, very good site investigation practices and sampling details must be incorporated and in some cases it may be necessary to use inclined boreholes to reach under the existing sites where inclined anchors may be located.

The drilling of an anchor hole should be a relatively straightforward operation provided the correct plant is available. An important point to recognise is that a range of ground conditions may be encountered by one anchor hole and it is essential that the correct drilling machinery is used to cause the minimum of damage and disturbance to the ground in which the anchor is to be formed. Care must be taken in the use of vibratory driving and drilling techniques as, in some ground conditions, settlement of loose sand and fill strata may occur with serious damage to property and services. Useful guidance is given by Hanna[4] and Littlejohn and Bruce.[13]

For closely spaced anchor holes considerable care must be given to tolerances of drilling and it is generally agreed that it is possible to drill holes to tolerances of 1 in 50.

Tendon fabrication is normally a site operation and points that must be borne in mind are to ensure that all tendon material is protected against site damage and corrosion, that no splashes from welding occur and that the protective sheathing is not dragged across abrasive surfaces. Care must be given to cleaning the strand to be used in a fixed anchor length, whilst positive fixing of spacers must be provided, not only in the fixed but also in the free anchor zone. During installation the tendon should be carefully inspected for damage or bad fabrication. The design of a grout mix must ensure adequate pumpability, low bleed characteristics, a low water–cement ratio and a slight expansion on setting, if the necessary strength is to be developed in bond, both at the grout tendon and the grout–soil or grout–rock interfaces.[4] To achieve these, good site practices should be followed whereby the materials are measured by weight, mixing machinery is such that a uniform mix is obtained, and all grouting equipment and pumps are kept clean and well-maintained. Thus the same practices should apply to the grouting of anchors as are used on conventional grouting schemes. To quantify this, further measures of the strength capacity of the grout should be obtained from each anchor. This may consist of taking several pairs of 50 mm cubes for crushing at 3, 7, 14 and 28 days for each anchor constructed.

The success of any anchor depends on the proper grouting of the tendon and here either single stage or multiple stage grouting can be used. With the single stage method the complete anchor hole is grouted in one operation. With the double stage method the fixed anchor zone is grouted first, the tendon tested and stressed, then a second stage of grout injection is applied to provide corrosion protection to the free anchor length. With all grouting work the borehole must be thoroughly cleaned by flushing from the bottom upwards and at no stage should the end of the grout pipe be lifted up above the surface of the grout or the level of the grout in the storage tank drop to enable air to be drawn into the system. Also, the control of grouting pressures is prudent to ensure no damage to adjacent property or nearby anchors occurs from the fracturing of the ground. These safeguards can be achieved by rigorous site discipline being enforced and proper and detailed records of every operation being taken. There are a number of grout quality measures that can be taken on site and details are given in references 13 and 17–20.

Load capacity and construction costs limit the use of ground anchors to clays of stiff and hard consistency, sands, gravels and rocks, and at present anchors are generally not used in soft to firm clay and organic soils. Also it is important to remember that in difficult soil conditions only proven anchor systems should be used. This is particularly important where the ground properties change during the anchor hole boring process.

Other points of importance are the occasional necessity to install additional anchors between already constructed anchors. This normally arises where one has to upgrade the load carrying capacity of a block of anchors, or perhaps one or more anchors within a cluster fail to carry their proof load adequately. Under such circumstances care must be exercised to ensure that the additional construction work does not distress the already stressed anchors.

7.5 TESTING OF GROUND ANCHORS

The most common method of anchor stressing is by means of a direct pull using either multistrand jacks, where the tendons are stressed simultaneously, or using monojacking, where each strand is stressed in turn. In general single strand stressing is advantageous with tendons of up to five strands. There are many difficulties with this system and Mitchell[21] points out that the load should be applied in increments to each strand in turn in an attempt to prevent non-uniform load distributions developing throughout the tendon. Thus the multistrand stressing approach is to be preferred for medium to large capacity anchors. Whilst this does not prevent non-uniformity of load in individual strands with long anchors the problem is not serious. It should be appreciated that when an anchor is loaded the top bearing plate which is being stressed against may move and consequently it is unreliable to measure the gross extension of the tendon. The interpretation of a load extension relationship requires a knowledge of the mechanism of load transfer involved, as well as the characteristics of the steel tendon and much useful guidance is given in papers by Fenoux and Portier,[22] Portier,[16] Littlejohn and Bruce[13] and Mitchell.[21]

The subject of anchor testing is important and here one may be required to test the tendon components, as well as the completed ground anchor. In general, test certificates will be supplied by the suppliers of the steel components in the form of load extension curves, as well as ultimate strength values. Thus the major requirement is to assess the quality of the completed anchor. Here there are several Standards and Codes available, which include those for Australia,[6] Czechoslovakia,[7] France,[8] Germany,[9] South Africa,[10] Switzerland[23] and the United States.[24] The user of anchors is particularly recommended to examine the French[8] and German Codes.[9] Anchor testing may be subdivided into three categories. Firstly, the development of a new anchoring system where the objective is to ensure that the anchor system performs as planned. Such tests are not often carried out and here it is important to test at least three production anchors to about 1·5 times the permissible working load or to failure and then to dig out the anchors for inspection to ensure that the dimensions as specified have been achieved under field conditions. The German Code DIN 4125[9] provides an excellent outline guide to how such tests should be carried out and the type of information that should be obtained.

The second category of test is somewhat similar to a prépile testing programme and its purpose is to confirm a design load value or to arrive at the ultimate carrying capacity of an anchor system in a particular ground condition. Thus the number of tests is controlled by the ground conditions encountered as well as the inclination of the anchors, and the French Code of Practice[8] relates the minimum number of tests to the number of anchors falling within each ground condition and anchor inclination. It is usual to load test the anchor in increments up to at least 1·5 times the working load and an interpretation of this diagram provides a check on the free anchor length, as well as giving data on the inelastic movements of the fixed anchor zone.[4] In addition it is quite common practice to hold the prestress load for a period of 24 hours or more and to measure the load loss.

The third category of test is the load testing of each individual anchor during construction. With temporary anchors it is normal to load test to about 1·2 times the working load with a small percentage of the anchors tested to up to 1·5 times the working load. In the case of permanent anchors it is usual to stress each anchor to about 1·5 times the working load. Care must be taken to ensure that the tendon steel is not overstressed.

In rare circumstances it may be necessary to stress a cluster of anchors especially if they are closely spaced. Guidance on the testing of anchors is given in the recent CIRIA Report No 65.[4]

7.6   OVERALL STABILITY

At present the greatest use of ground anchors is for the support of retaining wall structures in the solution of deep excavation problems, and these types of structure fall into three main categories: massive, flexible and rigid. While such a wall is being constructed the installation of rows of ground anchors results in the structure being progressively fixed and thus the lateral deformations are limited to such an extent that failure within the retained ground mass is unlikely. I believe, therefore, that this problem is somewhat analogous to the braced excavation and the earth pressure distribution changes with construction progress. The envelope to all these earth pressure distributions approximates to the rectangular shape as originally suggested by Terzaghi and Peck.[25] However, the magnitude of the earth pressure is governed by the extent of the lateral yield which can take place. Normally a k_a earth pressure coefficient can be used, but many designers have tended to use the value between the at-rest value k_0 and the active value k_a in attempts to control the extent of lateral movements.

Once the distribution of earth pressure and its magnitude have been arrived at, it is possible to work out the arrangements for the horizontal and vertical spacing of the anchors. The simplest method follows the procedures developed for strut spacing determination, although other methods are available such as the empirical design techniques developed by James and Jack.[26]

In all calculations the horizontal component of the anchor force is used to balance the earth pressure assumed. This means that the flatter the anchor inclination the more efficient the load carrying capacity of the anchor becomes. Also the flatter the anchor inclination the smaller the vertical component of force which the wall must carry. It should be appreciated that the design of the individual anchors does not necessarily give an acceptable design with respect to lateral and vertical movements and in attempts to control these movements the overall stability of the anchor/wall/ground system has to be considered. As an excavation is formed in front of a wall and rows of inclined anchors are placed and prestressed, vertical force is developed in the wall; with excavation progress, bearing capacity at wall tip level becomes important and must be checked by methods such as those of Meyerhof.[27] Care must be taken to ensure that wall base failure cannot occur and this may be very important with walls founded on bedrock. In addition, checks should be made on the heave of excavations in clay and the methods of Stille[28] are recommended.

Most methods of overall stability checking are based on a limited state whereby a failure surface is assumed and the forces causing disturbance are compared with the forces resisting disturbance to give an overall factor of safety. One of the most commonly used methods

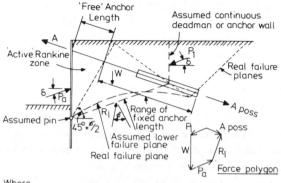

Where

   P_1 – earth pressure against imaginary continuous 'deadman' or anchor wall.

   W – total weight of assumed sliding soil mass

   P_a – total active earth pressure on wall

   δ – angle of friction between soil and wall

   R_1 – resultant reaction force against lower failure plane

   ϕ – angle of shearing resistance of soil

A poss – anchor force which is divided by an appropriate factor of safety to give anchor force required

FIG. 8.   Failure mechanism—Kranz method of analysis.[29]

is that due to Kranz[29] which is illustrated in Fig. 8. Many modifications to this method have occurred such as those by Broms[30] and useful guidance is given in the French Code of Practice on the Kranz method of overall analysis. Other methods of analysis are in use, such as a Coulomb wedge, but the most promising and logical solution to this problem is to model the wall construction excavation sequence using a finite element technique. This is a relatively expensive operation and there are a number of difficulties, particularly the modelling of soil behaviour, the modelling of wall–soil interface and the modelling of the method of excavation, as well as the idealisation of the anchors in the ground. One of the most useful programmes is that developed by Clough and Tsui,[31] which has enabled a number of parametric studies to be carried out concerning the yield of anchors.

   There are many examples of anchor-supported retaining walls and most of the literature is given in reference 4. In addition, excellent information is given in reference 18.

   There are many other applications of ground anchors and various mechanisms are assumed for checking overall stability of anchors and anchor groups. In the case of a dry dock floor slab which is anchored against hydrostatic uplift the most common assumption is that the carrying capacity of an individual anchor is controlled by the weight of a volume of soil or rock above it. This is usually taken as a cone of apex angle 2β (Fig. 9) with the vertex of the cone at the mid-point of fixed anchor length for a tension anchor and at the bottom for a compression anchor. Where anchors are closely spaced these cones interact and as a result the carrying capacity of the average anchor in the cluster may be arrived at. The value of the angle β is controlled by the frictional properties of the ground and normally is about $2/3\phi'$ in sands and about 45° in clays and rocks. An appropriate factor of safety has to be used since such methods are far from precise and values as high as 3 are in use. Much of this information is reviewed by Hobst and Zajic.[3]

In the stabilisation of hillsides subjected to creep and instability, the principle is to secure the moving mass of ground to intact ground beyond the surface of slip. This may necessitate a large number of rows of anchorages of considerable length, in some cases greater than 30 m. Generally the anchors are of modest capacity, in the order of 50 tons. With such methods it is important to provide adequate

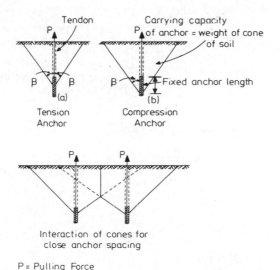

P = Pulling Force

FIG. 9. Mechanism of failure for checking overall stability of vertical anchors.

drainage to control and reduce the pore-water pressures within the unstable mass of ground. Thus the effective stress on the surface of failure can be increased considerably. The analysis of such a solution is somewhat inexact and the more common methods are to use conventional slip circle and wedge analyses simulating the anchor forces by point loads on the ground surface. Figure 10 shows an anchoring scheme for a slope executed in steps.

With all anchored structures attempts are made to ensure that each anchor is load tested and that it has been properly constructed. Despite such controls it is still possible that anchors may be damaged after construction. For this reason it is usual to design the structures on the assumption that one of the anchors may fail, but that overall failure of the structure does not occur to any of the adjacent anchors or that excessive yield of any of the structural components does not take place. When an anchor fails a load redistribution occurs and small additional movements of the structure result. This general problem has been examined by Stille[28] with respect to the failure of anchors supporting sheet pile walls and he has examined four sites where one or several anchor failures occurred. He has shown that failure of an anchor is not a serious problem if each anchor has been tested as detailed earlier and if sound and reliable methods of structural analysis have been used.

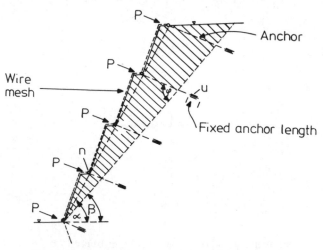

FIG. 10.   Anchoring system for a steep slope.[3]

Nearby construction activity must always be considered and this is particularly important with respect to changes in prestressed load caused by blasting. A most useful review of this problem is given by Littlejohn et al.[32] where they studied the influence of blasting adjacent to highly loaded prestressed anchors and they showed that the working load can be maintained.

Like all civil engineering techniques, anchoring methods are of great value to the geotechnical engineer and will provide economical solutions provided correct design and construction considerations are given. It is for this reason that detailed check lists should be drawn up for each scheme to ensure that not only the components of the anchor system, but also the complete anchor as constructed and the consequences of anchor use, are fully appreciated. Useful guidance is given in reference 4 on the minimum checks that should be made for factors such as corrosion protection, tendon design, site investigation information, anchor hole drilling, tendon manufacture, grouting operations, anchor stressing and performance evaluation of the anchored structure.

REFERENCES

1. *Encyclopaedia of Anchoring*, A. B. Chance Co., Bulletin 424, 1969.
2. DROUHIN, M. (1935). Consolidation du barrage des Cheurfas par tirants métalliques mis en tension. Fermeture partielle de la brèche de L'oeud Fergoug, *Annales des Ponts et Chaussées*, Paris, **105**, 257–272.
3. HOBST, L. and ZAJIC, J. (1977). *Anchoring in Rock, Developments in Geotechnical Engineering 13*, Elsevier Scientific Publishing Co., Amsterdam, p. 390.
4. HANNA, T. H. (1977). Design and construction of ground anchors, *Construction Industry Research and Information Association, Report No. 65*, July 1977, p. 64.
5. TSUI, Y. (1974). A fundamental study of tied-back wall behaviour, Ph.D Thesis, Duke University, Durham, North Carolina.
6. *Prestressed concrete code*, Standards Association of Australia, Australian Standard CA35, 1973.
7. *Draft standard for prestressed rock anchors*, Research Institute of Civil Engineering, Bratislava, 1976.
8. BUREAU SECURITAS (1972). *Recommendations concernant la conception, le calcul, l'execution et le controle des tirants d'ancrage*, Recommendations TA72, 1972, Paris.
9. DEUTSCHES INSTITUT FÜR NORMEN (1972, 1974). *Soil and rock anchors; bonded anchors for temporary use in loose ground—dimensioning, structural design and testing*, DIN 4125, Sheet 1, 1972, Sheet 2, 1974.

10. *Lateral support in surface excavations*, Code of Practice, South African Institution of Civil Engineers, 1972.
11. LITTLEJOHN, G. S. (1970). Soil anchors, *Proc. Ground Engineering Conference*, Institution of Civil Engineers, London, 1970, 30–44.
12. OSTERMAYER, H. (1974). Construction, carrying behaviour and creep characteristics of ground anchors, *Proc. Diaphragm Walls and Anchorages Conference*, Institution of Civil Engineers, London, 1974, 141–151.
13. LITTLEJOHN, G. S. and Bruce, D. A. (1975, 1976). Rock anchors: state of the art, *Ground Engineering*, **8**(3), 25–32; **8**(4), 41–48; **8**(5), 34, 35; **8**(6), 34–45; **9**(2), 20–29; **9**(3), 55–60; **9**(4), 33–44.
14. ANTILL, J. M. (1967). Relaxation characteristics of prestressing tendons, *Civil Engineering Transactions, Institution of Engineers of Australia*, **7**(2), 151–159.
15. BANNISTER, J. H. L. (1959). Characteristics of strand prestressing steel, *Structural Engineer*, **37**(3), 79–96.
16. PORTIER, J. L. (1974). Protection of tie-backs against corrosion, *Proc. Technical Session on Prestressed Concrete Foundations and Ground Anchors, 7th FIP Congress*, New York, 1974, 39–53.
17. *Proceedings, Diaphragm Walls and Anchorages Conference*, Institution of Civil Engineers, London, 1974, 223 pp.
18. *A Review of Diaphragm Walls*, Inst. Civ. Engrs., London, 1977, 148 pp.
19. NEVILLE, A. M. (1972). *Properties of Concrete*, 2nd edition, Pitmans, London.
20. POWERS, T. C. (1968). *The Properties of Fresh Concrete*, John Wiley & Sons, New York.
21. MITCHELL, J. M. (1976). Some experiences with ground anchors in London, *Proc. Diaphragm Walls and Anchorages Conference*, Institution of Civil Engineers, London, 129–133.
22. FENOUX, G. and PORTIER, J. L. (1972). La mise en précontrainte des tirants, *Travaux*, **54**(449/450), 33–43.
23. *Recommendations on prestressed ground anchors*, FIP Sub-committee on Prestressed Ground Anchors, Berne, 1973.
24. *Tentative recommendations for prestressed rock and soil anchors*, Prestressed Concrete Institute, Chicago, 1974.
25. TERZAGHI, K. and PECK, R. B. (1967). *Soil Mechanics in Engineering Practice*, 2nd edition, John Wiley & Sons, New York.
26. JAMES, E. L. and JACK, B. J. (1974). A design study of diaphragms walls, *Proc. Diaphragm Walls and Anchorages Conference*, Institution of Civil Engineers, London, 1974, pp. 41–49.
27. MEYERHOF, G. G. (1953). The bearing capacity of foundations under eccentric and inclined loads, *Proc. 3rd International Conference on Soil Mechanics and Foundation Engineering*, Zurich, 1953, **1**, 440–445.
28. STILLE, H. (1976). *Behaviour of Anchored Sheet Pile Walls*, Dept. of Soil and Rock Mechanics, Royal Institute of Technology, Stockholm.
29. KRANZ, E. (1953). *Uber die Verankerung von Spundwande*, 2nd edition, Wilhelm Ernst and Son, Berlin.

30. BROMS, B. B. (1968). Swedish tie-back anchor system for sheet pile walls, *Proc. 3rd Budapest Conference on Soil Mechanics and Foundation Engineering*, 1968, 391–403.
31. CLOUGH, G. W. and TSUI, Y. (1974). A study of the performance of anchored walls, *Proc. American Society of Civil Engineers, Journal of the Geotechnical Division*, December 1974, **100**(GT12), 1259–1273.
32. LITTLEJOHN, G. S., NORTON, P. J. and TURNER, M. J. (1977). A study of rock slope reinforcement at Westfield open pit and the effect of blasting on prestressed anchors, *Proc. Conference on Rock Engineering*, University of Newcastle upon Tyne, 1977, 293–310.

Chapter 8

THE DESIGN OF PILED FOUNDATIONS

R. W. COOKE

The Building Research Station, Garston, Herts, UK

SUMMARY

The foundation designer needs to understand how the performance of foundations is influenced by the properties of the supporting soil and, particularly in the case of piled foundations, how these properties are modified either permanently or for limited periods by the installation of the foundation. The settlement of a piled structure depends largely on the manner in which stresses are transferred away from the area loaded by the piles and on the immediate and longer term values of the soil properties in the stressed zone. Recent research has been concentrated on these aspects of piled foundation behaviour and this chapter attempts to highlight developments that are already influencing design or are likely to do so within the next few years.

8.1 INTRODUCTION

During the last two decades significant advances in piling technology have been made, and the numbers and sizes of piles employed in multi-storey building construction and motorway and off-shore structures have continued to increase. Advances in pile design during this period have been less spectacular, and empirical approaches based on experience seem likely to remain for the foreseeable future. For the design of piles in granular soils, recent published data have endorsed and extended the empirical factors used in conjunction with standard penetration test results, but little is known of the effects on bearing capacity of the soil compaction due to pile driving. Thus, driving

275

formulae are still employed although, from an analysis of 45 loading tests on driven piles, Tavernas and Audy[1] concluded that bearing capacities computed by five well-known formulae had no useful correlation to the actual bearing capacity.

Until more information on the process of load transfer becomes available, friction mobilisation factors (α), highly dependent on the quality of pile formation, must continue to be used for the design of bored piles in stiff clays. As the dimensions of large bored piles increase, accurate predictions of settlement become more desirable, yet loading tests remain the best method of assessing the performance of individual piles. Prediction of the settlement of free-standing pile groups and piled rafts continues to be based on extrapolation from the results of small model tests or on idealised theoretical studies, and very few observations have been made to check the validity of these approaches. However, recent research has contributed to improved understanding of pile behaviour. This has probably been concentrated in three main areas:

(a)   the effects of pile installation on the properties of the pile–soil system;

(b)   the mechanism of load transfer from piles to the soil;

(c)   the estimation of settlement.

Some of the more important developments in these areas will be discussed in this chapter, and an attempt will be made to relate them to existing or potential design approaches for axially loaded piled foundations in sands and clays. Space does not permit consideration of the increasingly important problems of the design of piles in soft rocks. This subject was very thoroughly covered in the 'Symposium in Print' (*Piles in Weak Rock*).[2]

8.2   PILES IN SAND

8.2.1.   Driven Piles

The process of driving a pile into sand displaces the particles and changes the density of the sand for some distance radially around the shaft and vertically beneath the base. The bearing capacity of a driven pile in sand depends very largely on the mean density of this disturbed zone. Loose sands may be densified considerably by the installation of driven piles and it is frequently beneficial to drive to large penetrations into them. On the other hand, driving piles deep into dense sands is liable to cause dilation of the sand with a consequential reduction in bearing capacity and probably heave of the ground surface.

Studies to estimate the size of the compacted zone around a driven pile have been undertaken by Meyerhof[3] who assumed that driving a pile of circular cross-section into loose sand would increase the angle of internal friction close to the shaft and immediately beneath the base from 30 to 35°. Meyerhof concluded that the diameter of the compacted zone around the shaft could be of the order of six times the shaft diameter. He also found that the zone would extend to a depth below the base of about five diameters and that its diameter a short distance below the base would be seven or eight times the shaft diameter. A five-degree increase in the angle of friction could approximately double both the base and shaft resistances of a driven pile in loose sand.

Radiography can be employed to show particle displacements and compaction in sand beds in the laboratory, and Robinsky and Morrison[4] used the technique to compare the behaviour of straight-shafted and tapered piles. For loose sand the dimensions of the compacted zone beneath the pile base are much as Meyerhof predicted. However, the process of compaction below the base is accompanied by sand movements adjacent to the shaft that tend to decrease the density there and to nullify some of the benefits resulting from the primary compaction around the shaft. This accounts for the low load transfer from the shaft widely associated with straight-shafted piles in sand. The same loosening process appears to take place close to the shaft of tapered piles, but the reduction in density is partially compensated by the wedging action of the taper so that much more of the load transfer is from the shaft than from the base. Robinsky and Morrison found that a tapered pile had an ultimate load capacity 40% greater than a straight-shafted pile of similar volume.

Because of the high level of compaction attained beneath the base when piles are driven into loose and medium sands, traditional bearing capacity theories (i.e. those of Terzaghi,[5] Meyerhof,[6] and Berezantzev et al.[7]) in which uniform density is assumed, cannot model the failure mechanism with accuracy. However, values of the bearing capacity factor N_q, given by Berezantzev et al. show a significant decrease with increasing length–diameter ratio L/d, that agrees with observations and which, for values of ϕ less than 35°, are more conservative than the corresponding values given by Terzaghi and by Meyerhof. Many textbooks now show curves of the variation of N_q with ϕ given by the solution of Berezantzev et al. for a range of values of L/d. Norlund[8] has reported the usefulness of this solution for interpreting pile test data.

Vesić[9] showed good agreement between the bearing capacity factors of Berezantzev et al. and the results of model pile tests. Vesić assumed that the effective sand density was the mean of the values measured prior to, and after, driving. The model piles, which were instrumented so that the development of the shaft and base resistances could be observed independently were driven by a laboratory drop hammer to various depths in sand beds having a range of relative densities. Neither the unit base resistance nor the unit shaft friction increased linearly with depth, but both approached constant values at a critical depth depending only on the relative density of the sand. In loose sand, constant values were attained at L/d ratios of about 10 and, in dense sand, of about 30. The existence of a critical depth related to sand density has since been demonstrated by numerous investigators including Kerisel,[10] Hanna and Tan,[11] and Tavernas.[12]

Field tests made by Vesić[13] on a 15 m long, 450 mm diameter pipe pile formed of five instrumented segments, confirmed his earlier model results. The pile was driven and tested by 3-m stages in deep deposits of medium dense sand. Both the unit base and unit shaft resistances increased approximately linearly for penetration less than 10 diameters, but became constant after depths greater than 20 diameters.

Because the ultimate bearing capacity of the bases of most piles of common dimensions $(20 < L/d < 40)$ is usually larger than the total friction, designers frequently fail to take advantage of the considerable frictional resistance that is available under working conditions. At small displacements friction develops more rapidly than support at the base, and Vesić's results (Fig. 1) showed that at a load factor of 3, when the L/d ratio of the segmental pile was 27, approximately 60% of the total resistance was mobilised on the shaft. In comparison, at the ultimate load, the proportion of the total load developed by friction had dropped to 40%. The maximum load transfer moved down the shaft as the L/d ratio was increased by adding pile segments, and Fig. 2 shows the distribution of unit frictional resistance at ultimate load for five values of L/d. For the larger values of L/d the

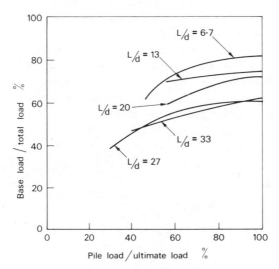

FIG. 1. Relative magnitudes of base and shaft load at various proportions of the ultimate load for piles in sand having a range of L/d ratios. (After Vesić.[13])

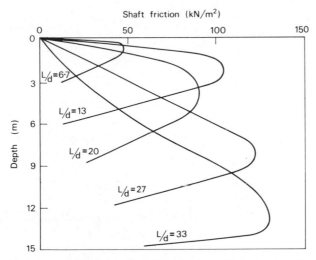

FIG. 2. Variation of load transfer from the shafts of piles in sand with increasing L/d ratio. (After Vesić.[13])

mean unit resistance over the whole shaft was close to one half of the maximum value. Gregersen *et al.*[14] also studied the distribution of friction on the shaft in instrumented pile tests in loose homogeneous sands in Norway. For straight-shafted piles having an L/d ratio of 57, Gregersen *et al.* found that 75% of the ultimate resistance was developed in friction and that when the lower half of the shaft was tapered this proportion increased to 85%.

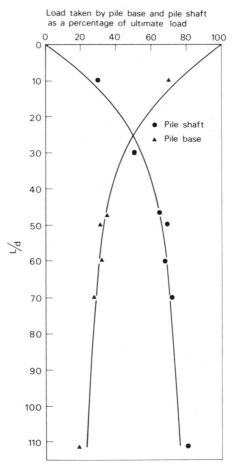

FIG. 3.   The proportions of ultimate shaft and base resistance for long piles in sand. (After Hanna and Tan.[11])

When it is necessary to found heavy structures on very deep deposits of loose sand, steel sections and jointed precast concrete piles are frequently used. The performance of piles having much larger L/d ratios is then of importance to designers. Hanna and Tan[11] made model studies of piles 25 mm in diameter having values of L/d up to 110 in a laboratory bed of compacted dry sand. Although the piles were not driven but were buried in either 'free' or 'locked' conditions as sand was poured around them, test results confirmed that unit base resistances are not proportional to penetration. At an L/d ratio of 25 the ultimate shaft and base components of resistance were equal and at larger values of L/d friction exceeded the bearing capacity of the base. Figure 3 shows the proportions of the two components at ultimate load as L/d was increased from zero to 110.

Uncertainty is often expressed as to whether an effective plug forms when open-ended steel sections are used as piles in granular soils or whether base resistance develops only over the net section. In the case of H-section piles and open box and tube sections, Tomlinson[15] has said that the friction mobilised between the flanges and on the inner surfaces may not be sufficient to develop useful resistance at the base. A test of an instrumented steel H-pile driven in 3-m lengths to a total depth of 21 m into medium dense sand was reported by Tavernas.[12] A precast concrete jointed pile was also driven for comparison and tested at similar depths. During the early stages of driving the H-pile developed a plug of dense soil between the flanges and behaved in a similar manner to the concrete pile. At greater depths the increased driving resistance dislocated the plug and the pile penetrated as a small displacement pile causing only limited densification of the surrounding soil. Under loading, as distinct from driving conditions, however, the bonds between the H-pile flanges and the soil were re-established and base resistance was developed over the gross area similar in magnitude to that mobilised by the precast concrete pile. The increase in bearing capacity with depth and the driving energy necessary to install both the steel H-pile and the precast jointed pile are shown in Fig. 4.

8.2.2 Bored Piles
Bored piles are seldom used in deep granular soils, at least in the UK, because of the high load capacities which can be provided by precast jointed piles formed of high-strength concrete. Taking out a borehole causes a reduction in the horizontal earth pressure, and De Beer[16] has

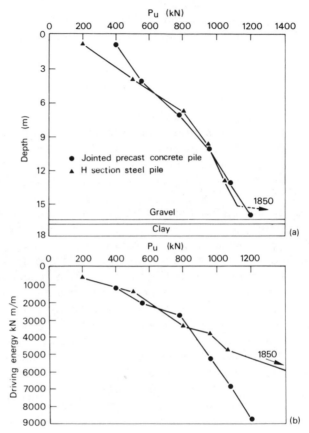

FIG. 4. The variation of ultimate bearing capacity (a) with depth and (b) with driving energy for a jointed precast pile and an H-section steel pile. (After Tavernas.[12])

pointed out that, from theoretical considerations alone, the base resistance of a bored pile in sand having an angle of internal friction ϕ of 30° should be approximately one third of that of a driven pile. Pouring high-slump concrete does little to restore the original state of stress in the soil and radial shrinkage associated with curing the concrete will reduce the effective pressure on the shaft surface. Concrete shrinkage in the vertical direction may reduce the bearing pressure at the pile base.

The effects of these phenomena on the bearing capacity and settlement of instrumented bored piles have been investigated by

Touma and Reese[17] at two predominantly sandy sites in Texas, USA. Additional tests of instrumented bored piles at sites on mixed layers of sand and clay have been reported by Reese et al.[18]

These tests showed that in homogeneous sand the maximum stress transfer from the pile shaft occurred just above the pile base. The distribution of stress along the shaft was similar to that shown in Fig. 2 for driven piles having L/d greater than 20. A clay crust over the sand increases the horizontal earth pressure around the upper shaft and thus increases the mean frictional resistance. Touma and Reese suggested that if the mean unit friction $\bar{f}_s$ is expressed in the form:

$$\bar{f}_s = K\bar{p}' \tan \delta \qquad (1)$$

where $\bar{p}'$ is the mean effective overburden pressure over the length of the shaft, δ is the angle of friction between the soil and the pile material, and K is the mean earth pressure coefficient, then values of K decreasing from about 0·7 for piles less than 7·5 m long to 0·5 for piles longer than 12 m may be used in design.

8.2.3 Design of Piles in Granular Soils

Recent research, some of which has been outlined above, has shown that at small displacements considerable friction can be developed but that its distribution along the pile shaft is complex. Equation (1) can only be used in the design of both bored and driven piles in cohesionless soils if the variation with depth of the earth pressure coefficient K is known with some certainty, and the choice of suitable values is difficult even for depths less than the critical depth. From an examination of published records, Meyerhof[19] concluded that K decreases from near the top of a pile, where it may approach the passive earth pressure coefficient, to the base, where it may be less than the initial earth pressure coefficient K_0. For short-driven, jacked and bored piles above the critical depth Meyerhof has suggested the values shown in Fig. 5. Because K is so dependent on the method of forming the piles, the nature of the deposits, and the material and shape of the pile, empirical assessments of the unit frictional resistance f_s are widely used.

Meyerhof[20] proposed:

$$\bar{f}_s = \frac{\bar{q}_c}{200} \qquad (2)$$

where $\bar{q}_c$ is the average cone resistance within the depth penetrated by the pile, measured in the static cone test. Thorburn and MacVicar[21]

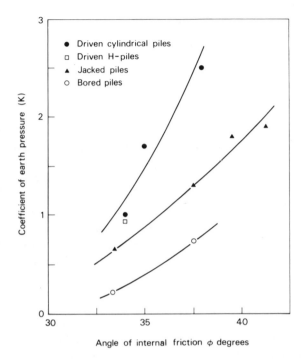

FIG. 5.   Earth pressure coefficients for estimating the shaft friction of piles
above the critical depth in sand. (After Meyerhof.[19])

used this formula for sands and recommended $\bar{f}_s = \bar{q}_c/150$ for silts.
Support for both the above formulae for initial design purposes has
been given by Rodin *et al.*[22] and by Thorburn.[23]

Where only Standard Penetration Test (SPT) results are available,
Thorburn and MacVicar[21] recommended:

$$\bar{f}_s = \frac{\bar{N}}{50} \text{ ton/ft}^2 \text{ for sand} \tag{3}$$

and

$$\bar{f}_s = \frac{\bar{N}}{60} \text{ ton/ft}^2 \text{ for silt} \tag{4}$$

where $\bar{N}$ is the average standard penetration resistance (blows per ft).
Caution must be exercised with empirical formulae of this type and a
minimum safety factor of 2·5 is recommended for calculating allowable working loads.

Since there is no compaction of the soil around the shafts of bored piles in cohesionless soils, Meyerhof[19] has suggested that for these piles the unit frictional resistance for design purposes be taken as one half of that given for sand by eqn. (3).

The unit base resistance, q_u, of a driven pile may be taken as equal to q_c, but the pile should penetrate at least 8 diameters into the stratum, and sand should be present to a depth of at least 3 diameters beneath the base. Where SPT N-values have been obtained at pile base level, $q_u = 2 \cdot 5 N$ (ton/ft^2) for silts, $4N$ for sands and $6N$ for gravels have been recommended by Thorburn[24] and endorsed by Rodin et al.[22] Meyerhof[19] has suggested that the base bearing capacity of a bored pile should be taken as roughly one third of the value given by the above formulae for driven piles.

8.3 DRIVEN PILES IN CLAY

8.3.1 The Effects of Pile Installation on the Soil

When a pile is driven into deep beds of clay, clay is displaced to accommodate the pile. Some of the clay close to the pile axis is carried downward, but the majority of the soil movements are outward and upward so that the volume heave at ground level is comparable with the embedded volume of the pile. These soil displacements are generally accompanied by increases in the pore-water pressure, and the resistance to penetration during driving or the ultimate load measured in a loading test carried out shortly after installation may be significantly less than that observed some time later, after the excess pore pressures have dissipated.

Attempts have been made to assess the extent of the disturbed zone around a driven pile by measuring changes in the *in situ* shear strength of the soil. Using the vane test in soft silty clay in Sweden, Orrje and Broms[25] showed that the undrained shear strength decreased by up to 40% within 1·5 pile diameters of the shaft surface.

Measurement of soil movement and pore-water pressure change were made by Adams and Hanna[26] as a circular group of 750 steel H-piles was driven through a clayey, sandy silt overlying sandy till and bedrock. The piles were driven from the centre of the group outwards and the high pore pressures which developed at the centre of the group were dissipated as driving progressed. All the piles were installed within 80 days, and during this time the pore pressure at the centre pile fell to one third of its maximum value. Soil displacements were concentrated within the foundation plan area, the heave volume being approximately equal to the volume displaced by the piles.

Direct measurements of the soil displacements around a single driven pile are few, since the displacements can be so small as to approach the limit of accuracy of the gauges used. However, Cooke and Price[27] measured radial displacements by probes from a trench as a tubular steel pile was jacked into London clay. The observations made at approximately mid-pile depth showed that when the pile base had penetrated below the probes the displacements decreased linearly with radial distance from the shaft surface (Fig. 6) and became negligible at about 4 pile diameters from the axis.

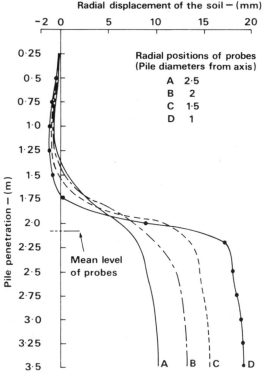

FIG. 6.   Radial displacements of the soil close to the shaft of a pile 0·17 m in diameter, jacked into stiff clay.

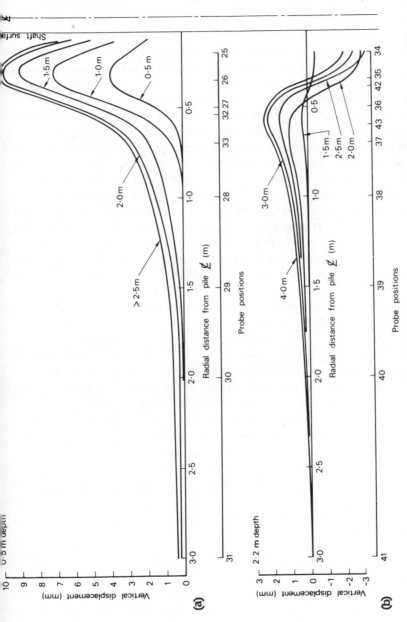

Fig. 7. Vertical displacements of the soil in the plane containing a pile 0·17 m in diameter, jacked into stiff clay (a) 0·5 m below ground surface, (b) at mid-pile depth 2·2 m. (Figures in metres indicate depth of pile point below ground surface.)

Precise measurements of the vertical displacements of the soil as piles were jacked into London clay have been reported by Cooke et al.[28] Observations of heave close to the ground surface (Fig. 7(a)) showed that the maximum vertical movements occurred at a radius of about 2 pile diameters from the pile axis and that these were reached by the time the pile had penetrated 12 diameters. At mid-pile level (Fig. 7(b)), clay close to the shaft surface was carried downward as the pile penetrated and upward movements were greater at somewhat larger radial distances than the maximum heave at the surface.

Redriving of piles in cohesive soils is frequently necessary because soil disturbance due to the installation of piles causes those already in position to lift. Cole[29] observed the uplift of piles driven at various spacings and load-tested some of these piles to detect any deterioration in performance. Where uplift is associated with heave of the main supporting stratum the effect on pile performance is likely to be small, but, to be on the safe side, Cole recommended that piles should be redriven whenever the uplift due to driving each adjacent pile exceeds 3 mm. The cheapest way to eliminate the adverse effects of uplift may be to redrive all piles within 8 diameters of each other whenever the soil conditions indicate some degree of risk.

In addition to affecting the support at the pile base, soil movements due to driving additional piles nearby can cause tensile forces to be developed in the pile shaft. Measurements by Cooke et al.[28] showed that piles already in position could be lifted by about one third of the heave occurring at the ground surface at the same radial distance. In piles spaced 3 diameters apart, the tensile forces could rise to about 20% of frictional resistance of the shaft and, at double the spacing, to about half this. During the final stages of installing each additional pile, these tensile forces reduced rapidly, and when full penetration of later piles was reached the original pattern of residual compressive forces in the piles placed previously was largely reinstated. Figure 8 shows the changing pattern of forces in the shaft of an existing pile, pile A, when a second pile was jacked to the same depth 3 pile diameters from it.

8.3.2 The Development of Resistance with Time

Whilst the extent of the remoulding and the magnitudes of soil displacements and pore-water pressures occurring in cohesive soils cannot be defined with any certainty it is clear that the bearing capacity of driven piles increases with time as pore pressures dis-

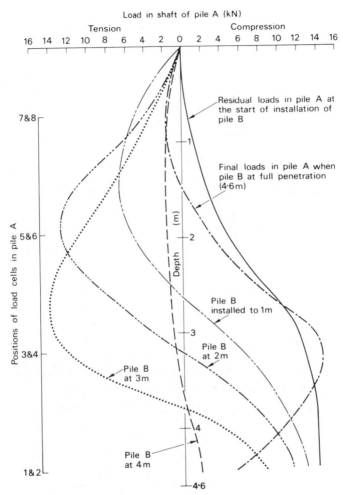

FIG. 8.   The forces in a pile under zero applied load as a second pile was jacked into the ground 3 pile diameters from it.

sipate. In soft clays the increase in strength is quite dramatic. Seed and Reese,[30] using an instrumented tubular steel pile in San Francisco Bay mud, showed that the bearing capacity attained in a test 20 days after installation was 50% greater than that measured 3 days after driving.

A similar result was reported by Thorburn and MacVicar[21] who tested a concrete shell pile on six occasions between 2 and 66 days

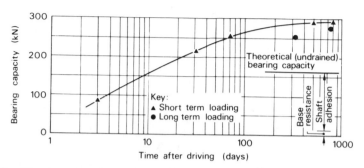

FIG. 9.   The increase in bearing capacity with time (log scale) of a pile in normally consolidated clay. (After Eide *et al.*[31])

after installation in a laminated clayey silt in the Clyde Valley. The bearing capacity as measured by the constant rate of penetration test at the end of the period was more than 50% greater than that measured two days after driving the pile.

A special rig was developed by Eide *et al.*[31] to apply constant loads to a pile for long periods, simulating the construction and life of a pile-supported structure. Short- and long-term tests were made on a timber pile in a deep deposit of normally consolidated marine clay in Norway, two successive long-term tests together occupying a period of nearly 2 years. In the long-term tests constant loads were applied until the settlement–time curves indicated that the pore pressures had largely dissipated and that primary consolidation was complete. Additional load increments were then applied. The maximum loads reached in the tests carried out by Eide *et al.* are plotted against log time in Fig. 9. The bearing capacity of the pile calculated from the undrained strength of the clay was 45% lower than the measured values shown in Fig. 9 for both the final long- and short-term tests. Eide *et al.* attribute the high bearing capacities measured in this and other long-term tests of piles in sensitive Norwegian clays to the formation of a clay annulus surrounding each shaft of such rigidity that failure occurs outside the shaft–clay interface.

8.3.3 The Forces in the Pile

When a pile is driven into stiff clay the resistance to penetration of the base increases at a rate only slightly greater than the rate of increase in shear strength with depth of the clay. This is shown on the left of Fig. 10 for the base of the instrumented pile jacked into

London clay by Cooke et al.[28] In this figure the distribution of load in the pile shaft at each level of penetration is also shown. The transfer of stress from the shaft to the soil derived from each load distribution curve, and shown on the right of Fig. 10, increased significantly down the embedded length of the shaft. However as the pile penetration was increased, the stress transfer from the upper part of the shaft decreased at a greater rate than it increased near the base. As a result, the mean value of the stress transfer over the shaft fell from about 27 kN/m^2 to 22 kN/m^2, i.e. by nearly 20%, as the embedded length was increased from 1 m to 4·5 m.

After installation in stiff clay, residual forces remain in a jacked or driven pile, and negative friction on the upper shaft balances a vertical force acting on the base. An example of the residual pile loading under zero applied load is shown by the solid line in Fig. 8. Redistribution of shaft loading and stress transfer occurs with the passage of time. In loading tests made some months after installation of the pile, Cooke et al.[28] observed a large increase in the friction developed at working load over the upper part of the shaft.

8.3.4 Effective Stress Design

Since it has been demonstrated that the bearing capacity of piles in clay increases with time as pore pressures set up in the clay by the installation and loading of the piles are dissipated, it follows that pile design should be based on drained values of the soil parameters. Effective stress design methods have been available for piles in normally consolidated clays but Burland[32] broadened their application to other clays. Burland assumed that by the time significant structural loads are applied to a pile the effective cohesion will be so small it can be ignored. The unit frictional resistance at any level (f_s) is then given by:

$$f_s = Kp' \tan \delta \qquad (5)$$

where p' is the effective overburden pressure at the level considered and the other symbols are as previously defined. The earth pressure coefficient K will depend on the type of soil, its stress history and the method of forming the pile. Burland denoted the terms $K \tan \delta$ by β so that $f_s = \beta p'$.

8.3.5 Driven Piles in Soft Clays

For piles driven into soft normally consolidated clays it is reasonable to assume that K is greater than K_0, the coefficient of earth pressure

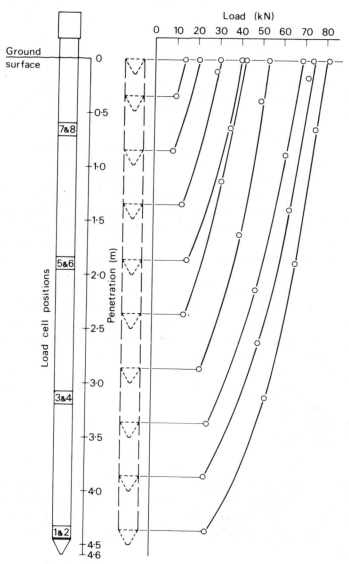

FIG. 10. The distribution of load in a pile being jacked into London clay.

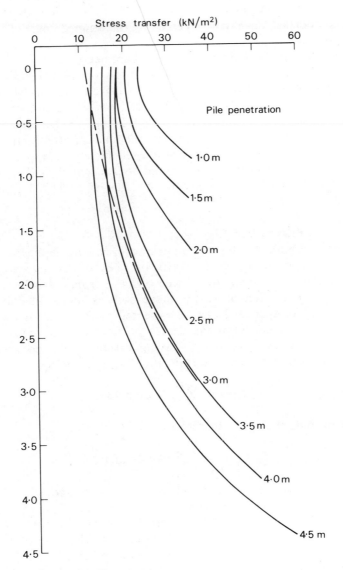

FIG. 10–*cont.*   The corresponding stress transfer from the shaft surface.

at rest. In addition, since failure under overload applied slowly probably occurs in the soil close to the shaft rather than at the clay–pile interface, δ may be taken as ϕ_d, the drained angle of friction of the soil.

For normally consolidated clay it has been shown that $K_0 = 1 - \sin \phi_d$. Hence $\beta = (1 - \sin \phi_d) \tan \phi_d$ probably represents a lower limit of the coefficient β for the design of driven piles in these clays. Values of ϕ_d will generally lie between 20° and 30° and over this range β, as expressed above, varies only from 0·24 to 0·29. Burland showed that in widely differing clays in which pile shaft friction corresponds to a very broad band of values of the coefficient α applied to the undrained shear strength, β is limited to a narrow band of values between 0·25 and 0·4. For design, a value of β of 0·3 is recommended.

8.3.6 Driven Piles in Stiff Clay

The approach outlined above is, in principle, equally applicable to the estimation of friction on the shafts of piles driven into overconsolidated clays. However, the choice of an appropriate value for the earth pressure coefficient K for these soils is more difficult since values of K_0 as high as 3 can be expected near the ground surface prior to pile driving. Summing values of f_s from eqn. (5) over the length, L, of a cylindrical pile having diameter, d, and assuming $K = K_0$, the ultimate frictional resistance, R_u, is given by:

$$R_u = \pi d \sum_0^L p' K_0 \tan \delta \Delta L \tag{6}$$

The mean shaft friction is therefore:

$$\bar{f}_s = \frac{R_u}{\pi d L} = \frac{1}{L} \sum_0^L p' K_0 \tan \delta \Delta L \tag{7}$$

Since the driving of piles into overconsolidated clay causes remoulding in the vicinity of the shaft, the angle δ should be the remoulded drained angle of friction. Burland determined the variation of $\bar{f}_s$ with depth given by eqn. (7) for the particular case of London clay, assuming that the remoulded drained angle of friction δ is 21·5° and that K_0 decreases from 3·5 at the surface to 2·0 at a depth of 22 m. This variation is shown in Fig. 11 compared with results of tests on driven piles reported by Tomlinson[34] and of jacked piles reported by Cooke et al.[28] The results given by Cooke et al.[28] are for piles only

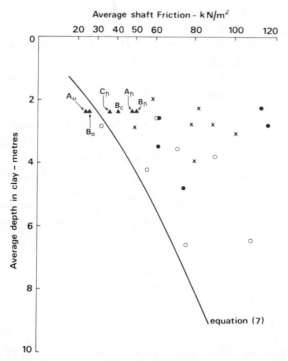

FIG. 11. Relationship between shaft friction and average pile depth for driven piles in London clay. (Tomlinson:[34] ●, driven through sand and gravel; ○, driven through soft clays or silts; x, no overlying strata. Cooke *et al.*:[28] A_0 and B_0, at end of installation; A_r, pulling test 32 months after installation; B_c, compressive test 28 months after installation; B_r, pulling test 29 months after installation; C_r, pulling test 22 months after installation.)

4·6 m long, but they show how, at installation, the piles failed to mobilise the frictional resistance predicted by eqn. (7). Nevertheless, in both compressive and pulling tests made at periods of the order of 2 years after installation the expected frictional resistance was amply exceeded. The curve given by eqn. (7), therefore, probably represents a lower limit of the unit frictional resistance for piles driven or jacked into stiff clays like London clay. Until sufficient tests have been made to assess the accuracy of the design approach outlined above it should be used in conjunction with traditional methods of design based on the undrained strength of the clay and empirical mobilisation factors.

8.3.7 The Effect of Superficial Soils Above the Clay

It is widely known that the frictional resistance measured in short-term tests of piles driven into stiff clays is less than the undrained shear strength of the clay around the shaft. Tomlinson[33] attributed the reduction in friction to the formation of a gap around the upper part of the shaft as a result of whipping of the pile under the hammer blow. In his more recent studies Tomlinson[34] has shown that the friction developed by a pile driven into stiff clays is further influenced by the characteristics of soils through which the pile passes before reaching the bearing stratum. The soils are carried down by the pile shaft to form a skin of limited depth in the stiff clay around the shaft. The skin is a sand–clay mix having considerable strength when piles are driven through sands and gravels and is of soft clay or silt having reduced frictional resistance when piles are driven through these soils. The effects are probably significant for penetrations of the stiff

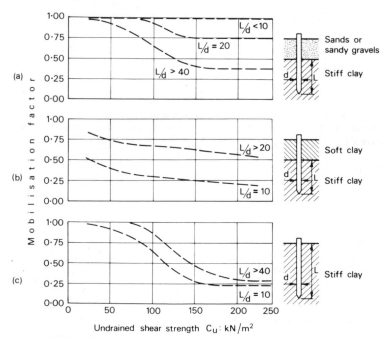

FIG. 12.   The variation of adhesion mobilisation factors with undrained shear strength for different pile penetrations. (a) piles driven through overlying sands or sandy gravels; (b) piles driven through overlying soft clay; (c) piles without different overlying strata. (After Tomlinson.[35])

clay of less than about 20 diameters. Figure 12 presents values for the friction mobilisation factor for the design of preformed driven piles given by Tomlinson[35] for a range of undrained shear strengths and three common soil situations. A minimum safety factor of 2·5 is recommended unless loading tests are carried out.

8.4 BORED PILES

8.4.1 Installation Aspects

The process of boring a vertical hole to form a cast *in situ* bored pile affects the properties of the clay around the pile in a manner as difficult to quantify as does the driving of a pre-formed or cast *in situ* driven pile. During the boring of a dry hole, horizontal stresses at the exposed surface fall to zero, and water migrates towards the surface so that swelling and softening start to occur. In overconsolidated clays, fissures may open, facilitating the movement of water towards less heavily stressed zones around the hole. Similar phenomena occur at the base of the shaft where the vertical stresses are reduced to zero. The extent of the swelling and softening depends on the properties of the clay and the length of time the hole is open before concrete is poured.

Recent studies[36] have shown that a large proportion of the limited number of defects encountered in the formation of bored cast in place piles are associated with the installation and removal of temporary casings. Of these defects probably the most important is the occurrence of overbreak cavities, frequently caused by boring ahead of the casing in unstable or water-bearing strata above the clay. When the casing is withdrawn during concreting large quantities of water can be released as shown in Fig. 13. Concrete slumps into the cavity and the water is trapped near the top of the overbreak. The concrete of the shaft, depending on its workability, is contaminated, segregated, or its cross-section area depleted. Concrete within the reinforcing cage is generally intact but reinforcing bars can be exposed and the cross-section of the pile shaft so reduced that its structural stability is impaired (Fig. 14).

Where very heavy loads are to be carried, it is frequently more economical to enlarge the bases of bored piles by underreaming than to extend the shafts or to provide rigid caps linking groups of smaller, straight-shafted piles. As a result of research (for example by Burland

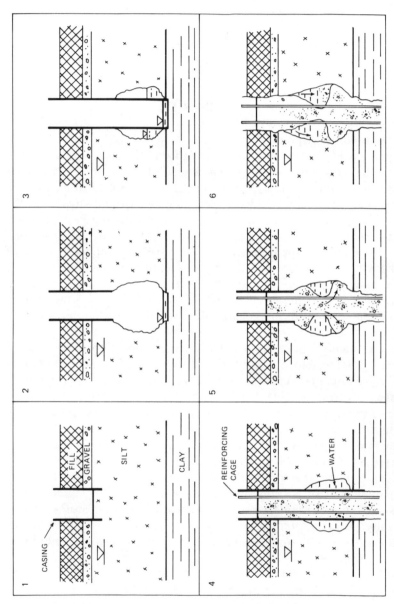

Fig. 13. The formation and effect of cavities in bored pile construction (by permission of the DoE and CIRIA Piling Development Group and Cementation Piling and Foundations Ltd).

FIG. 14. Exposed reinforcement and reduced shaft section due to slumping of concrete into a water-filled cavity (reproduced by permission of the DoE and CIRIA Piling Development Group).

et al.,[37] Whitaker and Cooke[38] and Burland and Cooke[39]) it is generally understood that large settlements are necessary to develop the available capacity of enlarged bases and that in many cases the resistance of the shaft will be completely mobilised when the pile is loaded under working conditions.

In addition to the loading and the pile dimensions, the elastic modulus E (or the shear modulus G) of the supporting soil has a profound effect on the settlement of a pile under working conditions. Marsland[40] has shown that the ratio of E to the undrained shear strength c_u determined from plate loading tests carried out in the bottom of shafts in London clay is dependent on the time taken to set up the tests. For very short times, E/c_u values of about 500 were measured, whereas for times longer than about 8 h E/c_u fell to between 100 and 200. The true *in situ* value of E/c_u for London clay is therefore probably at least 500 and lower values indicate boring disturbance, inadequate clearance of debris from the borehole base and swelling due to delays in concreting.

8.4.2 Bored Pile Design

The design of bored piles in stiff clays is widely based on the equation:

$$P_u = A_s\alpha\bar{c} + A_bN_cc_b \qquad (8)$$

where P_u is the ultimate bearing capacity of the pile,

A_s and A_b are the effective areas of the shaft surface and of the underside of the base,

$\bar{c}$ is the mean undrained shear strength of the clay over the depth of the shaft,

c_b is the undrained shear strength of the clay at the level of the pile base,

N_c is the base bearing capacity factor, and α is the friction mobilisation factor.

The coefficient α takes account of clay softening during pile construction, and variability in $\bar{c}$ due to the properties of the soil and sampling disturbance. For this reason the range of α-values is usually given as 0·3 to 0·6 but the value used for design must depend on the engineer's confidence in the soil sampling and testing and the quality of the pile construction. In their tests in London clay Whitaker and Cooke[38] separated the base and frictional components of resistance by means of load cells at the bottom of each shaft and observed mean frictional resistances corresponding to $\alpha = 0·44$, when calculated from the mean shear strength profile given by tests of small samples.

O'Neill and Reese[41] reported tests of piles with and without enlarged bases formed with an auger rig in stiff fissured Beaumont clay in Texas, USA. Strain gauge load cells incorporated in the shafts permitted accurate measurement of the development of friction. The mean peak values of friction corresponded to $\alpha = 0·5$. The results of these tests confirmed that when piles having enlarged bases are employed economically the friction in many cases is fully mobilised at the settlements necessary for the development of an acceptable working load at the base. The design of piles with enlarged bases must take this fact into account. Burland and Cooke[39] therefore suggested calculating the working load P_w in terms of two design criteria:

(a) friction pile behaviour, where the ultimate bearing capacity of the base, Q_u, is small compared with the ultimate frictional resistance, R_u, so that:

$$P_w \leqslant \frac{R_u + Q_u}{F} \qquad (9)$$

where the overall load factor F is normally 2; and

(b) that R_u is fully mobilised, but that overstressing of the soil beneath the base must be prevented. In this case:

$$P_w \leqslant R_u + \frac{Q_u}{F_b} \qquad (10)$$

Criterion (b) only applies when the base diameter is large compared with the shaft diameter, and F_b should not be less than 3. The

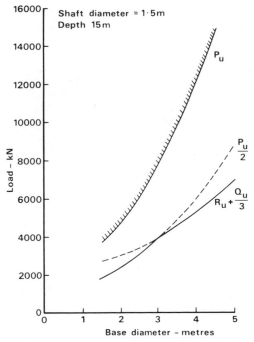

FIG. 15. Graphical representation of the application of the two design criteria for bored piles for a range of base diameters.

application of the two criteria to piles 1·5 m in diameter, 15 m long, having a range of base diameters, is shown graphically in Fig. 15. In order to limit settlements when very large bases are employed, it may be necessary to increase F_b. Methods of estimating settlement are discussed later.

Because of the difficulty of interpreting the undrained shear strength–depth profile, and of choosing an appropriate value for α, there is a need for a design approach based on the long-term behaviour of bored piles comparable with that already discussed for driven piles. Whilst K_0 may be as high as 3 near the ground surface it is unlikely that such a value is re-established in the clay around a bored pile shaft. Equation (7) may therefore be expected to represent an upper limit of the mean shaft friction $\bar{f}_s$ for bored piles. In Fig. 16, Burland[32] has plotted values of $\bar{f}_s$ observed in a number of tests of

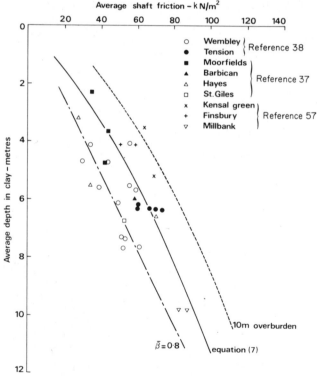

FIG. 16.   Relationship between shaft friction and average depth in the London clay for large bored piles.

large bored piles in London clay for comparison with the curve of eqn. (7). At some of the sites London clay did not extend to the ground surface, and Burland calculated the addition to $\bar{f}_s$ due to the effective vertical pressure of 10 m of dry overburden. The increased values of $\bar{f}_s$ is shown as a broken line in Fig. 16, and clearly represents the upper limit for bored piles in London clay. Burland has therefore suggested that the chain-dotted line in Fig. 16 given by:

$$\bar{\beta} = \frac{\bar{f}_s}{\bar{p}'} = K \tan \delta = 0{\cdot}8 \qquad (11)$$

be used as the basis of a conservative design for London clay, where $\bar{\beta}$, $\bar{f}_s$ and $\bar{p}'$ are the mean values of β, f_s and p' respectively.

8.5   THE SETTLEMENT OF PILES

The previous sections have been largely concerned with the effects of the installation process on the subsequent behaviour of piles and with the design of piles for bearing capacity, taking account of changing soil properties. The settlement of structures erected on piled foundations is also of importance to designers, although the settlement of individual piles when tested under working load is generally quite small.

In cohesive soils exhibiting broadly elastic properties the stresses and displacements at any point can be assumed to superimpose, and the settlements of large closely spaced groups will be many times that of an isolated pile. On the other hand, piles driven to end-bearing in sand can so increase the density by compaction that a new bearing stratum is, in effect, created, having much greater stiffness. Pile group settlements then remain small. Thus, it is normal practice to consider the overall behaviour of pile groups in sand although no approach having a satisfactory theoretical background is available and, in contrast, to extrapolate from the characteristics of a single pile for groups in clay. If the thickness of the bearing stratum is limited and any underlying soils are liable to consolidate under increased stress, the settlement due to this must, of course, be investigated.

8.5.1 Piles in Sand
The settlements of piles in sand follow the application of load so rapidly that most have occurred by the time the structure is erected

and before finishes susceptible to damage have been applied. For these reasons field and laboratory tests to provide data for the prediction of the settlements of piled structures on sands not overlying softer soils liable to consolidation may not be justified. However, since much of the load is transferred to the level of the pile bases, settlements can be crudely estimated by methods developed for shallow foundations, assuming that the piles and soil act as an equivalent pier. A survey of methods used for estimating the settlements of structures having shallow foundations on granular soils has been made by Sutherland.[42] Most of the methods are empirically based on the results of *in situ* tests. Meyerhof[19] has discussed the application of empirical relationships developed for shallow foundations to the prediction of the settlement of piled foundations, using an influence factor depending on the ratio of the effective depth to width of the pile group. The effective depth is usually assumed to be two thirds of the penetration into the bearing stratum for friction piles, whilst for end-bearing piles it is the depth of the pile bases.

The most satisfactory method of determining the settlement of a single pile in cohesionless soil is by carrying out a loading test. The ratio of the group settlement to the settlement of the test pile is unlikely to be large under working conditions, but it is affected by the initial state of compaction of the soil. Thus, in extrapolating from the observed settlement of a single pile to estimate that of a group, engineering judgement should be exercised. Leonards[43] tested a group of 10 small piles in the field and showed that the group settlement at working load was approximately equal to the mean settlement recorded in tests in individual piles. Where there is no compressible stratum below a pile group, Broms[44] has said that it is usual to assume the group settlement is the same as that of a single pile. The order of driving piles into granular soils affects the distribution of load between the piles[45] and this may influence the group settlement unless re-driving is carried out.

8.5.2 Piles in Clay

Piles in clay subjected to maintained loads may settle for long periods, and the settlement of closely spaced groups must be expected to be much greater than that of a single pile carrying the mean load of the group. The few published observations on piled structures on overconsolidated clays, however, have shown that the total long-term settlements are seldom as great as double the settlements measured at the end of construction. Morton and Au[46] have reported on eight major buildings in London, five of which were supported on rafts and three on large bored piles. The settlements at the end of construction were found to be about 60% of the maximum settlements, irrespective of the type of foundation.

8.5.3 Theoretical Studies

At loads up to about 50% of the ultimate, the load–settlement relationships for piles in stiff clays are frequently sensibly linear. Elastic analysis can therefore provide useful indications of pile behaviour for working conditions as long as appropriate values for the soil parameters are available. However, since the strengths and elastic moduli of clays generally increase with depth, and large variations can be given by different methods of measurement, the choice of suitable values is difficult. Further, the alterations to the soil properties that occur due to installation of the piles are almost impossible to quantify.

Poulos and Davis[47] analysed the effects of a range of pile dimensions and soil properties, assuming the pile was incompressible and the soil elastic. The results were presented in the form:

$$\rho = \frac{P}{LE} I_\rho \tag{12}$$

where ρ is the settlement corresponding to an applied load P,

L is the length of shaft,

E is Young's modulus for the soil, and

I_ρ is an influence factor for an incompressible cylindrical pile in an elastic medium overlying a rigid stratum.

Poulos and Davis[47] prepared charts showing the variation of I_ρ with the depth, h, of the rigid stratum, with the length–diameter ratio for the pile $L:d$, and with Poisson's ratio for the soils. For many common situations I_ρ varies from 1·2 to 2·0 as the $L:d$ ratio is increased from 10 to 50.

Unless the base is enlarged by underreaming, the analysis showed that the base load seldom exceeds 10% of the total load under working conditions. Thus, pile settlement is largely controlled by the mechanism of load transfer from the shaft. From this reasoning Cooke[48] was able to present a simple physical model of the stress transfer from a cylindrical shaft and to derive an expression for pile settlement of the same form as that developed by Poulos and Davis.

Cooke assumed that shear stresses are transmitted radially from the shaft through a series of annular elements as shown in Fig. 17, an element ABCD being deformed to A'B'C'D' while transferring the stress from the shaft surface to the element BEFC. The process is assumed to continue until at some point X (distance nd from the pile axis) the shear strains are so small that they can be ignored.

By integrating the shear strains, dv/dr, with respect to r from X to the shaft surface the shaft displacement v_s can be obtained. The settlement $\bar{v}_s$ (the mean value of v_s) is given by

$$\bar{v}_s = \frac{3R}{2\pi LE} \log_{exp}(2n) = \frac{R}{LE} I \tag{13}$$

where the shaft resistance $R = \pi\, dL\bar{f}_s$, $\bar{f}_s$ is the mean value of f_s, and the influence factor I is equal to $3/2\pi \log_{exp}(2n)$.

Cooke and Price[27] showed experimentally that the shear strains become insignificant for values of n greater than 10. If n is between 8 and 25, values of $3/2\pi \log_{exp}(2n)$ are between 1·3 and 1·9. This is within the range given by the more exact analysis of Poulos and Davis using the combined base and shaft loads P for piles having $L:d$ ratios between 10 and 50.

The effect of including pile compressibility in a theoretical analysis of pile behaviour is likely to be significant only for long slender piles. Mattes and Poulos[49] showed that, when compressibility is included, the settlement at the top of a pile increases while the displacement at the base is reduced.

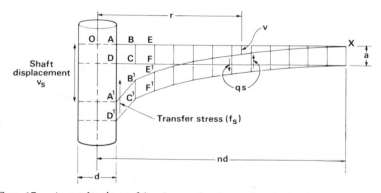

FIG. 17.   A mechanism of load transfer from the shaft of a friction pile.

8.5.4 Remoulding and Consolidation

Using finite element techniques, it is possible to take account of variations in soil properties, radially and with depth, associated with remoulding and softening or consolidation and stiffening. Balaam *et al.*[50] examined the effect on pile behaviour of a cylindrical disturbed zone having a radius r_1 around the pile and extending to depth r_1 below the pile base. The Young's modulus of the disturbed zone was assumed to vary linearly from a value E_r at the pile–soil interface, to the value E_s of the undisturbed soil at the periphery of the disturbed zone. In order to simulate both softening of the clay close to bored piles and consolidation around driven piles, values of E_r/E_s ranging from 0·2 to 5·0 were examined. The results shown in Fig. 18 are in the form of an influence factor I_R, the ratio of the pile settlement to the settlement of the pile in homogeneous soil, for various values of the ratio r_1 to the pile radius r_p, and for a range of values of E_r/E_s.

Very little experimental information has been published on the manner in which G or E in the soil around a pile is modified by the installation process or by consolidation when the pile is loaded. For

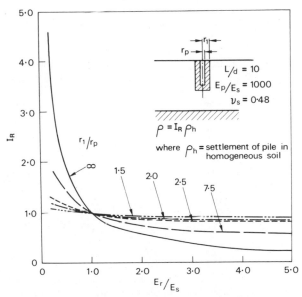

FIG. 18. The settlements of piles in zones of disturbed or consolidated clay relative to the settlement of a pile in homogeneous soil (after Balaam *et al.*[50]).

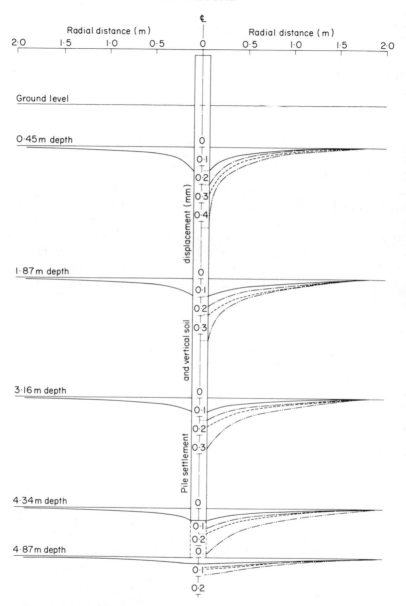

FIG. 19. Predicted and observed displacements of a friction pile shaft and the surrounding clay. *Left*: Finite element analysis for 20 kN head load. *Right*: Observations; Pile head load: ——— 20 kN, – · – · – 32 kN, ----- 40 kN, – · · – · · – 60 kN.

this reason Cooke and Price[51] developed instruments for measuring the small vertical displacements that occur in the soil around a pile loaded only up to working load. The instruments were used to measure vertical displacements at five levels around the pile jacked into London clay.[28] From the distribution of load in the pile shaft and the soil displacements shown in Fig. 19, values of G were calculated which were used as the basis for elastic finite element studies of the pile–soil system. In the loading test made 3 months after installation of the pile, G was found to be about one third of the undisturbed value adjacent to the pile near the ground surface, increasing to three quarters of the undisturbed value close to the pile base. Some reduction in G due to disturbance was detected to a radial distance of 3 diameters from the pile axis. Soil displacements computed with this shear modulus distribution are shown for an applied load of 20 kN, approximately half the design load, on the left side of Fig. 19.

8.5.5 Empirical Estimation of the Settlement of Piles in Clay

Many investigations have shown that the frictional resistance on bored piles in clay is mobilised at settlements of between 0·5 and 1% of the shaft diameter. Thus, for conservative design purposes the short term settlement ρ_w at working load of a bored pile without enlargement of the base may be taken as:

$$\rho_w = 0·01 \frac{P_w}{P_u} d \tag{14}$$

where P_w and P_u are the working load and ultimate bearing capacity of the pile. The working load for piles having enlarged bases, when friction is likely to be fully mobilised, is controlled by eqn. (10) and the actual load Q reaching the base is $P_w - R_u$. Plate loading tests and instrumented pile tests have shown that the load-settlement curve is sensibly linear to $Q/Q_u = 1/3$ and that the settlement ρ is related to Q/Q_u by:

$$\frac{\rho}{D} = K \left(\frac{Q}{Q_u} \right) \tag{15}$$

where D is the diameter of the enlarged base.

Burland and Cooke[39] derived values of K in terms of E and c_b and found that values of K obtained experimentally generally lie between 0·002 and 0·02, depending on the quality and speed of construction. Equation (15) may therefore be used for design, and $K = 0·02$ probably represents the upper limit of settlement. With good quality workmanship the settlement may be considerably less than this.

The settlement of driven piles is unlikely to be as large as that of bored piles of similar dimensions loaded under similar conditions. Cooke et al.[28] showed that at working loads, given by a safety factor of 2, jacked piles having $L : d$ greater than 10 settled less than 0·25% of the diameter in short-term tests in London clay. Confirmation of this result in tests of larger diameter piles at other sites should be obtained before the figure is widely used in design, but eqn. (14) probably represents an upper limit for driven piles.

8.5.6 Vertical Pile Groups in Clay

Curves for estimating the settlement of free-standing pile groups and piled foundations having caps cast on the clay surface from the results of model experiments have been provided by Whitaker.[45] These show that at normal spacings the settlement ratio (i.e. the ratio of the group settlement to the settlement of a single pile under working conditions) may be greater than 10 for square groups of more than 25 piles. The settlement ratio depends on the length and spacing of the piles and on whether the cap or raft transmits load directly to the main supporting stratum. A superposition approach to the estimation of group settlements, applicable to any number of piles at any spacing, has been presented by Poulos[52] for incompressible piles. The results of a more rigorous elastic analysis of compressible piles and pile groups made by Butterfield and Banerjee[53] differ from those given by Poulos by less than 5% for piles of normal length. Three-dimensional finite element analyses of small square and rectangular groups containing up to 15 piles have been made by Ottaviani.[54]

An alternative superposition approach using observed soil displacements rather than elastic analysis was suggested by Cooke[48] who assumed that any pile was displaced downward as a result of loading a nearby pile to the same extent as soil at the same radial distance. Since the observations presented in Fig. 19 had shown that significant soil displacements do not occur at radial distances greater than about 10 pile diameters, the number of piles affected by the settlements of other piles depends on the spacing but is severely limited.

Figure 20 shows the variation of settlement ratio with pile spacing given by this method for two small square groups and for an infinitely large group of free-standing piles. The results of some of Whitaker's model tests[58] are shown for comparison.

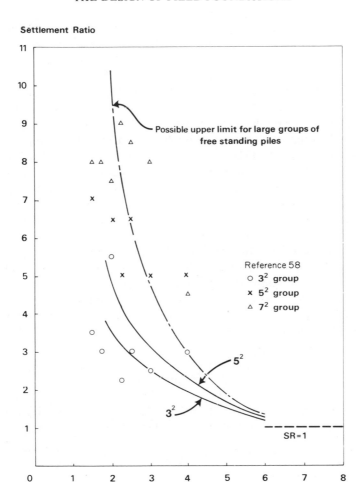

FIG. 20. The variation of settlement ratio with the spacing of the piles in a group.

In the design of piled foundations having a cap or raft cast on the clay surface, it is frequently assumed that all the support is provided by the piles. Butterfield and Banerjee[55] examined the effect of an elemental cap on the settlement of a single pile, and extended the study to include small pile clusters having a cap formed on the ground surface. The results of the study suggested that although the effect of the cap on a single pile is negligible a cap can reduce the settlement of a piled foundation by up to 15% depending on the group size and pile spacing.

Few reports of the behaviour of structures founded on piled rafts have been published. Some observations by Green and Hight[56] of the foundation loading and settlement of a 15-storey structure in London are therefore particularly valuable. The foundation consisted of a 1·5 m thick concrete raft formed on 400 bored piles 15 m long, 0·485 m in diameter, spaced at 1·5-m centres. A trial pile tested at the site settled about 0·6 mm under the mean pile load of 475 kN. At the end of construction the building had settled 33 mm, at which stage approximately 66% of the total load was being carried by the piles.

In their analysis of the settlement of a single pile Poulos and Davis[47] compared values of the influence factor I_ρ using both undrained and drained values for E and v. This approach suggested that nearly 90% of the total settlement occurs immediately and that only 10% is time-dependent. Poulos[52] extended the analysis to pile groups and showed that between 60% and 70% of the settlement occurs immediately. This analytical result is therefore in good agreement with the observations of Morton and Au referred to previously.[46]

REFERENCES

1. TAVERNAS, F. and AUDY, R. (1972). Limitations of the driving formulae for predicting the bearing capacities of piles in sands, *Canad. Geotech. J.*, **9**, 47–62.
2. INSTITUTION OF CIVIL ENGINEERS (1977). Symposium in print, *Géotechnique*, **26**(1), 1–147. Papers and discussion, *Piles in Weak Rock*, Inst. Civ. Engrs., London.
3. MEYERHOF, G. G. (1959). Compaction of sands and bearing capacity of piles, *J. Soil Mech. Fnd. Div., Proc. ASCE*, **85**(SM6), 1–30.
4. ROBINSKY, E. I. and MORRISON, C. F. (1964). Sand displacement and compaction around model friction piles, *Canad. Geotech. J*, **1**(2), 81–93.
5. TERZAGHI, K. (1943). *Theoretical Soil Mechanics*, John Wiley & Sons, Inc, New York.
6. MEYERHOF, G. G. (1951). The ultimate bearing capacity of foundations, *Géotechnique*, **2**(4), 301–332.
7. BEREZANTZEV, V. G., KHRISOFOROV, V. S. and GOLUBKOV, V. N. (1961). Load bearing capacity and deformation of piled foundations, *Proc. 5th Int. Conf. SMFE*, Paris, 1961, **2**, 11–15.
8. NORLUND, R. L. (1963). Bearing capacity of piles in cohesionless soils, *J. Soil Mech. and Fnd. Div., Proc. ASCE*, **89**(SM3), 1–35.

9. VESIĆ, A. S. (1964). Investigations of bearing capacity of piles in sand, *Proc. Conf. on Deep Foundations*, Mexico City, 1964, **1**, 197–224.
10. KERISEL, J. (1964). Deep foundations—basic experimental facts, *Proc. Conf. on Deep Foundations*, Mexico City, 1964, **1**, 5–44.
11. HANNA, T. H. and TAN, R. H. S. (1973). The behaviour of long piles under compressive loads in sand, *Canad. Geotech. J.*, **10**, 311–340.
12. TAVERNAS, F. A. (1971). Load test results on friction piles in sand, *Canad. Geotech. J.*, **8**, 7–22.
13. VESIĆ, A. S. (1970). Tests on instrumented piles, Ogeechee River site, *J. Soil Mech. Fnd. Div.*, *Proc. ASCE*, **96**(SM2), 561–584.
14. GREGERSEN, O. S., AAS, G. and DIBIAGIO, E. (1973). Load tests on friction piles in loose sand, *Proc. 8th Int. Conf. SMFE*, Moscow, 1973, **2.1**, 109–117.
15. TOMLINSON, M. J. (1975). *Foundation Design and Construction*, Pitman, London.
16. DE BEER, E. E. (1964). Some considerations concerning the point bearing capacity of bored piles, *Proc. Symp. Bearing Capacity of Piles*, Roorkee, 1964, 178–204.
17. TOUMA, F. T. and REESE, L. C. (1974). Behaviour of bored piles in sand. *J. Geotech. Eng. Div.*, *Proc. ASCE*, **100**(GT7), 749–761.
18. REESE, L. C., TOUMA, F. T. and O'NEILL, M. W. (1976). Behaviour of drilled piers under axial loading, *J. Geotech. Eng. Div.*, *Proc. ASCE*, **102**(GT5), 493–510.
19. MEYERHOF, G. G. (1976). Bearing capacity and settlement of pile foundations, *J. Geotech. Eng. Div.*, *Proc. ASCE*, **102**(GT3), 195–228.
20. MEYERHOF, G. G. (1956). Penetration tests and bearing capacity of cohesionless soils, *J. Soil Mech. Fnd. Div.*, *Proc. ASCE*, **82**(SM1), 1–19.
21. THORBURN, S. and MACVICAR, R. S. L. (1971). Pile load tests to failure in the Clyde alluvium, *Proc. Behaviour of Piles Conf.*, Inst. Civ. Engrs., London, 1971, 1–7.
22. RODIN, S., CORBETT, B. O., SHERWOOD, D. E. and THORBURN, S. (1974). State-of-the-Art Report on Penetration testing in the United Kingdom, *Proc. Europ. Symp. on Penetration Testing*, Stockholm, 1974, 139–146.
23. THORBURN, S. (1976). The static penetration test and the ultimate resistances of driven piles in fine-grained non-cohesive soils, *The Structural Engineer*, **54**(6), 205–211.
24. THORBURN, S. (1971). Discussion, *Behaviour of Piles Conf.*, Inst. Civ. Engrs., London, 1971, 54.
25. ORRJE, O. and BROMS, B. (1967). Effects of pile driving on soil properties, *J. Soil Mech. Fnd. Div.*, *Proc. ASCE*, **SM5**, 59–73.
26. ADAMS, J. I. and HANNA, T. H. (1971). Ground movements due to pile driving, *Behaviour of Piles Conf.*, Inst. Civ. Engrs., London, 1971, 127–133.
27. COOKE, R. W. and PRICE, G. (1973). Strains and displacements around friction piles, *Proc. 8th Int. Conf. SMFE*, Moscow, 1973, **2**(1), 53–60. (Also *BRE Current Paper* CP 28/73).
28. COOKE, R. W., PRICE, G. and TARR, K. To be published.

29. COLE, K. W. (1972). Uplift of piles due to driving displacement, *Civ. Engng. and Publ. Wks. Review*, March 1972, 263–269.
30. SEED, H. B. and REESE, L. C. (1955). The action of soft clay along friction piles, *Trans. Am. Soc. Civ. Engrs.*, **122**, 731–754.
31. EIDE, O., HUTCHINSON, J. N. and LANDVA, A. (1961). Short- and long-term test loading of a friction pile in clay, *Proc 5th Int. Conf. SMFE*, Paris, 1961, **2**, 45–53.
32. BURLAND, J. B. (1973). Shaft friction of piles in clay—a simple fundamental approach, *Ground Engineering*, **6**(3), 30, 32, 37, 38, 41, 42.
33. TOMLINSON, M. J. (1957). The adhesion of piles driven in clay soils, *Proc. 4th Int. Conf. SMFE*, London, 1957, **2**, 66–71.
34. TOMLINSON, M. J. (1971). Some effects of pile driving on skin friction, *Behaviour of Piles Conf.*, Inst. Civ. Engrs., London, 1971, 107–114.
35. TOMLINSON, M. J. (1971). Author's reply to the discussion, *Behaviour of Piles Conf.*, Inst. Civ. Engrs., London, 1971, 149–150.
36. THORBURN, S. and THORBURN, J. Q. (1977). Review of problems associated with the construction of cast-in-place concrete piles, *DoE and CIRIA Piling Development Group, Report PG2, CIRIA, London*.
37. BURLAND, J. B., BUTLER, F. G. and DUNICAN, P. (1966). The behaviour and design of large diameter bored piles in stiff clay, *Symp. on Large Bored Piles*, Inst. Civ. Engrs., London, 1966, 51–71.
38. WHITAKER, T. and COOKE, R. W. (1966). An investigation of the shaft and base resistances of large bored piles in London clay, *Symp. on Large Bored Piles*, Inst. Civ. Engrs., London, 1966, 7–49.
39. BURLAND, J. B. and COOKE, R. W. (1974). The design of bored piles in stiff clay, *Ground Engineering*, **7**(4), 28–30, 33–35. (Also *BRE Current Paper* CP 99/74).
40. MARSLAND, A. (1971). Laboratory and *in situ* measurements of the deformation moduli of London clay, *Proc. Symp. The Interaction of Structure and Foundation*, University of Birmingham, 1971, 7–17. (Also *BRE Current Paper* CP 24/73).
41. O'NEILL, M. W. and REESE, L. C. (1972). Behaviour of bored piles in Beaumont clay, *J. Soil Mech. Fnd. Div., Proc. ASCE*, **SM2**, 195–213.
42. SUTHERLAND, H. B. (1975). Granular materials, Review Paper Session 1: *Conf. on the Settlement of Structures*, Cambridge, 1975, 473–499. Pentech Press, London.
43. LEONARDS, G. A. (1972). Settlement of pile foundations in granular soil, *Proc. Conf. Earth and Earth Retaining Structures*, Purdue, USA, 1975, **1**(2), 1169–1184.
44. BROMS, B. (1976). Pile foundations—pile groups, *6th European Conf. SMFE*, Vienna, 1976, 103–132.
45. WHITAKER, T. (1976). *The Design of Piled Foundations*, Pergamon Press, Oxford.
46. MORTON, K. and AU, E. (1975). Settlement observations on eight structures in London, *Conf. on the Settlement of Structures*, Cambridge, 1975, Pentech Press, London, 183–203.
47. POULOS, H. G. and DAVIS, E. H. (1968). The settlement behaviour of single axially loaded incompressible piles and piers, *Géotechnique*, **18**(3), 315–371.

48. COOKE, R. W. (1974). The settlement of friction pile foundations, *Conf. on Tall Buildings*, Kuala Lumpur, 1974. (Also *BRE Current Paper* CP 12/75).
49. MATTES, N. S. and POULOS, H. G. (1969). Settlement of single compressible piles, *J. Soil Mech. Found Div., Proc. ASCE*, **95**, 189–207.
50. BALAAM, N. P., POULOS, H. G. and BOOKER, J. R. (1975). Finite element analysis of the effects of installation on pile load: settlement behaviour. *Geotechnical Engineering*, **6**, 33–48.
51. COOKE, R. W. and PRICE, G. (1973). Horizontal inclinometers for the measurement of vertical displacement in the soil around experimental foundations, *Proc. Symp. Field Instrumentation in Geotechnical Engineering*, London, 1973, 112–125. (Also *BRE Current Paper* CP 26/73).
52. POULOS, H. G. (1968). Analysis of the settlement of pile groups, *Géotechnique*, **18**(4), 449–471.
53. BUTTERFIELD, R. and BANERJEE, P. K. (1971). The elastic analysis of compressible piles and pile groups, *Géotechnique*, **21**(1), 43–60.
54. OTTAVIANI, M. (1975). Three-dimensional finite element analysis of vertically loaded pile groups, *Géotechnique*, **25**(2), 159–174.
55. BUTTERFIELD, R. and BANERJEE, P. K. (1971). The problem of pile group–pile cap interaction, *Géotechnique*, **21**(2), 135–142.
56. GREEN, P. and HIGHT, D. (1976). The instrumentation of Dashwood House, London, *CIRIA Technical Note No 78*, CIRIA, London.
57. SKEMPTON, A. W. (1959). Cast *in situ* bored piles in London clay, *Géotechnique*, **9**(4), 153–173.
58. WHITAKER, T. (1957). Experiments with model piles in groups, *Géotechnique*, **7**(4), 147–167.

Chapter 9

ANALYSIS OF AXIALLY AND LATERALLY LOADED PILE GROUPS

P. K. Banerjee

Department of Civil and Structural Engineering, University College, Cardiff, UK

SUMMARY

A completely general method for analysing the working load responses of pile groups embedded in a medium whose Young's modulus increases linearly with depth is described in this chapter. The method utilises a simplified formulation of the indirect boundary element method for representing the soil domain. The piles themselves have been idealised as linear structural members. It has been clearly demonstrated that, within working load restraints, this analysis may be used with confidence to yield meaningful results for practical pile group problems.

9.1 INTRODUCTION

A designer is interested in the following aspects of the behaviour of pile groups: (a) evaluation of the collapse load; (b) calculation of the immediate and long-term settlement so that he can select a suitable factor of safety in his design; and (c) the distribution of the loads and moments in piles so that he can provide adequate reinforcement in the piles.

Numerous attempts have been made by various investigators to solve the collapse problems. These range from purely theoretical studies to small and full scale experiments. Various approximations to

317

and simplified assumptions about the manner in which pile groups behave are fundamental to these studies and it is therefore difficult to extrapolate these results to all full scale foundations.

Fortunately, however, evaluation of the collapse loads of such foundations (except foundations on isolated single piles) are usually of secondary importance because relatively large factors of safety are needed to keep settlements within acceptable limits.

There are a number of methods which can deal with the settlement and the load distribution problems. These can be broadly classified into the following categories:

(a) Three-dimensional frame analysis.
(b) The modulus of subgrade reaction method.
(c) The pile–soil interaction method.
(d) The finite element method.
(e) The boundary element method.

The three-dimensional frame analysis method treats the piles as beam-columns and ignores the presence of the soil. The pile heads are assumed to be rigidly connected to a rigid cap and the ends are assumed to be fixed or pinned in a three-dimensional space. Although such an idealisation may appear to be realistic for end-bearing pile groups, driven or bored through a very soft subsoil, it has been seen from experimental studies that it leads to inaccurate estimates of displacements and moments even for such a special case.

The modulus of subgrade reaction method is essentially the same as the frame analysis method described above except that the presence of soil in the direction normal to the pile axis is included by treating the soil as an ideally elastic Winkler medium. The medium is defined by a single elastic constant (k), commonly called the modulus of subgrade reaction. Such a method can be applied to any end bearing pile group, provided the approximate value of k is assigned to the medium. This method, however, cannot be applied to floating pile groups because a Winkler medium cannot transmit shear which accounts for about 60% to 70% of the load transferred from the piles to the surrounding soil in these situations. It is possible to extend this method to deal with such problems by using two sets of spring constants, k_1 to transmit the loading intensity acting normal to the pile axis, and k_2 to transmit the shear stress along the pile shaft surface. To date, however, no such analysis has been reported.

The main advantage of this method is that it is possible to represent the non-homogeneity of the soil medium approximately by varying the value of k along the length of the pile. In spite of the possible generalisation and extension of this method its main disadvantage lies in the assessment of the value of the modulus of subgrade reaction (k). There have been at least 60 scientific publications on this subject alone over the last two decades. A careful study of these revealed that the value of the modulus of subgrade reaction that can be used to analyse the behaviour of pile groups depends on the following major parameters:

(a)   Young's modulus and Poisson's ratio of the soil or, alternatively, the basic deformation moduli representing the properties of the soil medium alone.

(b)   The length to diameter ratio of piles.

(c)   The diameter of piles.

(d)   The compressibility (EA) and the flexibility (EI) of piles.

(e)   The height of the cap above ground.

(f)   The spacing of piles.

(g)   The angle of rake, including the sign.

(h)   The number of piles in the group.

(i)   The relative magnitudes of the ratio of the vertical loads to that of the horizontal loads.

(j)   Whether the piles are end-bearing or floating within the soil layer.

In addition, the method of construction employed, e.g. boring or driving, the order in which piles are being installed and the separation of the soil from the pile near the ground level will alter the value of the modulus of subgrade reaction. Thus, it would appear from the foregoing that it is difficult for an engineer to evaluate appropriate values of k for his particular problem.

The pile–soil interaction method developed by Poulos and his associates[12,18,22–25] is based on the assumption that the piles themselves are linearly elastic and embedded in another linearly elastic three-dimensional solid. By using Mindlin's solution for a point load they obtained the interaction factors for a two-pile problem. A simple superposition technique was then utilised to obtain the solutions for a general pile group. The interaction factor method is quite satisfactory provided that only the load distributions at the pile heads are required, and that all piles in the group are of same diameter and length.

Therefore, the method can be applied only to symmetrical vertical pile groups where the interaction between the applied loading is negligible. Moreover, it is not possible to extend such analyses to solve problems involving soil inhomogeneity.

The finite element method can of course successfully solve the present problem. However, a recent attempt by Ottaviani[21] seems to indicate that the cost of analysing a realistic practical pile group problem by using this method could be prohibitively expensive. Indeed, for problems involving low surface area to volume ratios, such as those found in three-dimensional problems of foundation engineering, the finite element method, which employs a volume discretisation scheme, is not the most efficient numerical method.

The boundary element method or the boundary integral equation method, employs a surface discretisation scheme enveloping each homogeneous zone of a body. In this method, a particular solution of the governing differential equation of the problem is chosen, and, by distributing this solution over the surface of the given domain, a general solution is developed in terms of a boundary density function. For this general solution to satisfy the boundary conditions the density function must satisfy an integral equation on the boundary. The method is completely general and has attained a very high degree of performance, and, therefore, popularity. An indirect formulation of the method has been applied to problems of steady state and transient ground water flow, elasticity and elasto-plasticity by Banerjee, Butterfield and their associates.[1-8,14] Application of the direct formulation of the method to problems of elasto-dynamics, visco-elasticity and elasto-plasticity is described by Cruse and Rizzo.[11]

The application of a completely general algorithm to some pile group problems in which both the soil and the piles were idealised as three-dimensional elastic solids is described elsewhere.[1-4] Although such algorithms are considerably more efficient than the finite element method (the computing time being about one eighth of that required for similar problems analysed by the finite element method), it is still not feasible to analyse realistic practical pile group problems by using this completely general algorithm.

In what follows we will retain the basic features of the boundary element method and develop a simplified integral representation for the soil domain and combine this with a three-dimensional frame analysis. In the first part of the analysis the soil is assumed to be homogeneous, but later extensions of the method to non-homogeneous soils are described. It will be seen that the method can be applied to any combination of vertical load, horizontal load and moments.

While it must be emphasised that the idealisation used differs in many respects from the behaviour of real soil, it is, particularly in clayey soil, a much more satisfactory approximation to a real soil mass than is a Winkler medium. This is clearly demonstrated by comparing the results of this analysis with a series of model and full scale tests on piles and pile groups. It can be seen from these comparisons that the effects of the parameters (b) to (j) have already been included in the analysis. The user has to select the basic deformation parameters for the soil. Other effects such as the method of construction, order of installation, separation of the soil from the pile near ground level, etc. cannot be included in the present analysis.

9.2   THE METHOD OF ANALYSIS

9.2.1 A Brief Description of the Boundary Element Method

As mentioned earlier, the indirect formulation of the boundary element method involves the selection of an elementary singular solution of the governing differential equation of the problem. Then, by distributing this singular solution in terms of an arbitrary density function (traction vectors) over the surface of a given domain a general solution in terms of the arbitrary function is developed.[1]

For three-dimensional problems the elementary solutions may be chosen to be, for example:

(a)   Kelvin's solution for a point load within an infinite solid.[2]
(b)   Mindlin's solution for a point load within a semi-infinite solid.[19]
(c)   Boussinesq's solution for a point load acting on the surface of a half-space.[19]

Of these, Kelvin's solution provides the formulation for a solid of any shape, Mindlin's and Boussinesq's solutions provide the formulation for solids bounded by a plane. Whatever singular solution is chosen, the boundary element formulation closely follows the steps formally described in the paragraphs which follow.

The basic elementary solution is a function of the position of the point of application of load B and the field point A with respect to a cartesian co-ordinate system. Thus, the three components of displacements $u_i(A)$ and nine components of stresses $\sigma_{ij}(A)$ (by symmetry six components) at a point A due to the three components of the forces $e_j(B)$ acting at B can be written as:[2]

$$u_i(A) = e_j(B)K_{ij}(A, B) \tag{1}$$

$$\sigma_{ij}(A) = e_k(B)T_{ijk}(A, B) \tag{2}$$

where $K_{ij}(A, B)$ and $T_{ijk}(A, B)$ are functions of the positions of A and B with respect to a cartesian co-ordinate system. The traction vector $p_i(A)$ on a surface through the point A having outward normal vector $n_j(A)$ can be obtained from:

$$p_i(A) = \sigma_{ij}(A) \cdot n_j(A) = e_k(B) \cdot T_{ijk}(A, B) \cdot n_j(A)$$

or

$$p_i(A) = e_j(B)\Gamma_{ij}(A, B) \tag{3}$$

where

$$\Gamma_{ij}(A, B) = T_{ikj}(A, B) \cdot n_k(A)$$

If we now distribute fictitious traction vectors $\phi_j(B)$ over the surface S, the displacements and tractions due to the total effects of these fictitious tractions can be obtained from eqns. (1) and (3) as (see Fig. 1):

$$u_i(A) = \int_s \phi_j(B)K_{ij}(A, B)\, ds \tag{4}$$

$$p_i(A) = \int_s \phi_j(B)\Gamma_{ij}(A, B)\, ds \tag{5}$$

Equations (4) and (5) are valid as long as the load point B and the field point A do not coincide. Therefore we need to investigate what happens when they do coincide.

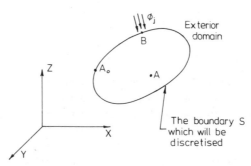

FIG. 1.   Any typical problem to be solved by the boundary element method.

For the elementary solutions stated above, functions K_{ij} and Γ_{ij} are singular whenever A and B coincide. The order of singularity of K_{ij} is of the order of $1/R$ and that for Γ_{ij} is of the order $1/R^2$, where R is the distance between A and B. It can be shown that as the point A approaches the point A_0 located on the surface, eqns. (4) and (5) can be written as:

$$u_i(A_0) = \int_s \phi_j(B)K_{ij}(A_0, B)\,\mathrm{d}s \tag{6}$$

$$p_i(A_0) = \alpha\phi_i(A_0) + \int_s \phi_j(B)\Gamma_{ij}(A_0, B)\,\mathrm{d}s \tag{7}$$

where α is a constant which depends on the nature of the boundary value problem, i.e. exterior or interior. For an interior problem, if the surface S at A_0 is flat, $\alpha = \frac{1}{2}$. Equation (7) can be seen as a statement of the equilibrium of the surface S. Equations (6) and (7) are the two integral equations which can be solved to satisfy any well-posed boundary conditions on S (see reference 1). Having obtained $\phi_j(B)$ from eqns. (6) and (7) the stresses and displacements at any interior point A can be obtained from eqns. (4) and (5).

9.2.2 Analysis of Pile Groups Embedded in Homogeneous Soils

With reference to the present problem which involves an unloaded ground surface, it is more convenient to adopt Mindlin's solution as a singular solution. If the displacements of the pile–soil interfaces (Fig. 2) are specified, the discretised form of eqn. (6) will provide the solution of the problem. It is also interesting to note that because the integral involving the displacement kernel functions $K_{ij}(A, B)$ is only weakly singular $(1/R)$, it exists in the ordinary sense over the surface and therefore, in the problem at hand, ϕ_j will be the real traction at the pile–soil interfaces.[2]

In order to couple these equations for displacements of the pile–soil interface for the soil domain with those for the pile domain, certain approximations are necessary. These arise from the fact that the piles have been idealised as linear structural elements for which the equilibrium and compatibility can be satisfied only at the centre line of the pile. In order to achieve this the value of ϕ_j was allowed to vary only along the pile length. Therefore a direct integration of K_{ij} over the circumference of each pile–soil interface can be carried out numerically. This approximation is quite consistent with the idealisation of

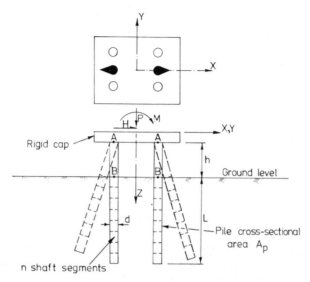

FIG. 2.   A typical pile group problem.

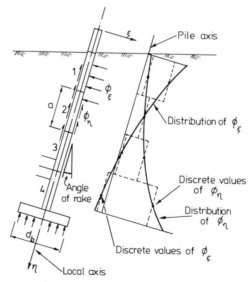

FIG. 3.   Discretisation of the pile–soil interface.

the piles as linear structural elements. Thus, if we divide each pile–soil interface into n cylindrical shaft segments (Fig. 3), and one circular base area, for a pile group containing N piles, eqn. (6) can be written for discretised pile–soil interfaces as:

$$u = K\Phi \tag{8}$$

where u and Φ are $3N(n+1) \times 1$ vectors and K is a fully populated $3N(n+1) \times 3N(n+1)$ matrix.

The order of the matrices involved may be further reduced by observing the symmetry of the group as well as that of the applied loading. For example:

(a) For piles which carry the same loads, values of ϕ_j are identical. Therefore a direct summation of the coefficients of K_{ij} can be carried out.

(b) If we restrict the applied loading to be a combination of a vertical load P, a horizontal load in the direction of the x axis H and a moment in the xz plane M, we can ignore the effects of the components of applied tractions in the y direction and the associated coefficients of the matrix K.

(c) Furthermore, if we assume the pile bases to be smooth, we can ignore the effects of the tangential components of ϕ_j over the base areas.

For raked piles, it is more convenient to introduce a local system of axes ξ and η, which are respectively perpendicular to and along the pile axis (as shown in Fig. 3). If we define U and Φ as the components of displacements and tractions in the direction of the local axes we can write eqn. (8) with respect to these components as

$$U = K_0 \Phi \tag{9}$$

where $K_0 = \lambda^T K \lambda$, λ = the direction cosine matrix, and $\lambda^T = \lambda$, transposed.

Explicit details of the various simple matrix operations needed to generate the system of equations represented by eqn. (9) are described in reference 6. If the displacements of the pile–soil interfaces are known then eqn. (9) can be solved directly. But these displacements can be specified only for perfectly rigid piles. For most practical situations piles cannot be considered as rigid, hence we must couple these equations with a corresponding set of equations for the pile domains.

The stresses Φ acting on the soil are equal and opposite to the stresses acting on the pile. Therefore, the equations for pile compressibility, with respect to the local axes, as indicated in Fig. 3, can be written as:[18]

$$\frac{\partial^2 u_\eta}{\partial \eta^2} = \frac{\pi d}{E_p A_p} \cdot \Phi_\eta \tag{10}$$

where d = the diameter of the pile, E_p = the Young's modulus of the pile material, and A_p = the cross-sectional area of the pile.

If we divide the pile shaft into n cylindrical elements, represent the base by a uniformly loaded circular disc, and express eqn. (10) by a finite difference approximation, we can write:

$$\lambda_0 D_0 U_\eta + B_0 = \Phi_\eta \tag{11}$$

where $\lambda_0 = E_p A_p/(\pi d a^2)$, $a = L/n$, and L = the embedded length of the pile.

$$
D_0 = \begin{bmatrix}
-3 & 1 & & & & & & \\
1 & -2 & 1 & & & & & \\
& 1 & -2 & 1 & & & \text{zeros} & \\
& & & \cdot & & & & \\
& & & & \cdot & & & \\
& \text{zeros} & & & & \cdot & & \\
& & & & 1 & -2 & 1 & \\
& & & & -0\cdot 2 & 2 & -5 & 3\cdot 2 \\
& & & & & 1\cdot 33f & 12f & -10\cdot 67f
\end{bmatrix}
$$

where $f = \pi d a/(4A_b)$, A_b = area of the base = $\pi d_b^2/4$, and U_η = the axial displacements of midpoints of the $n + 1$ elements of the shaft and the base.

$$
B_0 = \left\{ \begin{array}{c}
2\lambda_0 \delta_b \\
0 \\
0 \\
\cdot \\
\cdot \\
\cdot \\
\cdot
\end{array} \right\}, \quad (n + 1) \times 1 \text{ vector}
$$

where δ_b = the axial displacement of the pile at ground level.

In many practical situations it is customary to ignore the resistance offered by the top layer of soil thickness h (see Fig. 2). In such circumstances eqn. (11) can be modified to include the vertical displacement boundary condition at the pile head as:

$$\lambda_0 \mathbf{D}_0 \mathbf{U}_\eta + \beta \mathbf{B}_0 = \mathbf{S} \boldsymbol{\Phi}_\eta \qquad (12)$$

where $\beta = \delta_a/\delta_b$ and δ_a = the axial displacement of the pile head under consideration.

$$\mathbf{S} = \begin{bmatrix} 1+\dfrac{2l}{a} & \dfrac{2l}{a} & \dfrac{2l}{a} & \cdots & \dfrac{2l}{a} & \dfrac{ld_b^2}{2da^2} \\ & 1 & & & & \\ & & 1 & & \text{zeros} & \\ & & & \cdot & & \\ \text{zeros} & & & & \cdot & \\ & & & & & 1 \end{bmatrix}, \qquad \text{an } (n+1)^2 \text{ matrix}$$

and $l = h \sec \theta$.

Equation (12) can be further modified and written as

$$\mathbf{U}_\eta = \mathbf{D}_\eta \boldsymbol{\Phi}_\eta + \mathbf{B}_\eta \qquad (13)$$

where $\mathbf{D}_\eta = (1/\lambda_0)\mathbf{D}_0^{-1}\mathbf{S}$ and $\mathbf{B}_\eta = -(\beta/\lambda_0)\mathbf{D}_0^{-1}\mathbf{B}_0$.

The flexibility equation for the pile with respect to its local axes, as indicated in Fig. 3, can be written as:

$$E_p I_p \frac{\partial^4 U_\xi}{\partial \eta^4} = d \cdot \Phi_\xi \qquad (14)$$

where I_p = the second moment of area of the pile cross-section, d = the diameter of the pile, and $d \cdot \Phi_\xi$ = the transverse load per unit length of the pile.

Equation (14) can be solved for a pile which is rigidly attached to the pile cap with the following imposed boundary conditions:

(a) Transverse displacement of the cap = δ_ξ
(b) Rotation of the cap $= \theta_\xi$
(c) Moment at the base of the pile $= 0$
(d) Shear at the base of the pile $= 0$.

Thus, the transverse displacement of an element i due to transverse stress intensity Φ_ξ acting on an element j can be obtained from:

$$(U_\xi)_i = \delta_\xi - \theta_\xi\{(i - 0\cdot5)a + l\} - \frac{(\Phi_\xi)_j d}{E_p I_p}\left[\frac{a}{2}\{l + (j - 0\cdot5)a\}\right.$$
$$\times \{l + (i - 0\cdot5)a\}^2 - (a/6)\{l + (i - 0\cdot5)a\}^3 + (a^4/24)$$
$$\left.\times \langle(i - j + 0\cdot5)\rangle^4 - (a^4/24)\langle(i - j - 0\cdot5)\rangle^4\right] \tag{15}$$

where $\langle\ \rangle$ signifies McCauley brackets.

Equation (15), which is applicable to a fixed headed pile, can be written for all elements of the pile ($i = 1, 2 \ldots n$; $j = 1, 2 \ldots n$) as

$$\mathbf{U}_\xi = \mathbf{D}_\xi \boldsymbol{\Phi}_\xi + \mathbf{B}_\xi \tag{16}$$

where

$$\mathbf{B}_\xi = \begin{Bmatrix} \delta_\xi - \theta_\xi\{0\cdot5a + l\} \\ \delta_\xi - \theta_\xi\{1\cdot5a + l\} \\ \vdots \\ \delta_\xi - \theta_\xi\{(n - 0\cdot5)a + l\} \end{Bmatrix}$$

$\mathbf{D}_\xi = (n \times n)$ matrix formed by varying $i = 1, \ldots n$ and $j = 1, \ldots n$ in: $(d/E_p I_p) \times$ the appropriate expressions within the square brackets in eqn. (15).

Equations (13) and (16), which represent the compressibility and flexibility for the pile, must now be combined with eqn. (9) to yield the final solution.

We can combine eqns. (13) and (16) successively for each element of the pile and for every pile in the group and write them as:

$$\mathbf{U} = \mathbf{D}\boldsymbol{\Phi} + \mathbf{B} \tag{17}$$

which can be combined with the corresponding set of equations for the soil domain given by eqn. (9):

$$\mathbf{U} = \mathbf{K}_0\boldsymbol{\Phi} \tag{18}$$

By utilising the conditions of equilibrium and compatibility at the pile–soil interface, $\mathbf{U}$ can be eliminated and the following system of equations can be obtained:[6]

$$\mathbf{F}\boldsymbol{\Phi} = \mathbf{B} \tag{19}$$

where $\mathbf{F}$ is a fully populated $N(n + 1) + Nn$ square matrix. Equation (19) can be solved by a standard Gaussian elimination technique.

The analysis described above needs prescribed displacements of the pile cap. However, for most problems, it is the loading that is specified. Therefore, we need to establish the relationships between the displacements of the cap and the applied loading. This can be accomplished by successively applying unit vertical and horizontal displacements and rotations to the pile cap and calculating the vertical load, horizontal load and moments necessary to equilibrate the system of stresses developed (see Fig. 4).

Thus, an external loading system P, H and M, acting on the cap can be related to the displacements w, u and θ of the cap via

$$\begin{Bmatrix} P \\ H \\ M \end{Bmatrix} = \begin{bmatrix} S_{11} & S_{12} & S_{13} \\ S_{21} & S_{22} & S_{23} \\ S_{31} & S_{32} & S_{32} \end{bmatrix} \begin{Bmatrix} w \\ u \\ \theta \end{Bmatrix} \qquad (20)$$

The 3×3 matrix in eqn. (20) can be described as the global foundation stiffness matrix which may be used as boundary conditions to the analysis of superstructures. Equation (20) can be written as:

$$\begin{Bmatrix} w \\ u \\ \theta \end{Bmatrix} = \begin{bmatrix} f_{11} & f_{12} & f_{13} \\ f_{21} & f_{22} & f_{23} \\ f_{31} & f_{32} & f_{33} \end{bmatrix} \begin{Bmatrix} P \\ H \\ M \end{Bmatrix} \qquad (21)$$

where the matrix f can be described as the global flexibility matrix of the pile–soil system.

Having obtained the displacements from eqn. (21) the stresses resulting from applied unit boundary conditions can be scaled and the

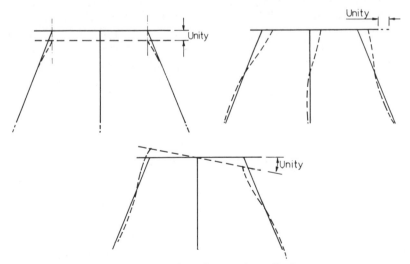

FIG. 4.   Application of unit boundary displacements.

various components can be algebraically added to give the final stresses due to any combination of P, H and M.[7]

9.2.3 Extension of the Above Method to Non-homogeneous Media

Although most naturally occurring soils are non-homogeneous, very few theoretical solutions of the problem of pile groups embedded in such soils have been attempted, primarily because of (a) the geometrical complexity of the problem, and (b) the definition of non-homogeneous soils. However, within the framework of the analysis presented above, and in boundary element methods generally, the geometrical complexity of a problem presents no special difficulty. Regarding the problem of the definition of non-homogeneous soil, the recent work of Gibson[16] has suggested a linear increase of Young's modulus with the depth, given by the equation:

$$E(z) = E(0) + mz \tag{22}$$

where $E(z)$ = Young's modulus at a depth z, $E(0)$ = Young's modulus at ground level, and m = an elastic constant.

With Poisson's ratio v remaining constant, this does provide a very satisfactory representation for analysing the linear part of the response of soil deposits where the effective stresses increase linearly with the depth. The main attraction of this model is the ease with which a non-homogeneous medium can be defined.

It is, of course, possible to solve this problem by approximating the 'Gibson soil' by a series of homogeneous layers and then by using the general version of the boundary element method algorithm. Indeed, such analyses for axially loaded pile groups are described by Banerjee[1] and Banerjee and Davies.[4] The main drawback of such a procedure is that the analysis is far too expensive in terms of computational costs. In what follows, therefore, we shall describe an approximate method, recently developed by Davies and Banerjee[14] for such problems.

The object of the present analysis is to develop a technique for calculating the coefficients of the pile–soil interaction matrix $\mathbf{K}$, i.e. we wish to generate the system of equations (see eqn. (8)):

$$\mathbf{K}\Phi = \mathbf{U}$$

where Φ is the applied tractions at the pile–soil interfaces and $\mathbf{U}$ is the pile–soil interface displacement vector, so that we can combine these equations with those obtained from the consideration of pile domains by using the three-dimensional frame analysis. The **K** matrix, of course, should now be derived for a Gibson soil. However, we can simplify our calculations to a considerable extent if we observe the following facts:

(a) The coefficients of the diagonal blocks of the matrix are the most dominant coefficients. These diagonal blocks contain the influence coefficients for displacements of points in the immediate neighbourhood of the applied pile–soil interface tractions.

(b) Any error in the magnitude of the coefficients in the off-diagonal blocks has less influence on the accuracy with which Φ is calculated. Indeed, this is the well-known St. Venant's principle in a slightly different form.

(c) If the value of the Poisson's ratio does not vary in a non-homogeneous medium, the stress distribution in a non-homogeneous soil is nearly identical to that in homogeneous soil. In fact, the stress distribution in Gibson soil is identical to that in homogeneous soils for $\nu = 0.5$, and differs negligibly for other values of the Poisson's ratio.

(d) In view of (c) above, the displacements in the neighbourhood of the point of application of the load can be calculated quite accurately if we replace that region by a homogeneous region having the Young's modulus E_s such that the strain energy density of the actual problem and of the idealised problem are identical. In simpler language, this simply means that work done by the two systems must be the same. Because the stresses resulting from the two systems are nearly identical, the weighted integral of the elastic compliances of the two systems taken over the volume must also be nearly equal.

In what follows we shall briefly describe two methods of numerical calculations for **K** based on the ideas expressed in the preceding paragraphs.

Suppose it is necessary to calculate the vertical and horizontal displacements of a point A at a depth Z below the ground level due to vertical and horizontal forces acting at B at a depth C below the surface of a Gibson soil. We can idealise the problem as being a two-layer one, where the point of application of the forces is considered to be at the junction between the two layers (see Fig. 5). The elastic constants E_1 and E_2, for the layers 1 and 2 respectively, can be obtained such that the displacements in the neighbourhood of the point of application of the load are identical to those of the Gibson

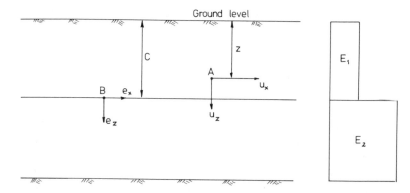

FIG. 5. A two-layer idealisation.

soil problem. This involves satisfying the strain energy criterion as discussed previously.

The problem shown in Fig. 5 has been recently solved by Davies and Banerjee[13] by utilising the integral transform technique developed in works of Chan *et al.*[9] and Muki[20] for such problems. From a series of computations it was found that the elastic moduli E_1 and E_2 can be chosen approximately as:

$$E_1 = E(0) + 0.5mC \tag{23}$$

and

$$E_2 = E(0) + 3.0mC \qquad \text{for } 5C < 2L$$

or

$$E_2 = E(0) + m(L + 0.5C) \qquad \text{for } 5C > 2L$$

where L is the embedded length of the pile.

It should be emphasised that eqn. (23) has been obtained specially for pile group problems, i.e. the idealisation used here is problem-dependent. Moreover, the values of E_1 and E_2 have to be calculated every time the locations of the unit point forces are altered; this means that different values of E_1 and E_2 are needed for calculating every column of the matrix **K**.

A novel method of solving the problem of a pile group embedded in a moderately non-homogeneous soil was proposed by Poulos.[25] In this analysis, the boundary element method is used in conjunction with Mindlin's solution for the homogeneous problem to model the Gibson medium, by modifying the **K** matrix of the system. In effect, the displacement at a point—given by Mindlin's solution—is modified by using the soil modulus at that point. No theoretical justification can be given for adopting this procedure though intuitively this method would introduce some approximate degree of non-homogeneity into the system. Clearly, in the extreme case of $E = mz$ displacements at the surface due to a point load within the medium would tend to infinity. However, assuming a moderate degree of non-homogeneity (e.g. $\chi = E(0)/E(L) \not< 0.5$—typically for over-consolidated clay) then this method should yield acceptable results despite the marked non-symmetry of the **K** matrix. It should be noted that the equivalent two-layer idealisation discussed earlier produces an approximately symmetric **K** matrix, thus satisfying the reciprocal work theorem of solid mechanics.

9.3   NUMERICAL ACCURACY

The solution method described for homogeneous soils and that employing the two-layer idealisation for Gibson soil were found to be in agreement (to within 10%) with the general boundary element method of analysis as well as with the finite element method of analysis carried out on a single pile. Perhaps it is worth while to point out that the cost of analysing the single pile problem by using the general boundary element method algorithm was about eight times, and by using the finite element method about 50 times, that of the present solution, thus demonstrating the usefulness of the present method for routine solution of practical pile group problems.

The method suggested by Poulos consistently provided larger displacements. For moderately homogeneous soils ($\chi = 0.5$), however, it gave solutions to within 10% of the present method. For highly non-homogeneous soils the upper elements of the piles were found to be far too flexible, which in turn decreased the overall stiffness of the system very considerably. Thus, although it seems unlikely that this formulation could provide meaningful results for sands and other soils with very low surface stiffness, it appears to be an attractive method of solving less extreme cases (typically $\chi \not< 0.5$).

Another test on accuracy of the solution can be carried out by observing the symmetry of the global stiffness and flexibility matrices described in eqns. (20) and (21). These were symmetric to within 6%

over the complete range of pile–soil compressibility ratios ($K = E_p/E(L)$) and flexibility ratios ($K_R = (E_p I_p/(E(L) \cdot L^4))$).

9.4   RESULTS OF THE ANALYSIS

For symmetric pile groups, the terms $S_{12}, S_{13}, S_{21}, S_{31}$ in the global stiffness matrix (see eqn. 20) are negligible, indicating the fact that there is negligible interaction between the axial loading and lateral loads and moments. For non-symmetrical pile groups, however, these are of significant magnitude. In what follows, a series of non-dimensional plots of results for single piles and symmetrical pile groups are described.

9.4.1 Behaviour of Axially Loaded Single Piles and Pile Groups
Figures 6 and 7 show the load-displacement behaviour and the proportions of total load carried by the bases for plain and under-

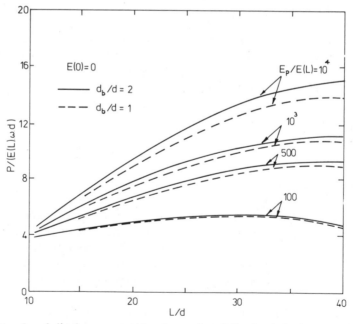

FIG. 6.   Load-displacement behaviour of axially loaded single plain and underreamed piles embedded in Gibson soil.

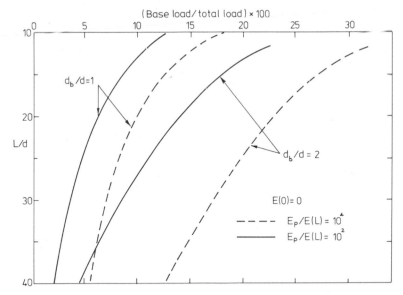

FIG. 7.   The percentage of total load carried by the pile base.

reamed piles. These results have been plotted for $v = 0.5$ and a ratio of depth of the soil layer to pile length of 2.0 throughout. Figure 6 shows the plot of non-dimensional stiffness $(P/E(L)wd)$ against the pile compressibility ratio $(E_p/E(L))$. These indicate approximately 30% reduction in stiffness compared with the corresponding solutions of Butterfield and Banerjee[7] for homogeneous soils.

Figure 7 shows the corresponding plots of the percentage of load carried by the pile base. It is interesting to note that, contrary to intuitive reasoning, the base carries only marginally higher loads than those reported by Butterfield and Banerjee for the corresponding homogeneous cases.

The distribution of the shear stresses at the pile–soil interface for an axially loaded plain pile embedded in Gibson soil is shown in Fig. 8. This shows that the pile compressibility has a dramatic effect on the shear stress distribution.

Figure 9 shows the interaction factors plotted against the spacing to diameter ratios for various pile compressibility ratios and length to diameter ratios. The interaction factor is defined as the ratio of the additional settlement due to the presence of a second pile in a two

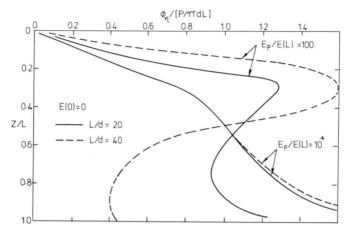

FIG. 8.   Distribution of the shear stress at the pile–soil interface.

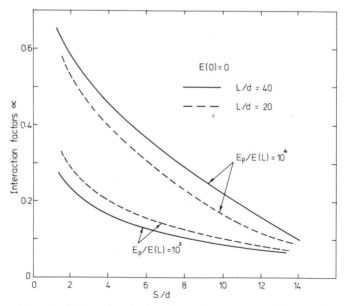

FIG. 9.   Interaction factors for pile groups in Gibson soil.

pile group each carrying a load P to that of an isolated single pile under a load P. It can be seen that these interaction factors are very much smaller than those reported for homogeneous soils.[7,12]

It is also possible to calculate the settlement of any axially loaded symmetrical pile group by utilising these interaction factors, e.g. the displacement of the ith pile in a general pile group is given by:

$$\omega_i = \omega_0 \sum_{j=1}^{N} \alpha_{ij} P_j \qquad (24)$$

where N = the number of piles in the group

$\alpha_{ij} = 1$ for $i = j$

$= \alpha$ from Fig. 9 for $i \neq j$

ω_0 = the settlement of the single pile due to unit load (obtainable from Fig. 6).

The system of eqn. (24) can then be solved for a fully rigid or fully flexible pile cap in the usual manner.

9.4.2 Behaviour of Laterally Loaded Single Piles and Groups

As discussed earlier, for symmetrical pile groups and single piles subjected to lateral loads and moments, we can write eqn. (21) as:

$$u = f_{22}H + f_{23}M$$
$$\theta = f_{32}H + f_{33}M$$

where, by reciprocal theorems, $f_{23} = f_{32}$.

We can write the above equations in non-dimensional form as:

$$u = I_H \frac{H}{E(L) \cdot L} + I_{HM} \frac{M}{E(L) \cdot L^2} \qquad (25)$$

$$\theta = I_{HM} \frac{H}{E(L) \cdot L^2} + I_\theta \frac{M}{E(L) \cdot L^3} \qquad (26)$$

These equations apply to pile caps which are free to rotate when H and M are applied to the group. It may also be necessary to analyse laterally loaded pile groups where the pile cap is fully restrained ($\theta = 0$) due to its connection to very stiff structural members of the superstructure which it supports. For such cases, the restraining moment M can be related to H from eqn. (26):

$$M = -\frac{I_{HM}}{I_\theta} \cdot HL$$

Substituting for M in eqn. (26) we have for restrained caps subjected to lateral loading:

$$u = (I_H - I_{HM} \cdot I_{HM}/I_\theta)\frac{H}{E(L) \cdot L} = I_{FH}\frac{H}{E(L) \cdot L} \qquad (27)$$

Figures 10 and 11 show the semi-log plots of these factors, I_H, I_{HM}, I_θ and I_{FH} for various values of pile–soil flexibility ratio K_R. The two most striking features to note are (a) the relative unimportance of the slenderness ratio L/d, and (b) the pronounced effects of variation of the flexibility parameter K_R on the displacement factors. It is also of interest to note that $K_R = 0\cdot1$ corresponds to rigid piles and $K_R = 10^{-4}$ corresponds to piles commonly encountered in practice, which are flexible piles.

Figure 12 shows the bending moments along the pile length due to applied horizontal load on an unrestrained pile cap, and Fig. 13 shows the maximum design moments on a laterally loaded pile with un-restrained pile cap. Once again, key features are (a) the relatively

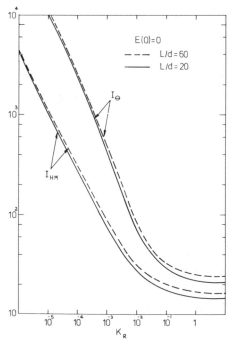

FIG. 10. The variation of I_{HM} and I_θ with the variation of the slenderness ratio and the flexibility factor.

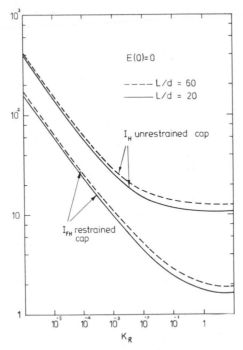

FIG. 11.   The variation of I_H and I_{FH} with the variation of the slenderness ratio and the flexibility factor.

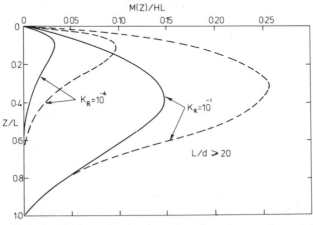

FIG. 12.   The distribution of moments in piles due to lateral loads (unrestrained cap). $---\chi = E(0)/E(L) = 0$; $———\chi = E(0)/E(L) = 1$.

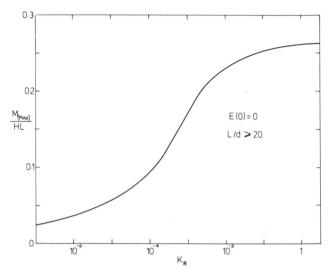

FIG. 13. The maximum bending moment in a pile due to lateral load on an unrestrained cap.

small effects of L/d, (b) the pronounced effects of K_R and the degree of non-homogeneity χ.

9.5 COMPARISONS WITH EXPERIMENTAL DATA

In order to investigate the usefulness and applicability of the method of analysis to practical problems of load distribution and load displacement behaviour of pile groups in real soil, a series of comparisons with reported test results was undertaken. Of these only two are described in Sections 9.5.1 and 9.5.2. Since the present analysis is intended to describe the working load behaviour, these comparisons were carried out at loads corresponding to half the ultimate load of the group. The original dimensional units used by the experimentors have been retained.

9.5.1 Comparisons with Cooke[10]
Cooke reported the results of full scale tests on axially loaded single piles and a 3×1 pile group embedded in London clay at Hendon. Tubular steel piles, 168 mm in diameter and 5 m long, equipped with

strain gauge load cells, were used. The load distribution in piles, as well as the vertical displacements at two different levels below the ground surface, were measured. Comparisons between the theoretical displacements and the experimental observations at these two levels for a 3×1 pile group with flexible and rigid pile caps are shown in Figs. 14 and 15 respectively. These theoretical results were obtained by using the general method of analysis described earlier using $v = 0.5$ and $E(0) = 20 \text{ MN/m}^2$ and $m = 15 \text{ MN/m}^2/\text{m}$.

These adopted parameters are also in agreement with those reported by Marsland,[17] who carried out some plate bearing tests on the same location.

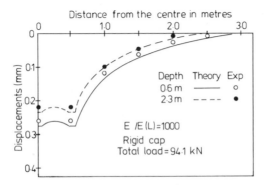

FIG. 14.   Comparisons with field test results of Cooke (flexible pile cap).[10]

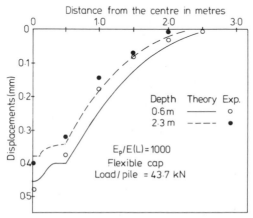

FIG. 15.   Comparisons with field test results of Cooke (rigid pile cap).[10]

9.5.2 Comparison with Feagin[15]

Feagin carried out a series of full-scale tests on groups of battered and vertical wooden piles fixed in concrete test monoliths subjected to lateral loads. Various configurations of piles were driven into the fine to coarse sand to determine the effect of group geometry on lateral stiffness.

Figure 16 shows the comparisons between the theoretical results and the test results for eight different group configurations. Details of

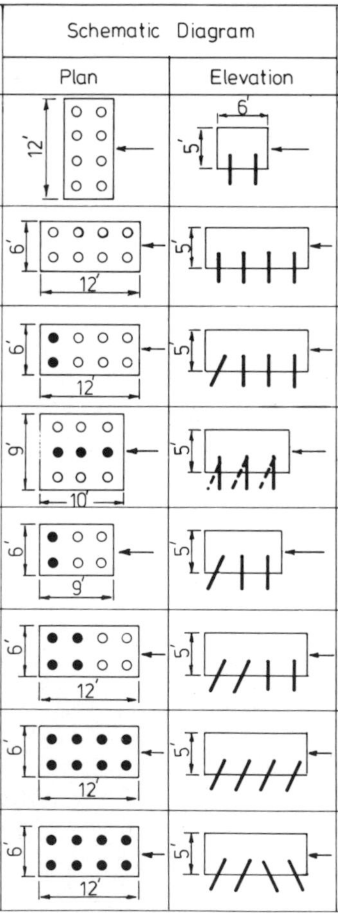

Test N°	Schematic Diagram		Lateral load for ¼" deflection		Lateral deflection due to vertical load of 20 T/pile	
	Plan	Elevation	Test	Theory	Test	Theory
1			4.8	4.8	0.0	0.0
2			5.8	5.3	0.0	0.0
3			7.0	7.3	0.04	0.06
4			7.1	6.8	0.06	0.08
5			7.3	8.1	0.05	0.07
6			9.0	8.4	0.07	0.11
7			9.0	8.2	0.21	0.27
8			15.8	11.7	0.0	0.0

o = Vertical   ● = Raked

FIG. 16.   Comparisons with full-scale tests by Feagin.[15]

the tests are as follows:

Depth of sand layer $\quad$ = 75 ft
Length of piles $\quad\quad$ = 32 ft
Young's modulus of timber = $2.9 \times 10^8 \, \text{lbf/ft}^2$
Mean diameter of pilc head $\;$ = 13 in
Mean diameter of pile base = 9 in
Angle of batter $\quad\quad\quad$ = 20°

The theoretical results were obtained using $E(0) = 0$ and $m = 70 \times 10^3 \, \text{lbf/ft}^3 (40 \, \text{lbf/in}^3)$. These results for lateral loading are in substantial agreement with the experimental results. The horizontal displacements due to vertical loads are not accurately specified by the author. This may have resulted in some divergence between the theoretical and experimental results.

9.6   DISCUSSION

The analysis of the experimental studies given above reveals encouraging correlations between soil properties and the deformation moduli and these are seen to be independent of geometric scale or loading conditions.

In particular, the rate of increase of modulus with depth for sands is consistently given by an average value of $40 \, \text{lbf/in}^3$ ranging from $25 \, \text{lbf/in}^3$ for loose sands to $70 \, \text{lbf/in}^3$ for dense sands.

Clays, however, may have substantial stiffness at ground level depending on the degree of over-consolidation. The maximum value of this parameter among those analysed was $1000 \, \text{lbf/in}^2$, though for highly over-consolidated clays, a value of $3000 \, \text{lbf/in}^2$ may be attained. Generally, soil moduli were found to be 100–150 times the measured shearing strengths, due consideration being accorded to the experimental techniques employed.

Of particular interest is the fact that these values of increase of modulus with depth are broadly of equal magnitude to the increase of subgrade reaction with depth proposed for sands by Terzaghi.[26]

The bending moments in laterally loaded single piles are generally very much higher than those in vertical groups (typically about 10–15% of those reported for single piles). For raked pile groups, however, these moments are of comparable magnitude to those for single piles.

The solutions using a Gibson model are generally better than those obtained using a homogeneous model, particularly for highly non-homogeneous soils, such as sands and normally consolidated clays, especially for the prediction of moments. For more homogeneous deposits—over-consolidated clays in particular—the simpler homogeneous model will suffice for most purposes especially since an appropriate soil modulus may be easily chosen in these cases.

9.7 CONCLUSIONS

The analysis of pile groups by means of a boundary element formulation has been shown to be capable of giving solutions that are satisfactory for practical engineering purposes. Comparisons between the current formulation which introduces the Gibson soil model into the pile–soil system and existing formulations which use homogeneous soil models reveal several distinct advantages in favour of the more complex model. These are, primarily, that more realistic values of interaction occur between piles in groups and that consistent relations between soil properties and soil moduli for a wide range of soil conditions are manifest in those data for which detailed analyses have been carried out.

These improvements on the existing formulations should provide sufficient impetus for use in practical situations from which data may be assembled to enable better assessment of soil moduli from other soil parameters, e.g. shearing strength, etc. Tentative correlations between these parameters are recommended in this chapter for sands and some clay soils.

The analysis is thus applicable to a wide range of pile-group problems, in particular, those most widely found in practice. Within working load restraints, the analysis may be used with confidence to yield meaningful results although the theoretical basis of the solution is approximate in some respects.

Previous studies of pile groups in Gibson soil (Banerjee and Davies[4]) have revealed the characteristics of these systems but have been limited both in their applicability and particularly by economic constraints. This formulation, however, is no greater than four times as expensive, in terms of computational time, as the corresponding solutions for the homogeneous case, and is thus of the order of 40–50 times faster than the equivalent finite element solution. In the case of moderately non-homogeneous soils, an approximate solution based on a modification of the Mindlin kernel for homogeneous soils will also yield moderately accurate results at very low cost.

ACKNOWLEDGEMENTS

The author is deeply indebted to Mr R. M. Driscoll and Mr T. G. Davies who carried out most of the computational work described here. The financial support of the work was provided by the Department of Transport, London.

REFERENCES

1. BANERJEE, P. K. (1976). Integral equation methods for analysis of piece-wise non-homogeneous, three-dimensional elastic solids of arbitrary shape, *Int. J. Mech. Sci.*, **18**, 293–303.
2. BANERJEE, P. K. (1976). Analysis of vertical pile groups embedded in non-homogeneous soil, *Proc. 6th European Conf. SMFE*, Vienna, 1976, pp. 345–350.
3. BANERJEE, P. K. and BUTTERFIELD, R. (1977). Boundary element methods in geomechanics, In, *Finite Element Methods in Geomechanics*, Gudehus, G. (Ed.), John Wiley & Sons, New York, pp. 529–70.
4. BANERJEE, P. K. and DAVIES, T. G. (1977). Analysis of pile groups embedded in Gibson soil, *Proc. 9th Int. Conf. SMFE*, Tokyo, 1977.
5. BANERJEE, P. K. and DRISCOLL, R. M. C. (1976). Three-dimensional analysis of raked pile groups, *Proc. Inst. Civ. Engrs.*, **61**, 653–671.
6. BANERJEE, P. K. and DRISCOLL, R. M. C. (1975). A program for the analysis of pile groups of any geometry subjected to any loading conditions, *HECB/B/7*, Department of the Environment, London.
7. BUTTERFIELD, R. and BANERJEE, P. K. (1971). The elastic analysis of compressible piles and pile groups, *Géotechnique*, **21**, 43–60.
8. BUTTERFIELD, R. and BANERJEE, P. K. (1971). The problem of pile cap and pile group interaction, *Géotechnique*, **21**, 135–142.
9. CHAN, K. S., KARASUDHI, P. and LEE, S. L. (1974). Force at a point in the interior of a layered elastic half space, *Int. J. Solids. Structs.* **10**, 1179–1199.
10. COOKE, R. W. (1975). The settlement of friction pile foundations, *Building Res. Establishment Report* CP/12/75, Department of the Environment, London.
11. CRUSE, T. A. and RIZZO, F. J. (1975). (Eds.), *Proc. ASME Conf. Boundary Integral Method*, Am. Soc. Mech. Engrs., New York.
12. DAVIS, E. H. and POULOS, H. G. (1968). Analysis of incompressible piles and piers, *Géotechnique*, **18**, 51.

13. DAVIES, T. G. and BANERJEE, P. K. (1978). Displacements due to a point force at the interface of a two-layer elastic half space, *Géotechnique*, **28**(1).
14. DAVIES, T. G. and BANERJEE, P. K. (1976). Analysis of pile groups embedded in non-homogeneous soils, *Report No SM/4/1976*, Department of Civil Engineering, University College, Cardiff.
15. FEAGIN, L. B. (1953). Lateral load tests on groups of battered and vertical piles, *Symposium on Lateral Load Tests on Piles*, Am. Soc. CE, New York, 12–20.
16. GIBSON, R. E. (1974). The analytical method in soil mechanics, *Géotechnique*, **24**, 113–140.
17. MARSLAND, A. (1971). Laboratory and *in situ* measurements of the deformation moduli of London clay, *Proc. Symp. Int. Struct. and Fdn.* Birmingham, 7–17.
18. MATTES, N. S. and POULOS, H. G. (1969). Settlement of a single compressible pile, *J. Soil Mech. Fnd. Div., Proc. ASCE*, **95**(SM1), 189–207.
19. MINDLIN, R. D. (1936). Forces at a point in the interior of a semi-infinite solid, *Physics*, **7**, 195–202.
20. MUKI, R. (1960). Asymmetric problems of the theory of elasticity for a semi-infinite solid and a thick plate, In *Progress in Solid Mechanics*, Sneddon, I. N. and Hill, R. (Eds.), Vol. 1, North-Holland Publishing Co, pp. 399–439.
21. OTTAVIANI, M. (1975). Three-dimensional finite element analysis of vertically loaded pile groups, *Géotechnique*, **25**, 159–174.
22. POULOS, H. G. (1968). Analysis of the settlement of pile groups, *Géotechnique*, **18**, 449–471.
23. POULOS, H. G. (1971a). Laterally loaded piles—single piles, *J. Am. Soc. CE*, **97**(SM5), 711–731.
24. POULOS, H. G. (1971b). Laterally loaded piles—pile groups, *J. Am. Soc. CE*, **97**(SM5), 733–751.
25. POULOS, H. G. (1973). Load deflection prediction for laterally loaded piles, *Aust. Geomech. J.*, G3(1).
26. TERZAGHI, K. (1955). Evaluation of coefficients of subgrade reaction, *Géotechnique*, **5**, 297.

Chapter 10

THE LONG-TERM STABILITY OF CUTTINGS AND NATURAL CLAY SLOPES

N. E. SIMONS and B. K. MENZIES

Department of Civil Engineering,
University of Surrey, Guildford, UK

SUMMARY

In this chapter, a summary is presented of the 'state of the art'
concerning the prediction of the long-term stability of cuttings and
natural slopes in clays. It is stressed that design must be based on an
effective stress approach using the relevant drained shear strength. A
brief classification of types of landslides is given and the measurement
of peak and residual drained shear strength is described. The im-
portant question of progressive failure is discussed. Several illustrative
case records are given. It is concluded that first-time slides in non-
fissured clays correspond to strengths very slightly below peak, and in
fissured clays to strengths well below peak, but generally above the
residual. The residual strength is relevant on pre-existing slip surfaces.

10.1 INTRODUCTION

The scientific study of earth and rock slopes has applications ranging
from problems in pure geomorphology to the prediction of slope
stability for civil engineering purposes and the design of remedial
measures where a landslide has destroyed or is threatening property,
communications or the lives of people. The study of instability in
natural slopes yields information essential to the engineering design
for long-term stability of man-made slopes which are part of construction works such as canals, cuts, embankments and earth dams.

TABLE 1
LONG-TERM FAILURES IN CUTS AND NATURAL SLOPES ANALYSED BY
THE TOTAL STRESS ($\phi = 0$) ANALYSIS (AFTER BISHOP AND BJERRUM[1])

Locality	Type of slope	Data of clay					Safety factor $\phi = 0$ analysis
		w %	w_L %	w_P %	I_P %	$\dfrac{w - w_P}{w_L - w_P}$	
1. *Overconsolidated, fissured clays*							
Toddington	Cutting	14	65	27	38	−0·34	20
Hook Norton	Cutting	22	63	33	30	−0·36	8
Folkestone	Nat. slope	20	65	28	37	−0·22	14
Hullavington	Cutting	19	57	24	33	−0·18	21
Salem, Virginia	Cutting	24	57	27	30	−0·10	3·2
Walthamstow	Cutting	—	—	—	—	—	3·8
Sevenoaks	Cutting	—	—	—	—	—	5
Jackfield	Nat. slope	20	45	20	25	0·00	4
Park Village	Cutting	30	86	30	56	0·00	4
Kensal Green	Cutting	28	81	28	53	0·00	3·8
Mill Lane	Cutting	—	—	—	—	—	3·1
Bearpaw, Canada	Nat. slope	28	110	20	90	0·09	6·3
English, Indiana	Cutting	24	50	20	30	0·13	5·0
SH 62, Indiana	Cutting	37	91	25	66	0·19	1·9
2. *Overconsolidated, intact clays*							
Tynemouth	Nat. slope	—	—	—	—	—	1·6
Frankton, NZ	Cutting	43	62	35	27	0·20	1·0
Lodalen	Cutting	31	36	18	18	0·72	1·01
3. *Normally consolidated clays*							
Munkedal	Nat. slope	55	60	25	35	0·85	0·85
Save	Nat. slope	—	—	—	—	—	0·80
Eau Brink cut	Cutting	63	55	29	26	1·02	1·02
Drammen	Nat. slope	31	30	19	11	1·09	0·60

In this chapter, the long-term stability condition is considered to apply when the pore pressures in a slope have reached an equilibrium value, i.e. they are no longer affected by construction operations. Under this class, therefore, falls the stability of natural slopes, and of cuttings when sufficient time has elapsed for the excess pore-water pressures set up during excavation to have dissipated and the water pressures in the slope are then governed by the prevailing ground water conditions. This is clearly a drained situation and it is to be stressed that any attempt to predict the stability of such slopes using the undrained shear strength as a basis for calculation (the $\phi = 0$ analysis) is bound in general to result in a completely unreliable calculated factor of safety. To illustrate this point, Table 1 shows calculated factors of safety obtained by the $\phi = 0$ analysis for long-term slope failures in different clays.[1] It can be seen that the factors of safety vary from 21 for overconsolidated clays, down to 0·6 for the normally consolidated Drammen clay. In an undrained test, an excess pore-water pressure is set up during shear so that the effective stress at failure is quite different from the effective stress acting on the failure surface in the field, and therefore a completely different strength is obtained.

In addition to keeping in mind the two classes of long-term stability problem, i.e. natural slopes and man-made cuttings, it is vital to take into account the type of clay involved, and a suitable classification appears to be:

(i) soft intact (i.e. non-fissured) clays,
(ii) stiff intact clays,
(iii) stiff fissured clays.

The majority of clays falls into one or other of these groups, although there are bound to be borderline cases and examples can also be found in a fourth group, namely soft fissured clays.

Finally, a distinction must be drawn between cases in which pre-existing slip surfaces exist, for whatever reason, and first time slide situations.

It is not possible to cover all aspects of this vast subject in detail, and the authors have therefore chosen to highlight a number of topics which are believed to be of practical importance.

10.2   GENERAL APPROACH

The general approach to estimating the long-term stability of cuttings and natural slopes is summarised in this section.

As pointed out by Skempton and Hutchinson,[2] a proper understanding is required of four interrelated groups of topics:

(i) recognition and classification of the various types of mass movement that can occur on slopes; their characteristic morphological features; their geological setting; their rates of displacement and the causes of failure;

(ii)   classification and precise description of the materials involved, including a close examination for evidence of pre-existing slip surfaces, and the quantitative measurement of their relevant properties;

(iii) analytical methods of calculating the stability of a slope;

(iv)   correlation between field observations and the results of the stability calculations based on measured soil properties.

The importance of this fourth topic cannot be overemphasised. No matter how competently the field, laboratory and analytical work has been carried out, reliable prediction of the long-term stability of a slope will, in general, only be obtained if suitable relevant case records of field behaviour are available.

The techniques of site investigation available are well-known and need not be discussed here, but some points of particular importance should be noted.

Much valuable information can be obtained from a detailed study of geological and Ordnance Survey maps. Aerial photographs also provide an important source of information and perhaps the value of this approach of investigation is not always fully appreciated. A thorough knowledge of the distribution of pore-water pressure in a slope is, of course, essential for a proper analysis of a long-term stability problem and, in addition to seasonal changes, any variation of pore-water pressure across the site or with depth must be measured. The distribution of pore-water pressure with depth will have a controlling influence on the depth of a potential slip surface.

The presence or absence of pre-existing slip surfaces must be ascertained and here continuous sampling or closely spaced alternating samples in adjacent bore holes, or visual inspection in deep test pits may provide crucial information. Reference may be made to Chandler[3] and Weeks.[4]

10.3   TYPES OF LANDSLIDES AND OTHER MASS MOVEMENTS

Skempton and Hutchinson[2] define the generic term landslide as embracing those down-slope movements of soil or rock masses which occur primarily as a result of shear failure at the boundaries of the

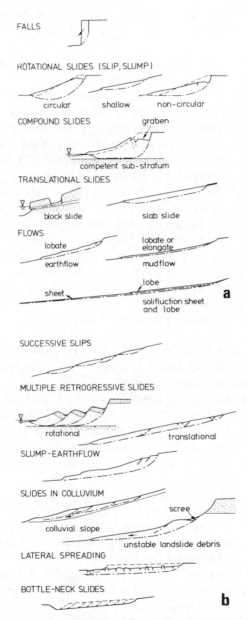

FIG. 1.   Basic types of landslide, after Skempton and Hutchinson.[2] (a) Movement on clay slopes. (b) Multiple and complex landslides.

moving mass. Some basic types of landslide on clay slopes are now summarised and are illustrated in Figs. 1(a) and (b).

Falls: Clay falls are typically short-term failures in the steep slopes of, for instance, artificial excavations or eroding river banks.

Rotational slides (slips, slumps): These occur characteristically in slopes of fairly uniform clay or shale. The curved surface of the failure, being concave upwards, imparts a back-tilt to the slipping mass which thus sinks at the rear and heaves at the toe.

Compound slides: These reflect the presence of a heterogeneity at moderate depth beneath a slope. In such cases the failure surface is formed of a combination of curved and planar elements and the slide movements have a part-rotational, part-translational character. Severe distortion and shearing accompany the sliding movements and the slide masses are correspondingly broken.

Translational slides: These generally result from the presence of a heterogeneity located at shallow depth beneath the slope. In such situations the failure surface tends to be relatively planar and to run roughly parallel to the slope of the ground. Slide movements are therefore predominantly translational and distortion of the sliding mass is small.

Earthflows: The term 'earthflow' is here confined to slow movements of softened, weathered debris which develop typically in material forming the toe of a slide.

Mudflows: These are glacier-like in form and have surface inclinations between 5° and 15°. As shown later in Table 3, rates of mudflow movement vary widely and can be highly seasonal.

Solifluction lobes and sheets: Shear surfaces are associated with certain fossil solifluction features in southern England and may be widespread on clay slopes.

Successive slips: Successive rotational slips consist of an assembly of individual shallow rotational slips. They are characteristic of the later stages of the free degradation process on slopes of overconsolidated, fissured clay which, for the London clay, occur on slopes of between about 13° and 8° inclination.

Multiple retrogressive slides: Multiple slides develop from single failures by the occurrence of further, retrogressive failures which interact to form a common basal surface.

Slump-earthflows: Slump-earthflows are a fairly common type of complex mass movement which occupies a position transitional between rotational slides, or slumps, and lobate mudflows. They develop typically in a rotational slide of considerable displacement, where the toe of the slipping mass is much broken by over-riding. In the presence of water, this debris then softens and develops into an earthflow and perhaps eventually into a mudflow.

Slides in colluvium: Although every slipped mass is, in a strict sense, colluvium, it is convenient in this context to use this word in the more restricted sense of material which is so shifted and weathered that individual slipped masses have become indistinguishable.

Spreading failures: Failures by sudden lateral spreading in clay slopes are a particular type of retrogressive translational slide. They are characterised by the gentle slopes involved, the broad front and the rapidity of the movements, which are usually completed in a few minutes, and the succession of graben and horst structures which are produced in the slide masses.

Quick clay slides: While in some circumstances quick clay slopes may fail in certain of the ways already mentioned, there is one type of landslide which is peculiar only to quick clays: the 'bottle-neck' type of retrogressive multiple rotational failure. Such slides generally begin with an initial rotational slip in the bank of a stream incised into quick clay deposits. The slipping mass is in part remoulded to the consistency of a liquid which runs out of the cavity, carrying flakes of the stiff weathered crust. The steep rear scarp is left unsupported and a further rotational slip takes place. This in turn becomes sufficiently remoulded to flow out into the stream bed, and retrogressive slips continue until a stable scarp is attained. The retrogression is extremely rapid and usually has a greater lateral extent in the deposits away from the stream than in the weathered, rather stronger material forming the bank, hence giving these slips their characteristic bottle-necked shape in plan. Such slides are common in the late- and postglacial marine clays of Norway and Eastern Canada.

Skempton and Hutchinson[2] further consider the rates of landslide movement, defining various categories as follows:

Creep: Here it is believed that there is a continuous gradation between the stationary and the moving material and hence no development of a shear surface. Creep may occur seasonally due to volume changes arising from swelling and drying and as a result of freeze–thaw action. Mass creep results only from gravity forces and is therefore of relatively constant rate. From drained laboratory tests on clays, continuing, long-term creep is known to take place at stresses that are only a fraction of their peak strength.

Pre-failure movements: From an engineering point of view great interest attaches to those movements which precede and lead up to the failure of a slope. These warn of the danger of sliding and may eventually form a basis for the prediction of failures. Some pre-failure movements are summarised in Table 2.

Movements during failure: The speed of landslides during failure is controlled chiefly by the nature of the clay in which shearing is taking place and by the shape and overall steepness of the failure surface. If the clay has a flat-topped or perfectly plastic stress–strain curve after failure, such as may be approximated in a till, the slide will experience no tendency to accelerate with increasing shear displacement and will move slowly down-slope until it reaches a stable position with a factor of safety close to 1·0. Conversely, if the shear resistance reduces appreciably once the peak strength is passed, the slide will accelerate and be carried past the stable position by its own momentum, coming to rest with a factor of safety higher than 1·0 on the residual strength.

Post-failure movements: After failure the slip surfaces in many clays will be at, or very close to, their residual strength, and further possible changes in shear parameters are likely to be small. A common feature of movements on such slip surfaces is their moderate or low speed. This applies whether the movements are brought about by seasonal pore-pressure changes or by some alteration in the loading of the slipped mass. The post-failure movements of several old slides are given in Table 3. Speeds of movement range from zero to 6 m per year. Movements of this type are particularly characteristic of slides in heavily overconsolidated clays, as the cases in Table 3 illustrate. Slides in normally consolidated or quick clays generally exhibit no post-failure movements. This is partly because their momentum during sliding has caused the slipped mass to over-ride strongly to a position with a fairly high factor of safety. Also in such clays the remoulded material in the slip surface has, once it has reconsolidated, a strength greater than that of the original undisturbed clay.

10.4 DRAINED STRENGTH OF CLAYS

It has been pointed out that the long-term stability of clay slopes must be analysed in terms of effective stress using the relevant drained

TABLE 2

EXAMPLES OF PRE-FAILURE MOVEMENTS (AFTER SKEMPTON AND HUTCHINSON[2])

Site	Description	Time before slip						Total movement before slip
		7 yrs	2 yrs	6 months	8 days	1 day	0	
		Average rate of movement over period of:						
		5 years	18 months	c. 6 months	7 days	1 day		
Kensal Green	Small retaining wall and slope	2 cm/year	9 cm/year	16 cm/year = 0.04 cm/day	—	—		35 cm
Ooigawa	Large retaining wall	—	—	5 cm/year* = 0.01 cm/day	1 cm/day	10 cm/day		> 20 cm
Dosan	Medium size landslide	—	—	—	3 cm/day	30 cm/day		> 40 cm
Gradot Ridge	Very large landslide	—	—	—	—	—		> 130 cm
Vajont	Extremely large landslide	—	70 cm/year	110 cm/year = 0.3 cm/day	6 cm/day	20 cm/day		250 cm

*Measured over 50 days only.

TABLE 3
EXAMPLES OF POST-FAILURE MOVEMENTS (AFTER SKEMPTON AND
HUTCHINSON[2])

Site	Slope angle	Period of observation	Average velocity	Max. daily velocity
California Loc 8*	9°	55 years	Almost imperceptible	—
Herne Bay, East Cliff	11°	4 years	2·5 cm/year	—
California Loc 10	10°	20 years	45 cm/year	—
Sarukuyoji	13°	8 months	300 cm/year	6 cm/day
Portuguese Bend	7·5°	3 years	600 cm/year	> 3 cm/day

*Re-activated by exceptional rainfall in 1958: moving at rate of 8 cm/year (observed for 8 months).

shear strength of the clay. The nature and determination of this strength are now considered.

10.4.1 Strain Softening in Soils

A number of investigations of slides in natural or man-made slopes have shown that the average shear stress along the failure surface on overconsolidated plastic clays and clay shales is considerably smaller than the peak shear strength measured in relevant shear tests in the laboratory. The failure of all such slopes, unless they have already failed in the past, is probably progressive.

As early as 1936, Terzaghi[5] postulated a mechanism by which the stiff fissured clays might grow progressively softer as a result of fissures and cracks associated with their structure. Although the concept of the residual strength has appeared in soil mechanics literature since 1937, the significance and relevance of it to the analysis of the stability of slopes in overconsolidated clays and clay shales was fully appreciated first by Skempton.[6] In his Rankine Lecture in 1964 Skempton presented strong evidence of correlations between residual shear strength obtained in the laboratory and average shear strength along slip surfaces in natural slopes in stiff fissured clays.

A number of papers have since been published which seem to

confirm the validity of the residual shear strength concept. Although the concept is relevant to all clays, it is of particular practical importance in overconsolidated clays and clay shales where the decrease in strength from peak to residual is large.

10.4.2 Residual Strength

If a specimen of clay is placed in a shearing apparatus and subjected to displacements at a very slow rate (drained conditions) it will initially show increasing resistance with increasing displacement. However, under a given effective pressure, there is a limit to the resistance the clay can offer, and this is termed the 'peak strength', s_f. With further displacement the resistance or strength of clay decreases. This process, which Skempton[6] refers to as 'strain softening', is not without limit because ultimately a constant resistance persists, regardless of the magnitude of displacement. This value of ultimate resistance is termed 'residual strength', s_r.

If several similar tests are conducted under different effective pressures, the peak and residual strengths when plotted against the effective normal pressure as shown in Fig. 2 will show a straight-line relationship, at least within a limited range of normal stress. Peak strength can therefore be expressed by

$$s_f = c' + \sigma' \tan \phi' \tag{1}$$

and the residual shear strength by

$$s_r = c'_r + \sigma' \tan \phi'_r \tag{2}$$

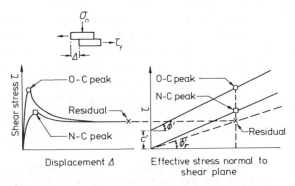

FIG. 2.   Peak and residual shear strength.

The value of c_r' is generally very small, but even so may exert a significant influence in the calculated factor of safety and depth of the corresponding slip surface.

Thus, in moving from peak to residual, the cohesion intercept approaches zero. During the same process the angle of shearing resistance can also decrease. During the shearing process, overconsolidated clays tend to expand, particularly after passing the peak. Thus, the loss of strength in passing from peak to residual is partly due to an increase in water content. A second factor that equally contributes in the post-peak reduction of the strength is the development of thin bands or domains in which the clay particles are orientated in the direction of shear, as noted by Skempton.[6]

In general, the difference between peak and residual strength depends on soil type and stress history and is most marked for heavily overconsolidated fissured clays. For normally consolidated clays this difference is generally small.* Thus, the concept of residual strength is of particular importance in the case of the long-term stability of slopes of overconsolidated fissured clay.

10.4.3 Factors Influencing Residual Strength

The difference in strength between peak and residual for overconsolidated clays increases with clay content and the degree of overconsolidation. Skempton[6] has shown that the residual strength decreases with clay fraction and that at a given effective stress the residual strength is practically independent of the past stress history. In fact, Kenney,[7] Bishop et al.[8] and Townsend and Gilbert[9] have shown that fully remoulded samples gave essentially the same residual strength as undisturbed samples of the same soil at the same normal stress. Furthermore, the residual shear strength has been found to be independent of the loading sequence (stress history) in multi-stage tests, because the same value ϕ_r' has been shown to exist, no matter if σ_n' is increased or decreased (see eqn. (2)).

Although it is generally believed that the amount of clay in the material controls the magnitude of the residual strength, Kenney[7] has investigated the influence of mineralogy on the residual strength and has shown that it is the type of clay mineral present that is the

---

*To the writers' knowledge, no long-term slip in a highly plastic normally consolidated intact clay has yet been analysed. For such a clay it is possible that a significant difference between peak and residual strength may exist.

governing factor. With the same ion concentration in the pore fluid, the residual friction angles reported by Kenney were about 4° for sodium montmorillonite, 10° for calcium montmorillonite, about 15° for kaolinite and from 16° to 24° for hydrous mica or illite.

The value of ϕ'_r has also been found by Kenney to depend on ion concentration in the pore fluid and to increase as salt concentration increases. In the case of sodium montmorillonite the residual friction angle increased from about 4° for negligible salt dissolved in the pore fluid to 10° with 30 g/litre sodium chlorite in the pore fluid. From his investigations on several natural soils, pure minerals and mineral mixtures, Kenney concludes that the residual shear strength is primarily dependent on mineral composition and, to a lesser degree, on the system chemistry and the effective normal stress, and that it is not directly related to plasticity or grain size of the soil. In general, according to Kenney, massive minerals such as quartz, feldspar and calcite exhibit high values of $\phi'_r > 30°$. For micaceous minerals (e.g. hydrous mica, illite) $15° < \phi'_r < 26°$. Soils containing montmorillonite exhibit low values of $\phi'_r < 10°$.

Although ϕ'_r and plasticity index (I_p) may not be related directly, subsequent investigations by Voight[10] have indicated that there is a definite statistical relationship between ϕ'_r and I_p, the general trend being that ϕ'_r decreases with increasing I_p. Correlations of residual shear friction angle with liquid limit have also been observed by Mitchell.[11]

The value of ϕ'_r may also exhibit a variation with the effective normal stress. For the brown London clay tested by Bishop et al.,[8] ϕ'_r varied from 14° at $\sigma'_n = 7 \, kN/m^2$ down to 8° at $\sigma'_n = 250 \, kN/m^2$, the increase in ϕ'_r below about 70 kN/m² being very marked. The general trend appears to be of ϕ'_r decreasing with increasing σ'_n. This may be attributed to the increased pressure at the interparticle contact points and the increased number of interparticle contacts per unit area on the slip surface as σ'_n increases. Mitchell[11] has also discussed the stress-dependency of ϕ'_r exhibited by some clays. Chandler[12] showed that the field values of ϕ'_r of Upper Lias clay are strongly stress-dependent, decreasing with increasing normal effective stress.

Non-linear Mohr–Coulomb failure envelopes have been observed for some clays, the curvature being more marked at low pressures. The cohesion intercept obtained by some investigators could be due to this curvature. Skempton and Petley[13] have shown that, in some cases, above a certain value of the normal stress, ϕ'_r can be considered independent of normal stress and a linear envelope can be fitted. Similar observations have been made by Townsend and Gilbert[9] for some clay shales.

The residual shear strength has been found to decrease very slightly with decreasing rates of shear. For most practical purposes it can be considered independent of the rate of shearing.[7,14-16]

10.4.4 Determination of Residual Strength

The residual shear strength is not only of practical importance in relation to the analysis of long-term stability of slopes, natural or man-made, but it may also be considered to be a fundamental property of the particular soil. Therefore, it is important in the laboratory to measure accurately residual strength. It is commonly measured by carrying out one or more of the following three types of tests:

   (i)   reversing shear box tests,
  (ii)   triaxial tests,
 (iii)   ring shear tests.

Reversing Shear Box Tests

In the case where a pre-existing surface is to be tested, large displacements have already reduced the strength to the residual value, and testing can be conveniently accomplished by employing either the shear box or the triaxial apparatus. During the past few years the direct shear box has been widely used and numerous data of values of residual strength have been reported.[6,7,17-19]

Skempton[6] determined the residual shear strength of soils by repeatedly shearing a specimen in a direct shear machine. After completing the first traverse, with a displacement of about 7·5 mm, the upper half of the shear box was pushed back to its original position and then pulled forward again, this process being repeated until the strength of the clay had dropped to a steady (residual) value.

Kenney,[7] who performed reversed direct shear tests on remoulded soil, followed a somewhat different procedure. The specimen with a moisture content exceeding the liquid limit was placed within a confining ring and between two circular carborundum plates to consolidate. The specimen had an initial thickness of about 2·5 mm and a diameter of 8 cm. When consolidation was complete, the confining ring was removed and the sample, with a thickness of about 1 mm, was sheared forwards and backwards with a travel of 2 to 2·5 mm each side of the centre.

Modified shear box devices have also subsequently been developed by many investigators.

The most serious drawback of the direct shear test in the measurement of the residual strength is that the laboratory conditions do not simulate the field conditions of a large relative displacement uninterrupted by changes in direction. Successive back and forth displacements may not be equivalent to a total displacement of the same amount in one direction. Although the effect of reversals is not exactly known in the direct shear test, it is believed to be accompanied by some degree of lack of perfect reorientation or disturbance of the previously orientated particles. Bishop et al.[8] noticed from direct shear tests on slip surfaces in blue London clay that ϕ'_r at the second forward travel is greater than that during and at the end of the first forward travel. Similarly, Cullen and Donald[19] have found that the residual strength for some of the soils tested is about 10% lower in the first forward travel compared with subsequent reversals. In such cases the lower value of residual strength at the end of the first travel has been accepted. Area correction problems may also arise, especially if the shear box travel is large.

The values of residual strength obtained in the direct shear box may be high compared with the corresponding values in the ring shear apparatus. It does not follow, however, that the estimation of the residual strength in the direct shear box always leads to significant errors.

Skempton[6] and Skempton and Petley[13] have shown that the measured residual strength in the reversing shear box correlates closely with the average mobilised strength calculated for a number of field failures in overconsolidated clays where movement has occurred along existing slip surfaces. Noble[20] has also measured residual strengths in the reversing shear box which were compatible with the observed behaviours of three landslides in the United States. In addition, Skempton and Petley[13] have demonstrated that in reversing shear tests performed on initially unsheared clays the residual strength was in good agreement with both shear box and triaxial tests on natural slip surfaces.

The residual strength obtained for Curaracha Shale by Bishop et al.[8] in the direct shear box and the ring shear apparatus was practically identical and they suggested that pre-cut samples in the direct shear box on hard materials, such as shales, give better estimates of ϕ'_r than for softer materials, such as clays, since the two halves of the box can be well-separated and 20 or 30 reversals can be imposed with less squeezing.

A series of tests on clay shales were conducted by Townsend and Gilbert[9] in the ring shear apparatus, the rotation shear apparatus and direct shear box (pre-cut samples). The results showed close agreement and Townsend and Gilbert concluded that the direct shear test can be conveniently used for hard overconsolidated shales.

The preparation of undisturbed samples for the direct shear test is easier than for any other type of test. Testing along discontinuities such as principal slip surfaces and joints may not lead to any errors due to reversal effects because the residual strength may be reached before the end of the first traverse. In such cases, due to the overall test simplicity, the direct shear test is preferable.

Triaxial Tests

The conventional triaxial cell has also been used for the measurement of the residual strength by a number of investigators. Chandler[12] measured the residual strength of Keuper Marl by cutting a shear plane in the sample at an angle of approximately $(45° + \phi'_r/2)$ to the horizontal and testing in the triaxial apparatus as suggested by Skempton.[6] Leussink and Muller-Kirchenbauer,[21] Skempton and Petley[13] and Webb,[22] among others, have published values of the residual strength obtained in the triaxial apparatus.

When employing the triaxial apparatus for the determination of the residual strength, special techniques and analyses must be used to take account of the following factors:

(i) horizontal thrust on the loading ram,
(ii) restraint of the rubber membrane,
(iii) change of the cross-sectional area.

After the peak strength has been reached in a triaxial test specimen the post-peak deformation is generally localised to a thin zone between the sliding blocks. The shear and normal stresses in this zone are functions of the vertical and horizontal loads on the end of the ram and the cell pressure. At the end faces of the test specimen, horizontal frictional forces can be mobilised due to end restraints. This could lead to erroneous results, especially for the deformation beyond the peak.

In order to maintain an even pressure along the failure plane, Chandler[12] employed a modified form of a triaxial cell with a loading cap freed to move laterally without tilting by the use of a number of ball bearings between the top cap and a special plate on the loading ram. Leussink and Muller-Kirchenbauer[21] minimised the horizontal load on the loading ram using a triaxial apparatus with a free moving pedestal. Bishop *et al.*[23] and Webb[22] considered the effects of the horizontal components of load on the measured test parameters.

The apparent increase in strength due to membrane restraint in triaxial tests where failure occurs on a single plane, has also been taken into account. The restraint provided by the membrane has been examined by using dummy specimens of plasticine (Chandler[12]), or perspex (Blight[24]).

The change in cross-sectional area resulting from movement along the shear plane has also been determined and incorporated in the analysis.

Polishing of the inclined cut-plane in triaxial tests by a flat spatula or glass plate produces quite a strong orientation of particles and the residual strength obtained by testing polished cut-plane triaxial test specimens has always been found to be lower than the values of residual strength measured in direct shear tests.[16]

Ring Shear Tests
Among the difficulties of obtaining the residual strength in the triaxial apparatus is that sufficient movement may not be obtained to achieve the residual stage on other than existing discontinuities or pre-cut planes. Herrmann and Wolfskill,[17] who also used triaxial tests which were continued to large strains and triaxial tests on specimens with a pre-cut inclined plane, concluded from their results that neither type of triaxial test was able even to begin to approach the residual state.

Skempton and Hutchinson[2] have noted that with some clays a true residual stage is reached only after large displacements (of the order of 1 m) and the residual strength obtained in reversal or cut-plane tests is considerably higher than this 'ultimate' residual strength obtained in the ring shear apparatus. La Gatta,[15] using data of previous investigators obtained from repeated reverse direct shear tests, has replotted the stress ratio (τ/σ'_n) versus the logarithm of the displacement and shown clearly that in many cases a constant residual strength was not reached and further displacement was necessary to establish the residual strength.

The ring and rotational shear tests (Fig. 3) are the only tests in which very large and uniform deformations can be obtained in the

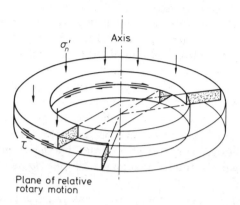

FIG. 3.   Ring shear test.

laboratory and have been used in soil mechanics for many years to investigate the shear strength of clays at large displacements.[25,26] Several designs of the apparatus and results have also been reported more recently by De Beer,[27] Sembelli and Ramirez,[28] La Gatta[15] and Bishop et al.[8]

The ring shear apparatus described by Bishop et al.[8] may be used to determine the full shear strength displacement of an annular soil specimen subjected to a constant normal stress, confined laterally and ultimately caused to rupture on a horizontal plane of relative motion. The apparatus may be considered as a conventional shear box extended round into a ring. Consequently large displacements (e.g. 1 m) may be obtained in one direction so that the residual strength may be accurately determined.

Tests in the ring shear apparatus give values of ϕ'_r generally lower than those obtained in the direct shear test or the triaxial test. The slip surface obtained after completing a ring shear test is generally more smooth and polished due to more complete orientation of particles.

Independent tests on blue London clay in the rotational Harvard apparatus by La Gatta[15] and in the ring shear apparatus described by Bishop et al.[8] gave essentially the same value of $\phi'_r = 9.3°$ $(c'_r = 0)$, whereas the average value of ϕ'_r obtained in the direct shear test was 3–4° higher.

The two advantages in determining the residual shear strength in a torsion or ring shear apparatus are that the cross-sectional area of the sample remains constant during testing and that the sample can be subjected to any uninterrupted displacement in one direction.

The reversal direct shear test may considerably overestimate the residual shear strength and generally give values of ϕ'_r higher than those determined in the ring shear tests. This is due to the disturbance of particle orientation and the change in direction of principal stresses in each reversal, so that attainment of the residual state is not achieved. Polished cut-plane triaxial tests give values of residual strength lower than the values measured in the direct shear box and may approach the values obtained in the ring shear tests.

It has been suggested that an accurate determination of residual shear strength can be made only by plotting τ/σ'_n against logarithm of displacement and taking as residual strength that value of τ/σ'_n which corresponds to zero slope of the curve.[15,16]

10.5   PROGRESSIVE FAILURE

In conventional analysis of slope stability problems, it is assumed that the peak shear strength of the soil is fully mobilised simultaneously along the whole length of the failure surface. Thus, the soil is treated as a rigid–plastic material and the actual soil stress–strain relationship does not enter into the method of analysis. The true stress–strain curve of a soil considerably deviates from that of a rigid plastic material and the ratio of strength to shear stress is not uniform along the entire length of a potential slip surface. In such cases the state of limiting equilibrium is associated with non-uniform mobilisation of shearing resistance and thus with progressive failure.

As progressive failure invalidates conventional stability limit analysis, it is important to specify the necessary conditions for a progressive failure to take place.

10.5.1 Necessary Conditions for Progressive Failure
Terzaghi and Peck[29] and Taylor[30] have associated progressive failure with non-uniform stress and strain conditions and redistribution of shear stress along a potential sliding surface.

If an element within a soil which possesses a strain-softening stress–strain curve is sheared beyond the peak failure strain it will lose part of its sustained stress. This part of stress must be shed to the neighbouring elements, which in turn may be brought past the peak by this additional stress and thus the process of progressive failure can initiate. Thus, local redistribution of stress can occur if the soil exhibits a brittle behaviour.

On the other hand, even if the soil stress–strain relationship exhibits a strong brittle behaviour, progressive failure cannot initiate if the stress and strain distribution within the soil mass is uniform.

Therefore, the development of a sliding surface by progressive failure is possible if the following three conditions are satisfied:[31]

(i) the soil exhibits a brittle behaviour with a marked decrease in strength after failure strain,

(ii) stress-concentrations take place,

(iii) the boundary conditions are such that differential strain may take place.

When all three conditions are satisfied, the likelihood of progressive failure is large and a potential slip surface may develop along which the average shear strength lies between peak and residual strength values.

10.5.2 The Influence of Normal Stress on the Post-peak Stress–Strain Behaviour

Drained triaxial test results on overconsolidated clays performed under different values of σ_3' have shown that the stress–strain curves can show plastic or brittle behaviour according to the magnitude of the confining pressure. Although variations due to clay type or the stress range used in the tests may exist, the general trends are, with increasing confining pressure:[23,32]

(i) the magnitude of the post-peak strength reduction decreases,

(ii) the rate of decrease in strength after passing the peak decreases,

(iii) the strain to reach the peak increases.

The reduction of strength in passing from peak to residual may be expressed by the 'Brittleness Index' I_B,[33] where:

$$I_B = \frac{s_f - s_r}{s_f} \tag{3}$$

This index depends on the normal pressure and generally decreases with increasing normal pressure.

Furthermore, the shearing displacement required to reach the residual strength strongly depends on the normal pressure, soil type and test conditions. Mitchell[11] has shown that shearing displacements of only 1 or 2 mm were necessary to reach the residual state for clay materials in contact with smooth steel or other polished hard surfaces. The required displacements for clay against clay were measured in several centimetres. Herrmann and Wolfskill[17] have found that the displacement required to reach the residual condition decreased considerably as the normal stress increased. Subsequent results in a rotary shear apparatus carried out by La Gatta[15] have not shown this trend. With blue London clay the displacement required to reach the residual state was found by La Gatta to be about 40 cm at a normal pressure of $\sigma'_n = 1$ and 2 kg/cm^2, about 3 cm at $\sigma'_n = 4 \text{ kg/cm}^2$ and 10 cm for $\sigma'_n = 8 \text{ kg/cm}^2$. Tests in the ring shear apparatus conducted by Garga[16] showed clearly that greater displacements are required to reach residual state when the sample is sheared under low effective stresses than those required under higher effective normal stresses. For blue London clay the displacements varied from about 50 cm at a normal stress of 42 kN/m^2 to approximately 12·5 cm at a normal stress of 280 kN/m^2.

Bishop et al.[8] have found that once the residual strength has been established under a given normal stress, subsequent rebound at lower normal stress requires further displacement to re-establish the residual strength. They point out that this effect should be allowed for when carrying out stability analyses of slides on pre-existing surfaces where re-initiation of slides takes place under a normal effective stress lower than that obtaining when the original surface was formed. The field value of ϕ'_r for clays such as brown London clay which show marked stress dependency in the re-initiated slide tends to be greater than that measured in the laboratory.

Apparently, the magnitude of the normal effective stress considerably influences the post-peak behaviour. This also emphasises the necessity for an accurate determination of the strength envelopes at the low stress level and the danger of extrapolating from test results at the high stress range. Determination of values of peak and residual strengths to be incorporated in the analysis of slope stability problems should be carried out in the laboratory under conditions simulating those in the field.

10.5.3 The Influence of Time
Results on the effect of rate of shearing on the peak drained strength are rather limited. Bishop and Henkel[34] reported drained triaxial tests

on remoulded Weald clay specimens in which the time to failure was varying between 1 day to 2 weeks. The tests showed a decrease in strength of about 5% per ten-fold increase in testing time. Tests on a normally consolidated, undisturbed marine clay carried out by Bjerrum *et al.*[35] showed that with times to failure of up to a month the drained shear strength was independent of the test duration when the latter was greater than 1 day. The authors suggested that the expected reduction of rheological component in this case was offset by an increase in true cohesion as a result of secondary consolidation as the time to failure increased. Constant stress level creep tests under drained conditions with duration up to $3\frac{1}{2}$ years were performed by Bishop and Lovenbury[36] on undisturbed brown London clay and a normally consolidated Pancone clay from Italy. Their results indicated that on an engineering time scale little decrease in strength from peak to residual can be accounted for by the time-dependent component of strength. Samples at stress levels below the residual strength were found to creep and there was no threshold value of stress below which time-dependent axial deformation did not take place.

There are some clays which may exhibit considerable drained strength reduction with time to failure. Bjerrum[37] suggested that the effect of time to failure on the peak drained strength must be considered in the case of plastic clays. Drained triaxial tests on St Vallier Clay, Canada, reported by Lo[32] showed that the drained shear strength decreased logarithmically with time to failure, amounting to a decrease in strength of about 12% per log cycle of time. The drained strength decrease with the logarithm of time was expressed by Lo using the equation suggested by Hvorslev:[38]

$$s_t = s_f - k \log_{10} \frac{t}{t_0} \tag{4}$$

where s_t = drained strength measured at time to failure t
  s_f = drained strength measured in time to failure t_0 in rapid conventional tests
  k = rate of decrease of strength per logarithmic cycle of time

In addition, Lo noted that the rate of post-peak reduction was changed with time to failure.

Consequently, for some clays, the time effect may constitute an important mechanism of progressive failure.

10.5.4 The Influence of Discontinuities

The majority of overconsolidated clays contain numerous discontinuities such as fissures, bedding planes, joints and faults. If, in addition, they have been sheared by landsliding or tectonic forces, shear zones will be formed containing minor shears and, usually, one or more principal slip surfaces.[39] Non-fissured, intact, overconsolidated clays such as boulder clays and clay tills are relatively rare. Since discontinuities represent local zones or surfaces of reduced shear strength which reduce the strength of the clay mass, it is expected that the stability of slopes in overconsolidated clays will be largely controlled by the strength along these discontinuities.

Terzaghi in 1936 gave the first explanation of the softening action of fissures in stiff clays with time. He pointed out the dangers of progressive failure if fissures and joints open out as a result of small movements consequent upon removal of lateral support when the excavation was made.

Skempton[6] has suggested that, in addition to allowing the clay to soften, the joints and fissures cause concentrations of shear stress which locally exceed the peak strength of the clay and lead to progressive failure. According to Skempton and La Rochelle[40] fissures can adversely influence the strength of overconsolidated clays as follows:

(i)   open fissures may form a portion of a failure surface across which no shear resistance can be mobilised,

(ii) closed fissures may form a portion of a failure surface on which only the residual strength can be mobilised,

(iii) fissures, whether open or closed, may adversely influence the stress within a slope, increasing the likelihood of progressive failure.

The shear strength along the different types of discontinuities generally depends on the amount of relative displacement which these planes of weakness have undergone. A tentative classification of discontinuities according to their occurrence and relative shear movement has been presented by Skempton and Petley.[13] According to this classification, principal displacement shears such as those found in landslides, faults and bedding-plane slips have undergone large displacements (more than 10 cm) and their surfaces appear polished. Minor shears such as Riedel, thrust and displacement shears of limited extent are described as non-planar and slickensided along which small displacements (less than 1 cm) have occurred. Joint surfaces, including systematic joints, displayed 'brittle-fracture' texture with little or no relative shear movement.

According to investigations by Skempton and Petley[13] and Skempton et al.[41] the strength along principal slip surfaces is at or near the residual. Along minor shears the strength may be appreciably higher than residual. On joint surfaces c' is small and ϕ' is approximately the same as at peak for intact clay indicating that the fracture which produced the joint virtually eliminated the cohesion but reduced the friction angle ϕ' by only a very slight amount. Movements of not more than 5 mm, however, are sufficient to bring the strength along the joint to the residual and to polish the joint.

10.5.5 The Influence of the Initial Stress State

It is well-established from studies in the field and the laboratory that in overconsolidated clays and clay shales the in situ horizontal stress may exceed the overburden pressure and in shallow depths the ratio of horizontal effective stress to vertical effective stress (K_0) may become large enough so that the soil approaches a state of passive failure. Skempton[42] used an indirect method to estimate the in situ stresses in the London clay at Bradwell and found that K_0 varied considerably with depth, increasing from a value of about 1·5 at a depth of 30 m to a value of 2·5 at 3 m.

Using finite element methods, Duncan and Dunlop[43] examined the effect of initial lateral stresses in excavated slopes. The soil was treated as a homogeneous linear elastic material. The process of excavation was simulated analytically in one step and the distribution of shear stresses was calculated. Two different soils were examined; a soil with $K_0 = 0·81$, representative of a normally consolidated clay, and a soil with $K_0 = 1·60$ (overconsolidated clay). The value of K_0 was found to greatly influence the magnitudes of the post-excavation shear stresses, which were much greater in the overconsolidated soil. The maximum shear stress in the region of the toe of the slope from which progressive failure was most likely to be initiated, was about ten times greater for the overconsolidated soil. The higher stresses were large enough so that failure could be expected and Duncan and Dunlop concluded that the high initial horizontal stresses in heavily overconsolidated clays and shales increase the likelihood of progressive failure in these materials. Lo and Lee[44] also showed the crucial dependence on the *in situ* effective stresses of analytical models incorporating strain-softening soil behaviour.

10.5.6 Thickness of Shear Zone

The displacement required to reach failure in slopes mainly depends on the strain which corresponds to the peak strength and on the thickness of failure zone. For a relatively thin failure zone the total movement before failure will be small whereas the required movement before failure occurs will be appreciably larger if the failure zone is relatively thick.

By analysing the measurements of horizontal movements obtained by a slope indicator, Gould[45] found that failures in landslides in overconsolidated clays in the California coast region occurred within a narrow zone of 0·6 m to 2 m in thickness. In the landslide of Jackfield, England, described by Henkel and Skempton[46] the failure zone was approximately 5 cm thick. The water content of the clay in this zone was 10% greater than in the adjacent material outside the failure zone. Skempton and Petley[13] have observed, in a large landslide in stiff fissured clay at Guildford, England, that the shear zone had a width of about 0·6 cm which contained numerous minor shears. The actual slip surface consisted of a band about $50 \, \mu$m wide in which the particles were strongly orientated. At Walton's Wood, England, they observed that the shear zone had a width of about 2 cm and the particles were strongly orientated within a band about 20–$30 \, \mu$m wide. The increase in water content in the shear zone was about 3%.

10.5.7 Mechanism of Progressive Failure

In a field failure, the average shear strength is the average value of the strengths of all the elements around the slip surface. This strength will lie between the peak and residual strengths. Skempton[6] has compared the average shear strength ($\bar{s}$) occurring at failure of several natural slopes and cuttings to the peak (s_f) and residual (s_r) shear strengths of specimens from the failure zone (surface) corresponding to the average effective normal stress, and defined the residual factor R by the equation

$$R = \frac{s_f - \bar{s}}{s_f - s_r} \tag{5}$$

If the average field shear strength equals the residual $R = 1·0$, and R is zero if the average field shear strength equals the peak shear strength. Skempton has found a residual factor $R = 0\cdot08$ for a landslide in a natural slope at Selset in a uniform non-fissured and unweathered clay, indicating that the average strength mobilised along the total failure surface was very close to the peak strength; and $R = 1$ for two landslides in natural slopes consisting of fissured, jointed and weathered clay (London clay and Coalport Beds respectively) indicating that the average strength mobilised at failure was very close to the residual value.

It is suggested that fissures and joints, apart from their weakening effect in the soil mass, act as stress concentrators in their edges which can overstress locally the soil beyond the peak strength and hence a progressive failure may be initiated. Although a slip may occur before the residual value is reached everywhere within the mass, continued sliding will cause the average strength to decrease toward that limiting value.

In clays without fissures or joints, however, the post-peak reduction is very small, or even negligible. Compacted clay fills as used in embankments and earth dams may belong in this category. If a failure has already taken place any subsequent movement on the existing slip surface will be controlled by the residual strength, no matter what type of clay is involved.

Bjerrum[31] has suggested a mechanism of progressive failure which is not associated with the presence of fissures in the clay. During the process of drained loading and after long periods of time, strain energy is stored in the soil mass. Depending upon the nature of the soil the strain energy may be stored or released upon unloading. If weak diagenetic bonds have been developed, the stored energy is soon released after unloading. If the bonds are strong the strain energy can only be released if the bonds are destroyed as a result of weathering during long periods of time. Bjerrum classified diagenetic bonds as weak, strong or permanent.

The rate of release of strain energy upon unloading is slower in soils with strong bonds, and if the bonds are permanent the energy may never be released. As the strain energy is released by the disintegration of diagenetic bonds, due to the process of weathering, progressive failure of a soil mass is initiated and continues retrogressively from the face of the slope.

Bjerrum[31] proposed a classification, shown in Table 4, of overconsolidated clays and shales on the basis of the likelihood of progressive failure.

TABLE 4
RELATIVE DANGER OF PROGRESSIVE FAILURE (AFTER BJERRUM[31])

Soil	Relative danger of progressive failure
Overconsolidated plastic clay with weak bonds	
Unweathered	High
Weathered	High
Overconsolidated plastic clay with strong bonds	
Unweathered	Low
Weathered	Very high
Overconsolidated clay with low plasticity	Very low

It can be seen from this classification that the danger of progressive failure is greater in the case of overconsolidated plastic clays possessing strong diagenetic bonds (e.g. shales) which have been subjected to weathering and the gradual release of the strain energy in their bonds.

Consequently, heavily consolidated plastic clays and shales have initially large lateral stresses and show a high tendency for lateral expansion. This could result in stress concentrations at the toe of an excavation or cut, local shear failure and gradual development of a continuous sliding surface.

Based on studies of a great number of case histories on overconsolidated clays, James[47] emphasised the necessity of relatively large deformations in the field to produce progressive failure and the reduction of strength near residual conditions. As a result, James suggested that a cutting in overconsolidated clay or clay shale designed with strength parameters $\phi' = \phi'_{peak}$ and $c' = 0$ can ensure long-term stability against first-time failures and will not be subject to progressive failure in the majority of cases encountered in practice. This suggestion is in accordance with the observations made by Henkel and Skempton,[46] Skempton and De Lory[48] and Skempton.[49]

As emphasised by Peck[50] and Bishop[51] a complete understanding of the problem of progressive failure would require a finite element solution for a strain-softening material. Lo and Lee[52] have presented a finite element solution for the determination of stresses and displacements in slopes of strain-softening soils. The various factors that influence the extent and propagation of the overstressed zone were investigated and typical results obtained with slope geometry and soil properties commonly encountered in practice indicated that the extent of the overstressed zone defined by the residual factor increases with the inclination and height of the slope and with the magnitude of the *in situ* stress as defined by the coefficient of earth pressure at rest.

10.6 METHODS OF SLOPE ANALYSIS

Currently, limit analysis is commonly used, assuming circular or non-circular trial slip surfaces. Such methods require the assumption of a shear strength. The results obtained are subject, to a varying extent, to the internal stress distribution which must be assumed within the sliding mass. With non-circular surfaces, the methods often do not take specific account of the deformations within the sliding mass which are necessary to allow movement. There is uncertainty concerning the depth of the tension crack which can form and the general admissibility of implied tensile stresses. Lastly, the strength to be assumed is ambiguous when the clay is brittle and progressive failure may occur, as discussed subsequently.

More recently, non-linear finite element techniques have been used to examine slope stability problems by Duncan and Dunlop,[43,53] Chang and Duncan,[54] Duncan,[55] Lu and Scott[56] and Lo and Lee.[52] As the rupture mechanism is developed from the soil properties and initial stresses assumed, these methods have great potential. Considerable difficulties exist, however, in formulating the soil stress–strain behaviour, in incorporating pore pressure changes and in handling progressive failure when strength drops with displacement and strain discontinuities occur. The application of finite element analyses to stability problems is consequently at an early stage.

Centrifuge model tests as an alternative to analytical methods are another recent development.[57] The formulation of soil properties for analysis is avoided and pore pressure changes can be examined. However, brittle failure cannot be reproduced,[58] and there are problems in simulating excavation, the correct stress history and the effects of discontinuities. Such tests may have their greatest value in providing checks for improved analytical methods.[59]

The back analysis of actual slope failures using limit analyses shows that the strength required for design cannot generally be taken directly from laboratory or *in situ* tests. Back analysis determines an average apparent strength, the value of which depends on the method of analysis adopted, on the accuracy with which the failure surface and the depth of tension crack are known, and on the pore pressures assumed. The back calculated strength includes the effects of progressive failure, anisotropy and time dependence. It is not a true material property and will seldom coincide with the strength measured in any particular test.

Ideally, design involves the determination of the true bulk clay properties, including the local effects of discontinuities, the determination of initial stresses in the ground and the use of a continuum boundary analysis which reproduces the correct slope behaviour during and after excavation. Recent developments in these areas are encouraging, but currently heavy reliance is placed on limit analysis and the results of back analyses of actual failures when these are available. New developments generally involve more sophisticated and expensive techniques of site investigation and analysis, and there is a parallel need to improve the simple methods suitable for routine design. The upgrading of these simple methods in the light of recent developments seems desirable. Perhaps correction factors of adequate reliability can be established, whereby the effects of progressive failure, anisotropy and loading rate can be allowed for in limit analyses based on simple site investigation tests.

As discussed previously, the long-term stability of clay slopes can only be satisfactorily investigated using an effective stress analysis and in many cases Bishop's slip circle method[60] is suitable. For the case of non-circular failure surfaces of any form, Janbu[61] has published a solution which is convenient to use.

Bishop and Morgenstern[62] have produced charts of stability coefficients from which a factor of safety for an effective stress analysis can be rapidly obtained. The general solutions are based on the assumption that the pore pressure u at any point is a simple proportion r_u of the overburden pressure γh. This proportion is the pore pressure ratio $r_u = u/\gamma h$ and is regarded as being constant throughout the cross-section. This is called a homogeneous pore pressure distribution.

For a simple soil profile and specified shear strength parameters, the factor of safety, F, varies linearly with the magnitude of the pore pressure expressed by the ratio r_u such that:

$$F = m - nr_u \qquad (6)$$

where m and n are termed 'stability coefficients' for the particular slope and soil properties. The use of pore pressure ratio, r_u, permits the results of stability analysis to be presented in dimensionless form.

While the use of stability coefficients provides a rapid method of calculating the factor of safety of a clay slope, it should be stressed that in many clay slopes, the value of r_u varies with the depth of the slip circle under consideration. Since the published charts do not indicate the position of a slip circle corresponding to a critical factor of safety, care must be taken to check that the pore pressure ratio, r_u, inserted into the design charts, is in fact correct for the critical slip circle implied.

10.7   CASE RECORDS

From the scientific and practical points of view the study of analytical case records is supremely important. Our knowledge of slope problems in clays can be assessed only by measuring the properties of clays under carefully controlled conditions of test, using these properties in a stability analysis, and seeing to what extent the results agree with field observations.

In the following, a number of selected case records have been summarised in order to give an appreciation of the present state of knowledge.

10.7.1   First-time Slides—Intact Clays

Drammen[63]

On the 6th January 1955 a rotational slide occurred in the north bank of the Drammen River, at the town of Drammen in Norway (Fig. 4). The slide was located in a soft intact marine clay of post-glacial age, covered by about 3 m of sand and granular fill. The clay has occasional extremely thin seams of silt and fine sand. Its index properties are typically $w = 35\%$, $w_L = 35\%$, $w_P = 18\%$, clay fraction = 38%, and sensitivity = 8. Piezometer readings showed that the pore pressures were hydrostatic. Beneath the ground surface the clay is normally consolidated. Below the slope the clay is very lightly overconsolidated as a result of removal of load by river erosion.

Average values of the peak strength parameters are: $c' = 2 \text{ kN/m}^2$, $\phi' = 32 \cdot 5°$, as measured in drained triaxial tests on vertical axis specimens taken with a piston sampler in boreholes.

The main cause of the slide was a gradual steepening of the slope by erosion at the toe, but the failure was triggered by placing a small amount of fill at the edge of the bank. At the time of the slide the

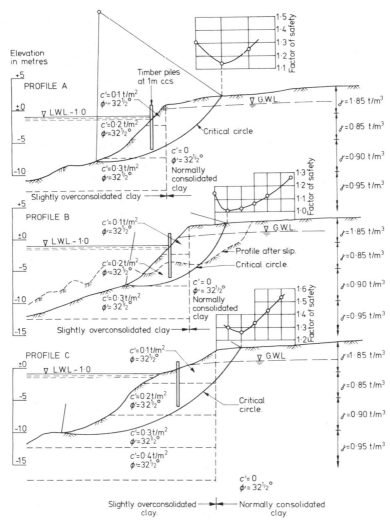

FIG. 4.   Drammen slide.[63]

water level in the river was about 1 m below normal. There were no obvious signs of impending movement beforehand.

Stability analyses using Bishop's effective stress method gave a minimum calculated factor of safety of 1·01 on a critical circle corresponding closely to the actual slip surface, so far as the latter could be determined from field observations.

This result suggests that the effects of rate of testing, progressive failure and anisotropy are more or less self-balancing in the Drammen slide; and in any case they would not be expected to exert a major influence on a long-term slide in soft intact clay, tested and analysed in terms of effective stress. No residual shear strength determinations have as yet been carried out on the Drammen clay.

Lodalen, Oslo[64]

A railway cutting, originally made in 1925, was widened in 1949. Five years later a slide occurred in the early morning of 6th October 1954. The sliding mass moved as an almost monolithic body, sinking about 5 m in the upper part and pushing forward about 10 m at the toe. Subsequent borings established the position of the slip surface at three points (Fig. 5) and, together with the back scarp, showed that the surface closely approximated to a circular arc. Pore pressure measurements by piezometers revealed a small upward component of ground water flow, presumably influenced by artesian pressures in the underlying rock.

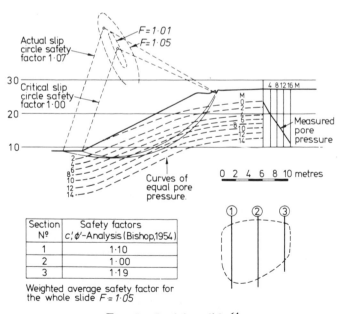

Section Nº	Safety factors c', ϕ'-Analysis (Bishop,1954)
1	1·10
2	1·00
3	1·19

Weighted average safety factor for the whole slide $F = 1·05$

FIG. 5.   Lodalen slide.[64]

The clay, of post-glacial age, is lightly overconsolidated with an intact structure, and has an undrained shear strength ranging from 40 to 60 kN/m^2. Average values of the index properties are: $w = 31\%$, $w_L = 36\%$, $w_P = 18\%$, clay fraction $= 40\%$, and sensitivity $= 3$. The peak strength parameters were determined in consolidated undrained triaxial tests, with pore pressure measurements, made on vertical axis specimens taken by a thin-wall piston sampler from boreholes. The scatter of results was extraordinarily small, with average values: $c' = 10$ kN/m^2, $\phi' = 27°$. The uppermost zone a few metres thick below original ground level constituted a 'drying crust' typical of Scandinavian clays, but the tension crack extended through this. Stability analyses using Bishop's method gave a minimum calculated factor of safety of 1·00. The corresponding critical slip surface differed slightly in position from the actual surface, and on the latter the calculated factor of safety was 1·07.

These figures are so close to unity that, as in the previous record from Drammen, the conclusion must be that the combined effect of various factors such as anisotropy and rate of testing is negligible.

Residual strength tests have not yet been made on the Lodalen clay, but as an upper limit we could safely assume: $c'_r = 0$, $\phi'_r = 27°$. With these parameters the factor of safety falls to 0·73. Thus it seems clear that progressive failure must have played a very small part as a cause of the slide. The delay of 5 years from excavation to failure is therefore probably associated with a slow decrease in effective stress following the removal of load from the slope in 1949.

Selset[65]
In the north Yorkshire Pennines, the River Lune, an upland tributary of the Tees, is eroding its valley through a thick deposit of clay till, probably of Weichselian age. At the section shown in Fig. 6 the river, when in flood, is cutting into the toe of a slope about 12·8 m high. When the site was first visited in 1955 clear evidence could be seen of a rotational landslide. Comparison of present topography with a map of 1856 showed that the rate of lateral movement of the river into the valley side was very slow.

Piezometers indicated a flow pattern rather similar to that at Lodalen, with a component of upward flow from the underlying bedrock.

The till consisted of stones and boulders set in a sandy clay matrix ($w = 12\%$, $w_L = 26\%$, $w_P = 13\%$, clay fraction $= 25\%$), forming a

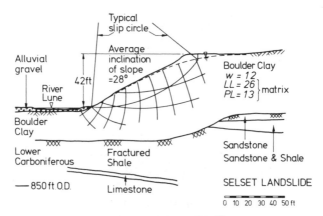

FIG. 6.   Selset slide.[65]

massive, stiff intact clay. Shear strength parameters of the matrix, measured in slow drained triaxial tests with a time to failure of up to 2 days, are: $c' = 8 \cdot 6 \, \text{kN/m}^2$, $\phi' = 30°$ (peak); $c'_r = 0$, $\phi'_r = 28°$ (residual). The stress–strain curves were gently rounded at the peak.

Stability analyses by Bishop's method with peak strengths gave a minimum calculated factor of safety = $1 \cdot 05$ (typical result within a range from $0 \cdot 99$ to $1 \cdot 14$ controlled by two limiting assumptions concerning flow net patterns). In contrast, the factor of safety using residual parameters was $0 \cdot 69$.

Thus, the actual strength around the slip surface at the time of failure must have been close to the peak strength as measured in the laboratory, and very much greater than the residual strength.

This conclusion is not unexpected. The strength of this type of clay would presumably be little influenced by rate of shearing and aniso-tropy may well be insignificant, while any substantial reduction from peak strength by progressive failure is most unlikely with the non-brittle, almost flat-topped stress–strain curves.

Vaughan and Walbancke[66] cite slope angles and heights of a number of other slopes in till under similar conditions, which give support to the conclusions derived from the detailed study of the Selset case.

10.7.2 First-time Slides—Stiff Fissured Clays

An outstanding analysis of the long-term stability of cuttings in the brown London clay has been given by Skempton.[67] Several cuttings constructed more than 40 years ago have been considered, as well as four, more recent, examples. The main conclusions resulting from this study are as follows:

(i) The shear strength parameters of the brown London clay relevant to the first-time long-term slides are $c' = 1 \text{ kN/m}^2$ and $\phi' = 20°$. These values approximate to the 'fully softened' values and are considerably lower than the peak strength, even as measured on large samples, so some progressive failure mechanism appears to be involved.

(ii) The residual strength is much smaller than this and corresponds to the strength mobilised *after* a slip has occurred, with displacements of the order of 1 to 2 m.

(iii) A characteristic feature of first-time slides in cuttings in London clay is that they generally occur many years after the cutting has been excavated. The principal reason for this delay is the very slow rate of pore pressure equilibration; a process which, in typical cuttings, is not completed until 40 or 50 years after excavation.

These conclusions drawn from the London clay have been broadly confirmed by Chandler[68] working with the Upper Lias clay, a very heavily overconsolidated clay of Jurassic age. It is generally less plastic than the London clay of the London area, and is brecciated to depths of about 10 m. Chandler[68] has analysed 12 first-time failures in this material and has shown that the strength at failure is well below the peak strength of conventional laboratory specimens, is close to the fully softened strength, and is significantly above the residual strength.

10.7.3 Slides on Pre-existing Slip Surfaces

Slip surfaces can be caused by landsliding, by solifluction, and by tectonic shearing. The movements may be continuing post-failure displacements or they may result from reactivation (for example, caused by excavation at the toe of a slope) but in both cases we are dealing with a condition of limiting equilibrium controlled by the residual strength along the slip surface. In this respect this set of records differs sharply from the first-time slides.

Sudbury Hill[6]

After the slide in 1949 in the London clay cutting at Sudbury Hill, no remedial measures were carried out and the slumped mass continued to move intermittently for several years. Small amounts of clay were removed from time to time at the toe (probably in the winter months) to prevent it encroaching on the railway track.

When the profile was surveyed in 1956 the overall displacement amounted to at least 2 m. The post-failure movements must therefore correspond to a factor of safety of unity on residual strength.

Using the Morgenstern–Price analysis the average effective normal stress and shear stress along the slip surface are $\sigma'_n = 28 \cdot 7 \, \text{kN/m}^2$, $\tau = 7 \cdot 7 \, \text{kN/m}^2$, respectively. If c'_r is made equal to zero the value of ϕ'_r corresponding to these stresses is $14 \cdot 4°$ (calculations on a circular slip surface by Skempton[6] give $\phi'_r = 14 \cdot 7°$ if $c'_r = 0$).

Tests on natural slip surfaces in brown London clay can be represented by $\phi'_r = 13°$ and a small cohesion intercept. For the London clay at Sudbury Hill, the tests gave $c'_r = 1 \cdot 44 \, \text{kN/m}^2$, but from the post-failure stability analysis a better value for c'_r is $0 \cdot 96 \, \text{kN/m}^2$, taking $\phi'_r = 13°$. These parameters give a calculated factor of safety on the non-circular slip surface equal to $1 \cdot 03$. A difference in c'_r of $0 \cdot 48 \, \text{kN/m}^2$ is inside the limits of experimental error.

Other classic slides of this type, which have been summarised by Skempton and Hutchinson[2] are:

(i)   Folkestone Warren
(ii) Walton's Wood
(iii) Sevenoaks Weald
(iv) River Beas Valley

All these cases confirm that, at failure, the mobilised strength was very close to the residual strength.

10.7.4 Natural Slopes in Stiff Fissured Clays

Jackfield[46]

In 1952 a landslide occurred at the village of Jackfield, Shropshire, on the River Severn 2 km downstream of Iron Bridge, destroying several houses and causing major dislocations in a railway and road. In this locality the Severn flows through a V-shaped valley (the so-called Iron Bridge gorge) which has been eroded largely, if not entirely, since the retreat of the main ice sheet of the Last Glaciation. Erosion is indeed still continuing and the sides of the valley are covered by a mosaic of landslides of varying ages.

It is possible that previous landslides may have taken place along at least a part of the present slip surface, but the slope must have been more or less stable for a long time before 1950, when warnings of instability were observed in the form of a broken water main serving cottages near the river bank. Towards the end of 1951 further movement was noted, and by February 1952 the road was becoming dangerous. During the next month or two the landslide developed alarmingly. Six houses were completely broken up, gas mains had to be relaid above ground, the railway could be maintained only by daily adjustments to the track and a minor road along the river had to be closed to traffic. By this time the maximum downhill displacement totalled 19 m.

The strata, consisting of very stiff clays and mudstones, alternating with marl-breccia and occasional coal seams, dip gently in a south-easterly direction with the strike running roughly parallel to the section of the landslide. The slide, however, was confined wholly within the zone of weathered, fissured clay extending to a depth of 6–8 m below the surface (Fig. 7). The slip surface ran parallel to the slope (which is inclined at 10°), at an average depth of 5·5 m. The length of the sliding mass, measured up the slope, amounted to about 170 m and in the winter of 1952–53 the ground water level reached the surface at a number of points, although on average it was located at a depth of 0·6 m.

An analysis of the forces acting on the slip surface showed that the normal effective stress and the shear stress were $\sigma'_n = 62 \text{ kN/m}^2$ and $\tau = 19 \text{ kN/m}^2$, respectively.

Drained shear tests on samples taken from depths between 4·6 m and 5·8 m, but not in the immediate vicinity of the slip plane, showed peak strength parameters of $c' = 10·5 \text{ kN/m}^2$ and $\phi' = 25°$. When these

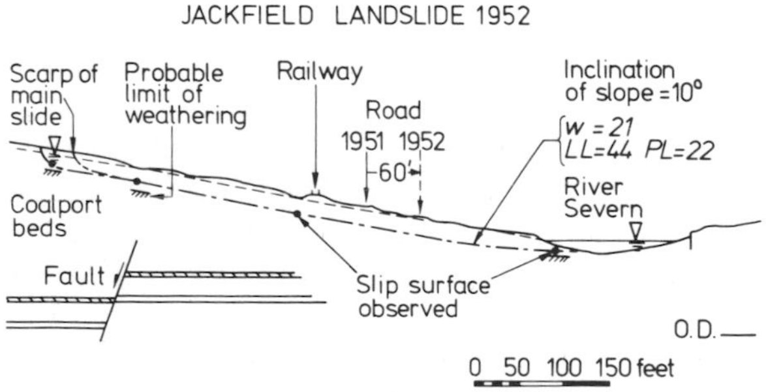

JACKFIELD LANDSLIDE 1952

FIG. 7.   Jackfield slide.[46]

tests were made, the significance of residual strengths was not clear. Fortunately, however, in most cases the observations were continued throughout the full travel of the shear box, and it is possible from the results to make an approximate estimate of the residual angle of shearing resistance, giving $\phi'_r = 19°$.

The peak and residual strengths corresponding to the average effective pressure of 62 kN/m^2 acting on the slip surface, are 39·5 and 20·4 kN/m^2 respectively. But, as previously mentioned, the average shear stress (and hence the average shear strength) along the slip surface at the time of failure was 19 kN/m^2.

It is therefore clear that when the landslide took place the strength of the clay was closely equal to its residual value. In fact, taking $\phi'_r = 19°$, it is found that the residual factor $R = 1·12$ but, when the approximate nature of ϕ'_r is taken into account, it is doubtful if the value of R is significantly different from 1·0.

Expressing the results in another way, had the peak strength been used in a stability analysis of the Jackfield slope, the calculated factor of safety would have been 2·06 (an error of more than 100%, since the true factor of safety was 1·0). On the other hand, using even the rather crude value of $\phi'_r = 19°$, the calculated factor of safety based on residual strength would differ by only 11% from the correct result.

Slopes in London Clay
Studies of natural slopes in London clay may conveniently be considered under two headings; coastal cliffs and inland slopes.

Coastal cliffs:[69] Under conditions of fairly strong marine erosion the cliffs are subject to rotational or compound sliding on deep or moderately deep slip surfaces. Where the rate of erosion is less severe the typical pattern is dominated by shallow slides and mudflows. The slopes of all these eroding cliffs are characteristically irregular, with average inclinations between about 15 and 30°.

At various places along the coast, sea defences have been constructed which prevent further erosion at the foot of the cliffs. If no further stabilisation works are carried out, such as drainage or regrading, the slopes then enter the phase of free degradation. Eight slopes in this category have been surveyed. Their inclinations range from 13 to 20° and they show clear evidence of instability, in the form of shallow rotational slides involving either the whole or part of the slope. These defended cliffs have been free from marine erosion for periods of about 30 to 150 years.

Where marshes have formed, particularly in estuaries, the sea has retreated from the old cliffs which, generally, have been left to flatten their slopes undisturbed by stabilisation measures. Surveys of ten of these freely degrading abandoned cliffs show inclinations of 8·5 to 13°. Slopes steeper than 9·5° are still unstable, and are characterised by successive shallow rotational slips; while the flatter slopes exhibit well-marked undulations which almost certainly represent the sub-dued remains of quiescent successive slips. These slopes, at 8·5 to 9·5°, may be regarded as being in a transitional state, approximating the condition of final equilibrium.

Inland slopes: Observations by Skempton and De Lory,[48] greatly extended by Hutchinson,[69] have shown that many inland slopes in London clay or soliflucted London clay are unstable even though they are not currently subject to stream erosion. Two clearly differentiated types of instability can be noted: successive slips and transitional slab slides. Shallow, markedly non-circular slides also occur which may be a variant form of the slab-like movements. In addition the undulations, previously mentioned, are common.

Transitional slab slides are found at inclinations ranging from 8 to 10°. Their shape suggests that failure is taking place on pre-existing solifluction shears running parallel to the surface.

Successive slips have been observed on slopes inclined at angles between 9·5 and 12°, with one exception at 8·5°. These slopes are so similar to the abandoned cliffs in their form of instability and range of inclination that we consider them to be closely equivalent, and infer that post-glacial erosion has removed the solifluction mantle leaving a slope essentially in the London clay.

Undulations occur at inclinations from 8·5 to 10·5°, the lower limit on these inland slopes being identical with that on the abandoned cliffs.

It seems, then, that while the minimum unstable angle is 8°, this is almost certainly associated with renewed movements on solifluction slip surfaces; and the angle of ultimate stability of London clay itself is around 9°.

It has also been observed that the maximum stable slope in London clay is about 10°. An overlap of 1° or so can easily be accounted for by modest differences in the position of ground water level as between one site and another.

There is no great difficulty in deriving a quantitative explanation of these field observations provided it is assumed that the strength of the clay has fallen to its residual value. This indeed will be the case on a solifluction slip surface, and with $c'_r = 1 \text{ kN/m}^2$ $\phi'_r = 13°$, movement can take place at an inclination of 8° if ground water is near the surface of the slope.

The slightly steeper inclination of 9° for un-soliflucted London clay presumably reflects the presence of a series of curved but inter-linking slip surfaces rather than a continuous planar shear.

10.8 SOME CONCLUSIONS DEDUCED FROM CASE RECORDS OF SLOPE FAILURES

The following conclusions appear to be valid at the present state of knowledge:

(i) After a slide has taken place, the strength on the slip surface is then equal to the residual value. The residual strength is associated with strong orientation of the clay particles and is represented by an angle of shearing resistance ϕ'_r, which in most clays is considerably smaller than the value of ϕ' at peak strength.

(ii) First-time slides in slopes in non-fissured clays correspond to strengths only very slightly less than peak.

(iii) First-time slides in fissured clays correspond to strengths well below peak, but generally above the residual.

(iv) Some form of progressive failure must be operative to take the clay past the peak. This is probably the result of a non-uniform ratio of stress to strength along the potential slip surface but also the fissures probably play an important role as stress concentrators and in leading to softening of the clay mass.

(v) The London clay, and probably many other stiff fissured clays, undergoes a loss in strength in cuttings tending towards the fully softened value. Just before a first-time slide occurs, there is a softened shear zone with many minor shears. It is possible that some overconsolidated clays may exhibit a more marked reduction in strength before a first-time slide takes place.

(vi)   In all clays, the residual strength will be reached after a continuous principal slip surface has developed and, in the field, this state appears to be attained typically after mass movements of the order of a metre or so.

(vii) The angle of ultimate stability of clay slopes is probably controlled by the residual strength and this may result from successive slipping or solifluction movements.

(viii) The residual strength obtains on pre-existing shear surfaces, whether these are the result of tectonic shearing or old landslides.

(ix)   First-time slides in cuttings in London clay generally occur many years after excavation; the main reason for this delay is the very slow rate of pore pressure equilibration which typically is not complete until 40 or 50 years after the cutting has been excavated.

ACKNOWLEDGEMENTS

In preparing this chapter, the authors have freely drawn on the published work of a number of investigators and references to the various papers considered have been given in the text. These sources are gratefully acknowledged.

The authors are also grateful for the valuable contributions of their colleague Mr N. A. Kalteziotis, to Julia Bentley who typed the script, and to Margaret Harris who drew the figures.

REFERENCES

1. BISHOP, A. W. and BJERRUM, L. (1960). The relevance of the triaxial test to the solution of stability problems. *ASCE Research Conf. on Shear Strength of Cohesive Soils*, Boulder, Colorado, 1960, 437–501.

2. SKEMPTON, A. W. and HUTCHINSON, J. N. (1969). Stability of natural slopes and embankment foundations, *Proc. 7th Int. Conf. Soil Mech. Fdn. Engng.*, State of the Art Volume, 291–340.

3. CHANDLER, R. J. (1976). The history and stability of two Lias Clay slopes in the upper Swash valley, Rutland, *Phil. Trans. Roy. Soc. London*, **A283**, 463–490.

4. WEEKS, A. G. (1970). The stability of the Lower Greensand Escarpment in Kent. Ph.D. Thesis, University of Surrey.

5. TERZAGHI, K. (1936). Stability of slopes of natural clay, *Proc. 1st Int. Conf. Soil Mech.*, Harvard, 1936, **1**, 161–165.

6. SKEMPTON, A. W. (1964). Long-term stability of clay slopes, *Géotechnique*, **14**, 77–101.
7. KENNEY, T. C. (1967). The influence of mineral composition on the residual strength of natural soils, *Proc. Geotech. Conf.*, Oslo, 1967, **1**, 123–130.
8. BISHOP, A. W., GREEN, G. E., GARGA, V. K., ANDRESEN, A. and BROWN, S. D. (1971). A new ring shear apparatus and its application to the measurement of residual strength, *Géotechnique*, **21**, 273–328.
9. TOWNSEND, F. C. and GILBERT, P. A. (1973). Tests to measure residual strengths in some clay shales, *Géotechnique*, **23**, 267–271.
10. VOIGHT, V. (1973). Correlation between Atterberg plasticity limits and residual shear strength of natural soils, *Géotechnique*, **23**, 265–267.
11. MITCHELL, J. K. (1976). *Fundamentals of Soil Behaviour*, John Wiley & Sons, Inc, New York.
12. CHANDLER, R. J. (1966). The measurement of residual strength in triaxial compression, *Géotechnique*, **16**, 181–186.
13. SKEMPTON, A. W. and PETLEY, D. J. (1967). The strength along structural discontinuities in stiff clays, *Proc. Geotech. Conf.*, Oslo, 1967, **2**, 29–46.
14. SKEMPTON, A. W. (1965). Discussion, *Proc. 6th Int. Conf. Soil Mech. Fdn. Engng.*, Montreal, 1965, **3**, 551–552.
15. LA GATTA, D. P. (1970). Residual strength of clays and clay shales by rotation shear tests, *Harvard Soil Mechanics Series, No 86*, Cambridge, Mass.
16. GARGA, V. A. (1970). Residual shear strength under large strains and the effect of sample size on the consolidation of fissured clay. Ph.D. Thesis, University of London.
17. HERRMANN, H. G. and WOLFSKILL, L. A. (1966). Residual shear strength of weak shales. *MIT Soil Mechanics Publication, No. 200*.
18. BISHOP, A. W. and LITTLE, A. L. (1967). The influence of the size and orientation of the sample on the apparent strength of London clay at Maldon, Essex, *Proc. Geotech. Conf.*, Oslo, 1967, **1**, 89–96.
19. CULLEN, R. M. and DONALD, I. B. (1971). Residual strength determination in direct shear, *Proc. 1st Australia–NZ Conf. on Geomechanics*, Melbourne, 1971, **1**, 1–9.
20. NOBLE, H. L. (1973). Residual strength and landslides in clay and shale, *J. Soil Mech. Fnd. Div.*, ASCE, **99**(SM9), 705–719.
21. LEUSSINK, H. and MULLER-KIRCHENBAUER, H. (1967). Determination of the shear strength behaviour of sliding planes caused by geological features, *Proc. Geotech. Conf.* Oslo, 1967, **1**, 131–137.
22. WEBB, D. L. (1969). Residual strength in conventional triaxial tests. *Proc. 7th Int. Conf. Soil Mech. Fnd. Eng.*, Mexico, 1969, **1**, 433–441.
23. BISHOP, A. W., WEBB, D. L. and LEWIN, P. I. (1965). Undisturbed samples of London clay from the Ashford Common shaft: strength–effective stress relationships, *Géotechnique*, **15**, 1–31.
24. BLIGHT, G. E. (1967). Observations on the shear testing of indurated fissured clays, *Proc. Geotech. Conf.*, Oslo, 1967, **1**, 97–102.

25. TIEDEMANN, B. (1937). Über die Schubfestigkeit bindiger Boden, *Bautechnik*, **15**, 433–435.
26. HAEFELI, R. (1938). Mechanische Eigenschaften von Lockergesteinen, *Schweiz Bauzeitung*, **111**, 321–325.
27. DE BEER, E. (1967). Clay strength characteristics of the Boom Clay, *Proc. Geotech. Conf.*, Oslo, 1967, **1**, 83–88.
28. SEMBELLI, P. and RAMIREZ, A. (1969). Measurement of residual strength of clays with a rotation shear machine, *Proc. 7th Int. Conf. Soil Mech. Fdn. Engng.*, Mexico, 1969, **3**, 528–529.
29. TERZAGHI, K. and PECK, R. B. (1948). *Soil Mechanics in Engineering Practice*, John Wiley & Sons, Inc, New York.
30. TAYLOR, D. W. (1948). *Fundamentals of Soil Mechanics*, John Wiley & Sons Inc, New York.
31. BJERRUM, L. (1967). Progressive failure in slopes of overconsolidated plastic clay and clay shales, *J. Soil Mech. Fdn. Engng. Div.*, ASCE, **93**(SM5), 3–49.
32. LO, K. Y. (1972). An approach to the problem of progressive failure, *Canad. Geotech. J.*, **9**(4), 407–429.
33. BISHOP, A. W. (1967). Progressive failure—with special reference to the mechanism causing it, *Proc. Geotech. Conf.*, Oslo, 1967, **2**, 142–150.
34. BISHOP, A. W. and HENKEL, D. J. (1962). *The Measurement of Soil Properties in the Triaxial Test*, Edward Arnold, London, p. 228.
35. BJERRUM, L., SIMONS, N. and TORBLAA, I. (1958). The effect of time on the shear strength of a soft clay, *Proc. Brussels Conf. Earth Pressure Prob.*, **1**, 148–158.
36. BISHOP, A. W. and LOVENBURY, H. T. (1969). Creep characteristics of two undisturbed clays, *Proc. 7th Int. Conf. Soil Mech.*, Mexico, 1969, **1**, 29–37.
37. BJERRUM, L. (1969). Discussion to Main Session 5, *Proc. 7th Int. Conf. Soil Mech. Fnd. Eng.*, Mexico, 1969, **3**, 410–412.
38. HVORSLEV, M. J. (1960). Physical components of the shear strength of saturated clays, *Proc. ASCE, Res. Conf. Shear Strength of Cohesive Soils*, Boulder, Colorado, 1960, 437–501.
39. SKEMPTON, A. W. (1966). Some observations on tectonic shear zones, *Proc. 1st Int. Conf. Rock Mech.*, Lisbon, 1966, **1**, 328–335.
40. SKEMPTON, A. W. and LA ROCHELLE, P. (1965). The Bradwell slip: a short-term failure in London clay, *Géotechnique*, **15**, 221–242.
41. SKEMPTON, A. W., SCHUSTER, R. L. and PETLEY, D. J. (1969). Joints and fissures in the London clay at Wraysbury and Edgeware, *Géotechnique*, **19**, 205–217.
42. SKEMPTON, A. W. (1961). Horizontal stresses in overconsolidated eocene clay, *Proc. 5th Int. Conf. Soil Mech. Fnd. Engng.*, Paris, **1**, 351–357.
43. DUNCAN, J. M. and DUNLOP, P. (1969). Slopes in stiff-fissured clays and shales, *J. Soil Mech. Fnd. Engng. Div.*, ASCE, **96**(SM2), 467–492.
44. LO, K. Y. and LEE, C. F. (1972). Discussion, *J. Soil Mech. Fnd. Engng. Div.*, ASCE, **98**(SM9), 981–983.

45. GOULD, J. P. (1960). A study on shear failure in certain tertiary marine sediments, *Proc. ASCE, Res. Conf.*, Boulder, Colorado, 1960, 615–641.
46. HENKEL, D. J. and SKEMPTON, A. W. (1955). A landslide at Jackfield, Shropshire, in a heavily overconsolidated clay, *Géotechnique*, 5, 131–137.
47. JAMES, P. M. (1971). The role of progressive failure in clay slopes, *Proc. 1st Australia–NZ Conf. on Geomechanics*, Melbourne, 1971, 1, 344–348.
48. SKEMPTON, A. W. and DE LORY, F. A. (1957). Stability of natural slopes in London clay. *Proc. 4th Int. Conf. Soil Mech.*, London, 1957, 2, 378–381.
49. SKEMPTON, A. W. (1970). First-time slides in overconsolidated clays, *Géotechnique*, 20, 320–324.
50. PECK, R. B. (1967). Stability of natural slopes, *J. Soil Mech. Fnd. Engng. Div.*, ASCE, 93(SM4), 403–417.
51. BISHOP, A. W. (1971). The influence of progressive failure on the choice of the method of stability analysis, *Géotechnique*, 21, 168–172.
52. LO, K. Y. and LEE, C. F. (1973). Stress analysis and slope stability in strain-softening materials, *Géotechnique*, 23, 1–11.
53. DUNLOP, P. and DUNCAN, J. M. (1970). Development of failure around excavated slopes, *J. Soil Mech. Fnd. Engng. Div.*, ASCE, 96(2), 471–493.
54. CHANG, C. U. and DUNCAN, J. M. (1970). Analysis of soil movements around a deep excavation, *J. Soil Mech. Fnd. Engng. Div.*, ASCE, 96(5), 1655–1681.
55. DUNCAN, J. M. (1972). Finite element analysis of stresses and movements in dams, excavations and slopes. *Proc. Symp. Applications of the Finite Element Method*, US Army Waterways Expt. Station, Vicksburg, 1972, 267–326.
56. LU, T. D. and SCOTT, R. F. (1972). The distribution of stresses and development of failure at the the toe of a slope and around the tip of a crack, *Proc. Symp. Applications of the Finite Element Method*, US Army Waterways Expt. Station, Vicksburg, 1972, 385–430.
57. LYNDON, A. and SCHOFIELD, A. N. (1976). Centrifugal model tests of the Lodalen landslide, *Proc. 29th Canadian Geotech. Conf.*, Vancouver, 1976, III, 25–43.
58. PALMER, A. C. and RICE, J. R. (1973). The growth of slip surfaces in the progressive failure of overconsolidated clay, *Proc. R. Soc.*, No. 1591, 527.
59. VAUGHAN, P. R. and CHANDLER, R. S. (1974). The design of cutting slopes in overconsolidated clay, Notes for *Informal Discussion, British Geotechnical Society*, p. 14.
60. BISHOP, A. W. (1955). The use of the slip circle in the stability analysis of slopes, *Géotechnique*, 5, 7–17.
61. JANBU, N. (1954). Stability analysis of slopes with dimensionless parameters, *Harvard Soil Mechanics Series No 46*, Cambridge, Mass., p. 81.
62. BISHOP, A. W. and MORGENSTERN, N. R. (1960). Stability coefficients for earth slopes, *Géotechnique*, 10, 129–150.
63. KJAERNSLI, B. and SIMONS, N. E. (1962). Stability investigations of the north bank of the Drammen River, *Géotechnique*, 12, 147–167.

64 SEVALDSON, R. A. (1956). The slide at Lodalen, October 6, 1954. *Géotechnique*, **6**, 167–182.
65. SKEMPTON, A. W. and BROWN, J. D. (1961). A landslide in boulder clay at Selset, Yorkshire, *Géotechnique*, **11**, 280–293.
66. VAUGHAN, P. R. and WALBANCKE, H. J. (1973). Pore pressure changes and the delayed failure of cutting slopes in overconsolidated clay, *Géotechnique*, **23**, 531–539.
67. SKEMPTON, A. W. (1977). Slope stability of cuttings in Brown London clay. Special Lecture, *9th Int. Conf. Soil Mech. Fnd. Eng.*, Tokyo, 1977.
68. CHANDLER, R. J. (1974). Lias clay: the long-term stability of cutting slopes, *Géotechnique*, **24**, 21–38.
69. HUTCHINSON, J. N. (1967). The free degradation of London clay cliffs, *Proc. Geotech. Conf.*, (1967). Oslo, **1**, 113–118.

Chapter 11

STABILITY OF EMBANKMENTS ON SOFT GROUND

B. K. MENZIES and N. E. SIMONS

*Department of Civil Engineering, University of Surrey,
Guildford, UK*

SUMMARY

*An outline is given of the current design approach associated with the
prediction of stability of embankments on soft ground. By way of a
fundamental introduction a brief resumé is given of the nature of
shear strength and the underlying assumptions implicit in the use of
shear test data in the prediction of field stability. This theme is
extended to include the role of analysis in forecasting stability. The
assessment of field strength is also considered, with particular
emphasis on the interpretation of shear vane test data. To illustrate
these factors the stabilities of two trial embankments on soft clay are
examined in some detail. In conclusion a summary is given of the
more important features considered relevant to the prediction of
stability on soft clays.*

11.1 INTRODUCTION

The prediction of settlements and stability of embankments on soft
ground is of particular interest to the geotechnical engineer because
of the large deformations associated with embankment loading of
normally consolidated and lightly overconsolidated clays. Due to
limitations of space, this chapter is concerned with giving an outline
of the current design approach to the prediction of stability only of
embankments on soft ground.

11.1.1 Short-term Stability

If a fine-grained saturated soil is rapidly loaded (e.g. by rapidly filling a large oil tank) in the short term the soil is effectively undrained due to the high viscous forces resisting pore-water flow within the soil. The excess pore pressure generated by the sudden application of load dissipates by drainage or consolidation over a period of time which may, in the case of clays, extend for tens or even hundreds of years. Hence the terms 'short' and 'sudden' are relative and a load application over several months during a construction period may be relatively rapid with the short-term or end-of-construction condition approximating to the undrained case.

In positive loading conditions like embankments and footings the subsequent consolidation under the influence of the increased load gives rise to increased strength and stability. The lowest strength and, therefore, the critical stability condition, hold at the end of construction. The critical strength is thus the undrained shear strength before post-construction consolidation.

One way of measuring this strength is to build up the load rapidly in a full-scale field test until the soil fails and this is sometimes done, particularly in earthworks, by means of trial embankments. Such full-scale testing can be costly and is appropriate only to large projects where the soil is uniform.

For conditions of variable soil and when large expenditure on soil testing is unlikely to effect economies in design, small-scale testing is more appropriate.

A given soil shear test specimen or zone may have many different measured strengths depending on a variety of factors. The factor which most significantly affects the measurement of shear strength is the mode of test, that is, whether the test is drained or undrained. Further significant factors which affect the measurement of shear strength, particularly undrained shear strength, are the type of test, that is, whether by direct shear box, triaxial compression, *in situ* shear vane, etc; the effects of orientation of the test specimen or test zone (the effect of anisotropy); time to failure; sampling disturbance and size of test specimen or test zone; and time between sampling and testing. These factors are now discussed with particular emphasis on undrained strength.

11.1.2 Strength Anisotropy

Soil strength anisotropy arises from the two interacting anisotropies of geometrical anisotropy, i.e. preferred particle packing; and of stress anisotropy. The geometrical anisotropy arises at deposition when the sedimented particles tend to orientate with their long axes horizontal, seeking packing positions of minimum potential energy. The horizontal layering or bedding which results is further established by subsequent deposition which increases the overburden pressure. The stress anisotropy arises because of a combination of stress history and the geometrical anisotropies of both the particles themselves and of the packed structure they form. The net effect is a clear strength and stress–strain anisotropy.

Consider, for example, the drained triaxial compression tests carried out on a dense dry rounded sand in a new cubical triaxial cell described by Arthur and Menzies.[1] The cubical specimen was stressed on all six faces with flat, water-filled, pressurised rubber bags. The test specimen was prepared by pouring sand through air into a tilted former. In this way, it was possible to vary the direction of the bedding of the particles which was normal to the direction of deposition. A clear stress–strain anisotropy was measured, the strain required to mobilise a given strength being greater for the bedding aligned in the vertical major principal stress direction than for the conventional case of the bedding aligned horizontally in the test specimen.

Parallel tests were carried out by Arthur and Phillips[2] in a conventional triaxial cell testing a prismatic specimen with lubricated ends. Remarkably similar anisotropic strengths were measured in the different apparatus.

Geometrical anisotropy, or fabric as it is sometimes called, not only gives rise to strength variations with orientation of the test axes but is also probably partly the cause of undrained strength variations between test type. Madhloom[3] carried out a series of undrained triaxial compression tests, triaxial extension tests and direct shear box tests using specimens of a soft, silty clay from King's Lynn, Norfolk. The soil was obtained by using a Geonor piston sampler. Samples were extruded in the laboratory and hand-trimmed to give test specimens in which the bedding was orientated at different angles to the specimen axes. It was found generally that for this type of clay the triaxial compression test indicated a strength intermediate between that indicated by the triaxial extension test and the direct shear box test. This was not the case, however, in tests on soft marine clays reported by Bjerrum[4] and given in Table 1. Here the direct shear box (and the corrected shear vane) indicated strengths intermediate between the triaxial extension and compression tests.

TABLE 1

COMPARISON BETWEEN THE RESULTS OF COMPRESSION AND EXTENSION TRIAXIAL TESTS, DIRECT SHEAR TESTS AND IN SITU VANE TESTS ON VARIOUS SOFT CLAYS, AFTER BJERRUM [4]

| Type of soil | Index properties (%) | | | | Triaxial test τ_f/p_0' | | Simple shear test τ_f/p_0' | Vane tests s_u/p_0' | |
	Water content, w	Liquid limit, w_L	Plastic limit, w_p	Plasticity index, I_p	Compression	Extension		Observed	Corrected for rate
Bangkok clay	140	150	65	85	0·70	0·40	0·41	0·59	0·47
King's Lynn clay	65	93	39	57	0·37	0·23	0·55	0·40	—
Matagami clay	90	85	38	47	0·61	0·45	0·39	0·46	0·40
Drammen plastic clay	52	61	32	29	0·40	0·15	0·30	0·36	0·30
Vaterland clay	35	42	26	16	0·32	0·09	0·26	0·22	0·20
Studentertunden	31	43	25	18	0·31	0·10	0·19	0·18	0·16
Drammen lean clay	30	33	22	11	0·34	0·09	0·22	0·24	0·21

11.1.3 Time to Undrained Failure

As demonstrated by Bjerrum et al.,[5] the greater the time to undrained failure, the lower will be undrained strength. It is therefore necessary to take this factor into account when using the results of in situ vane tests or undrained triaxial compression tests, with a failure time of the order of 10 min, to predict the short-term stability of cuttings and embankments, where the shear stresses leading to failure may be gradually applied over a period of many weeks of construction.

The greater the plasticity index of the clay, the greater is the reduction factor which should be applied to the results of the tests with small times to failure. Most of the reduction in undrained shear strength is because of an increase in the pore-water pressure as the time to failure increases, i.e. it is a pore pressure phenomenon.

A further factor to be considered is the elapsed time between taking up a sample, or opening a test pit, and performing strength tests. No thorough study of this aspect has been made to date for soft clays but it is apparent that the greater the elapsed time, for a stiff, fissured clay, the smaller is the measured strength. Marsland[6] noted that from loading tests made on 152-mm diameter plates at Ashford Common, strengths measured 4–8 h and 2·5 days after excavation were approximately 85% and 75%, respectively, of those measured 0·5 h after the excavation. Other evidence was provided by laboratory tests on 38-mm diameter specimens cut from block samples of fissured clay from Wraysbury, which were stored for different periods before testing. Strengths of specimens cut from blocks stored for about 150 days before testing were only about 75% of the strengths of specimens prepared from blocks within 5 days of excavation from the shaft. This could be attributed to a gradual extension of fissures within the specimens.

11.1.4 Sampling Disturbance

If the test specimen is disturbed by the sampling process, the measured undrained shear strength will generally be lower than the in situ value for a given test apparatus and procedure. Thin-walled piston samplers jacked into the ground cause very little disturbance and this technique, together with careful handing in the field, during transit, and in the laboratory, is believed to give reasonably reliable measurements of the undrained shear strength of clay. Hand-cut block samples of clay taken from open excavation may be used equally well.

A number of workers have investigated the effects of disturbance caused by physical disruption and also by stress change, and much of the work has been conveniently summarised by Davis and Poulos.[7] In the type of soft clays with low to medium sensitivity which are under consideration here, both of these sources of disturbance tend to reduce the shear strength, thus giving laboratory values below the field values. This would tend to lead to an underestimate of the stability in the field. Davis and Poulos suggest that the soil should be reconsolidated in the laboratory to the effective stress condition in the field, but in fact this would lead to lower voids ratios than in the field (for example, see Schmertman[8]) and perhaps to higher strengths than in the field, thus increasing the tendency which already exists to overestimate the stability.[9]

11.1.5 Selection of Appropriate Shear Strength Data

The selection of appropriate shear strength data for the prediction of field stability therefore requires some guiding principle unless empirical corrections based on experience are to be made. Apart from such corrections *the use of shear test data in stability analyses is appropriate only provided there is similitude between the shear test model, the analytical model and the field prototype.* The first steps, therefore, in ensuring similitude between test model and field prototype require that:

(i) prior to testing, the effective stresses and structural configuration of the test specimen or zone are identical to those *in situ* before testing disturbance, i.e. the soil specimen or zone is undisturbed:

(ii) the size of the test specimen or test zone is representative of the soil in the mass;

(iii) during the test the structural distortions and rates of distortion of the soil mass are similar to those which would arise in the field;

(iv) at failure in the test, the distortions and rates of distortion of shear surfaces are similar to those which would arise in a full scale failure.

Finally, of course, the principle of similitude requires a realistic analytical model in which the shear test data may be used to give an estimate of the stability of the field prototype. Of considerable importance in this respect is the phenomenon of progressive failure.*

The stability of foundations and earth-works in saturated fine-grained soil is time-dependent. This is because the average size of the interconnecting pores is so small that the displacement of pore-water is retarded by viscous forces. The resistance that a soil offers to water flow may be measured in terms of the soil permeability which is the velocity of flow through the soil under a unit hydraulic gradient.

Permeability is the largest quantitative difference between soils of different time-dependent stability.[11] A sand and a normally-consolidated clay, for example, may have similar effective stress shear strength parameters c' and $\tan \phi'$ but the permeability of the clay is several orders of magnitude lower. The stability of the clay is thus time-dependent, whereas the more permeable sand reacts to loading changes almost immediately.

If a saturated clay is loaded, such as may occur in soils supporting building foundations and earth embankments (Fig. 1), an overall increase in mean total stress occurs (Fig. 2(a)). In a fine grained soil like clay, the viscous resistance to pore-water expulsion prevents the soil structure from rapidly contracting. In the short-term loading condition, therefore, there is a change in effective stress due to shear strain only together with an increase in pore pressure (Fig. 2(b) and (c)). With time, this excess pore pressure is dissipated by drainage away from the area of increased pore pressure into the surrounding area of lower pore pressure unaffected by the construction. This flow

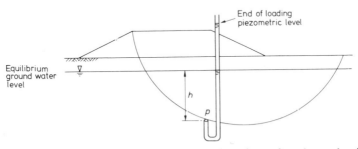

FIG. 1.   Pore pressure generated on a potential slip surface by embankment loading.

*There is some theoretical evidence[10] to suggest that this phenomenon may not be adequately modelled in laboratory shear tests on small scale specimens.

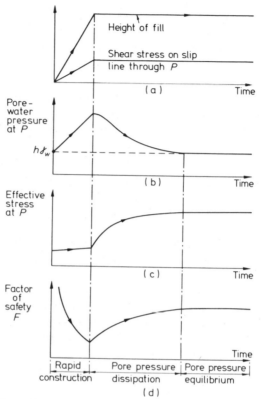

FIG. 2.   Variation with time of the shear stress, local pore pressure, local effective stress, and factor of safety for a saturated clay foundation beneath an embankment fill (after Bishop and Bjerrum).[11]

of pore-water causes a time-dependent reduction in volume in the zone of influence, the soil consolidating and the soil structure stiffening, giving rise to decreasing settlement and increasing strength. The minimum factor of safety thus occurs in the short-term undrained condition when the strength is lowest (Fig. 2(d)).

In this undrained condition the stressed zone does not immediately change its water content or its volume. The load increment does, however, distort the stressed zone. The effective stresses change along with the change in shape of the soil structure. Eventually the changes in structural configuration may no longer produce a stable condition and the consequent instability gives rise to a plastic mechanism or plastic flow and failure occurs.

The strength is determined by the local effective stresses at failure normal to the failure surfaces. These are conditioned by and generated from the structural configuration of the parent material (which is itself conditioned by the preloading *in situ* stresses) and its undrained reaction to deformation. A first step, therefore, in fulfilling the complex similitude requirements which accordingly arise is to ensure that the shear test is effectively undrained.

The undrained shear test may be used to give a direct measure of shear strength, namely the undrained shear strength s_u, or it may be used to give an indirect measure of shear strength, if the pore-water pressures are measured,* by providing c' and $\tan \phi'$. It is therefore possible to analyse the stability of the loaded soil by:

(i)   using the undrained shear strength s_u in a total stress analysis; or

(ii) using the effective stress shear strength parameters c' and $\tan \phi'$ in an effective stress analysis.

The use of (ii) requires an estimation of the end of construction pore pressures in the failure zone at failure, whereas the use of (i) requires no knowledge of the pore pressures whatsoever.

In general civil engineering works, the soil loading change is applied gradually during the construction period. The excess pore pressures generated by the loading are thus partially dissipated at the end of construction. The end of construction pore pressures and the increased *in situ* shear strength can be measured on site if the resulting increased economy of design warrants the field instrumentation and testing. On all but large projects this is rarely the case. In addition, the loading is localised, allowing the soil structure to strain laterally, the soil stresses dissipating and the principal stresses rotating within the zone of influence.

In the absence of sound field data on the end of construction shear strengths and pore pressures and in the face of analytical difficulties under local loading, an idealised soil model possessing none of these difficulties is usually invoked for design purposes. This consists of proposing that the end of the construction condition corresponds to the idealised case of the perfectly undrained condition. Here the soil

---

*In fact c' and $\tan \phi'$ may also be estimated from a drained test, the relationship between strength and effective stress being reasonably constant over a limited range of stress.

is considered to be fully saturated with incompressible water and is sufficiently rapidly loaded that, in the short term, it is completely undrained. Prefailure and failure distortions of the soil mass in the field are, *by implication if not in fact*, simulated by the test measuring the undrained shear strength. It follows that if the shear strength of the soil structure is determined under rapid loading conditions prior to construction, this undrained shear strength may be used for short-term design considerations. No knowledge of the pore pressures is required, the undrained shear strength being used in a total stress analysis (the so-called $\phi = 0$ analysis).

11.2 FACTORS AFFECTING THE EVALUATION OF FIELD STABILITY

11.2.1 Shear Test Similitude
Soil can only fail under conditions of local loading where the loading in the zone of influence distorts the soil mass as a whole. Beneath a rapidly constructed embankment, for example, the previously horizontal ground surface deflects, the zone of influence distorting without, at the instant of loading, changing its volume. The major principal stress direction is orientated vertically under the embankment. Towards the toe of the embankment the principal stress directions rotate until the major principal stress is horizontal. If model shear tests are to be used to predict the undrained shear strength in these varying stress and deformation zones, the tests must, ideally, simulate the actual field stress and deformation paths.

To approximate to similitude, Bjerrum[4] suggests using different modes of shear test to evaluate the undrained shear strength for different areas of the distorted soil. Thus the triaxial* compression test simulates distortion directly under the embankment where the shear surface is inclined near the major principal stress direction; the simple shear test simulates distortion where the shear surface is nearly horizontal, and the triaxial* extension test simulates distortion near the toe of the embankment where the shear surface is inclined near the minor principal stress direction (Fig. 3(a)).

In many respects, the ideal testing system is represented in Fig. 3(b), where the direct shear test specimens are subjected to the

---

*Plane–strain tests may be more appropriate here.

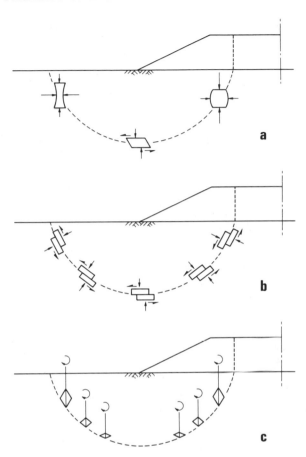

FIG. 3.   Possible shear test configurations for estimating shear strength of a soft clay beneath an embankment: (a) matching of *in situ* stresses;[4] (b) matching of *in situ* stresses and orientation of failure surface in direct shear; (c) measurement of an *in situ* strength anisotropy by diamond vanes at pre-construction *in situ* stresses.

post-construction *in situ* stresses. The diamond shear vane testing system represented in Fig. 3(c) is open to criticism because the shearing is *across* the surface of sliding.

The alternative to matching the shear test to the field failure zone on the basis of like soil distortions is to adopt a purely empirical approach.

TABLE 2

CORRELATION BETWEEN PLASTICITY INDEX AND FACTOR OF SAFETY PREDICTED FROM *IN SITU* SHEAR VANE MEASUREMENTS FOR EMBANKMENTS ON SOFT CLAY WHICH HAVE FAILED

No	Reference and location	Factor of safety, F	Plasticity index, I_p
1*	Parry and McLeod,[12] Launceston	1·65	108
2*	Eide and Holmberg,[13] Bangkok, A	1·61	85
3*	Eide and Holmberg,[13] Bangkok, B	1·46	85
4*	Golder and Palmer,[14] Scrapsgate	1·52	82
5*	Pilot,[15] Lanester	1·38	72
6	Eide,[16] Bangkok	1·5	60
7†	Peterson et al.,[17] Seven Sisters	1·5	59
8*	Pilot,[15] Saint André de Cubzac	1·4	47
9*	Dascal et al.,[18] Matagami	1·53	47
10*	Pilot,[15] Pornic	1·2	45
11	Serota,[19] Escravos Mole	1·1	40
12	Roy,[20] Somerset	1.2	36
13	Brent Knoll, (see section 11.3.1)	1.37	36
14*	Wilkes,[21] King's Lynn	1·1	35
15*	Lo and Stermac,[22] New Liskeard	1·05	33
16*	Pilot,[15] Palavas	1·30	32
17†	Stamatopoulos and Kotzias,[23] Thessalonika	$\simeq 1\cdot0$	30
18†	La Rochelle et al.,[24] Saint-Alban	1·3	25
19	Flaate and Preber,[25] Jarlsburg	1·1	25
20	Flaate and Preber,[25] Aulielava	0.92	23
21	Flaate and Preber,[25] Nesset	0·88	22
22	Flaate and Preber,[25] Ås	0·80	20
23	Flaate and Preber,[25] Presterødbakken	0·82	17
24*	Pilot,[15] Narbonne	0·96	16
25*	Ladd,[26] Portsmouth	0·84	16
26*	Haupt and Olson,[27] Fair Haven	0.99	16
27	Flaate and Preber,[25] Skjeggerød	0·73	11
28	Flaate and Preber,[25] Tjernsmyr	0·87	8
29	Flaate and Preber,[25] Falkenstein	0·89	8

*Quoted by Bjerrum.[4]
†Estimated average values.

By comparing the stability of embankments which had failed with the predicted stabilities based on *in situ* vane measurements, Bjerrum[4] derived an empirical correction factor based on the plasticity index to enable the strength measurements made by conventional rectangular vanes to be factored to give a realistic forecast of field stability, using a traditional limit analysis based on a rigid–plastic shear stress displacement relationship which does not vary with orientation. The correction factor therefore should specifically compensate for the effects of progressive failure and anisotropy. It also generally accommodates the effects of testing rate, size of test zone and any other effect which may cause a lack of similitude between the vane shear test model and the field prototype.

Bjerrum[4] considered 14 case histories when formulating his correction factor μ_B. As shown in Table 2, the authors have added a further 15 case histories. The data is plotted in Fig. 4 where the straight line has been fitted to the data using linear regression by the method of least squares. The coefficient of determination is $r^2 = 0.76$. Noting that

$$\mu_B = (s_u)_{\text{field}}/(s_u)_{\text{vane}} = 1/(F)_{\text{vane}} \qquad (1)$$

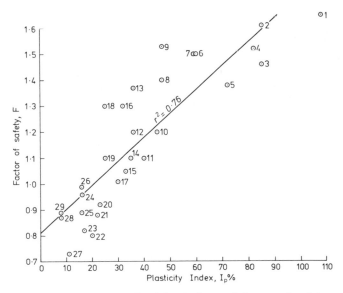

FIG. 4. Correlation between plasticity index and factor of safety predicted from shear vane measurements of strength for failed embankments on soft clays.

enables μ_β to be plotted against $I_p\%$ in Fig. 5 where a hyperbola is fitted giving $r^2 = 0.78$.

An *in situ* shear vane, rotated about a vertical axis, shears the soil on a circumscribing vertical cylinder and will indicate components of strength derived from shearing in the horizontal plane (shearing mode 1) and from shearing horizontally in the vertical plane (shearing mode 2). A two-dimensional field failure, in which the shear surface is partly a horizontal cylinder (Fig. 6), will mobilise strength components derived from shearing mode 1 and from shearing vertic-

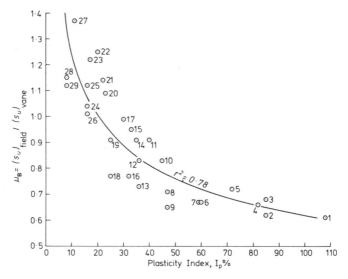

FIG. 5.   Correlation between plasticity index and empirical shear vane correction factor μ_B for failed embankments on soft clay.

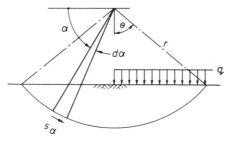

FIG. 6.   Simplified bearing capacity configuration.

ally in the vertical planc (shearing mode 3). A three-dimensional field failure in which the shear surface is conchoidal will mobilise strength components derived from shearing modes 1, 2, and 3. It may therefore be concluded from considerations of anisotropy alone that the conventional *in situ* shear vane is unlikely to measure strength that is directly appropriate to any field stability condition.

Bjerrum's correction factor, correlated as it is with plasticity index, i.e. with data obtained from tests on remoulded soils, is linked to a parameter that cannot measure inherent anisotropy. This may account for some of the scatter in the data used by Bjerrum in obtaining the relationship between his correction factor and plasticity index. The best-fit curve which resulted may well have eliminated the unrelated effect of anisotropy, in which case the correction factor does not correct for anisotropy but may well accommodate the effects of testing rate and progressive failure. If this is so then Bjerrum's correction factor must be applied to correct for testing rate and progressive failure and a separate further correction factor must be applied to correct specifically for anisotropy. In order to make some estimate of this correction factor, Menzies[28] considered the simplified bearing capacity configuration given in Fig. 6.

The soil was taken to be weightless and to fail on a circular arc. To simplify the analysis further the centre of rotation was taken above the edge of the uniformly loaded strip. The analysis is similar to that given by Raymond.[29] It was assumed that the difference between the vertical undrained shear strength s_{uv} and the horizontal undrained shear strength s_{uh} may be distributed as the square of the direction cosine,[30-33] giving the undrained shear strength in any direction α as

$$s_{u\alpha} = s_{uh}(1 + (R - 1)\cos^2 \alpha) \qquad (2)$$

where

$$R = s_{uv}/s_{uh} \qquad (3)$$

is the degree of undrained strength anisotropy.

A factor was obtained which corrects for the influence of strength anisotropy on conventional shear vane measurements used to predict field bearing capacity

$$\mu_A = \frac{((R + 1)2\theta - (R - 1)\sin 2\theta)/\sin^2 \theta}{2{\cdot}37(2R + 1/3)} \qquad (4)$$

where θ is given by

$$\tan \theta = 2\theta k \qquad (5)$$

where

$$k = \frac{(R+1) - ((R-1)\sin 2\theta)/2\theta}{(R+1) - (R-1)\cos 2\theta} \qquad (6)$$

The relationship between μ_A and R is plotted in Fig. 7.

In order that the conventional shear vane strength $(s_u)_{\text{vane}}$ may be used in a traditional limit analysis, i.e. assuming the soil has a rigid–plastic shear strength–displacement relationship which does not vary with direction, it is suggested that $(s_u)_{\text{vane}}$ should be corrected to give the field strength $(s_u)_{\text{field}}$ as follows:

$$(s_u)_{\text{field}} = \mu_A\mu_B(s_u)_{\text{vane}} \qquad (7)$$

where μ_A is the correction factor for strength anisotropy given in Fig. 7 and where μ_B is Bjerrum's correction factor for the effects of testing rate and progressive failure given in Fig. 5. This equation is similar to that proposed by Bjerrum.[4]

The correction factor μ_A has been derived using a simplified analysis which includes the assumption that shearing modes 2 and 3 give the same strengths, which is most probably not the case. The correction factor μ_A is therefore an approximation. Nevertheless, it is a useful indication of the effect of strength anisotropy on traditional predictions of field bearing capacity based on conventional shear vane measurements of strength. The degree of anisotropy R may be estimated from undrained compression tests on specimens sampled at

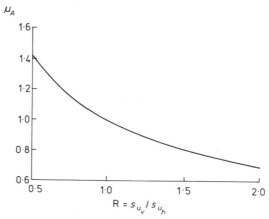

FIG. 7.   Variation of anisotropy correction factor μ_A with degree of strength anisotropy R.

differing orientations,[32] from undrained direct shear tests on specimens trimmed from block samples to various angles with respect to the bedding[34] or more simply by using diamond-shaped shear vanes *in situ*,[33,35] as described in Section 11.3.1.

11.2.2 Analytical Similitude

In the foregoing section the term 'field strength' refers simply to a number which gives a realistic estimate of field stability when used in a traditional limit analysis, assuming a rigid–plastic shear stress–displacement relationship which does not vary with direction. The limit analytical model does not therefore allow for the effect of progressive failure.[36]

Most naturally occurring soils are strain softening. Irrespective of whether the shear test is drained or undrained, they possess a shear stress–deformation relationship which is characterised by a peak followed by a reduction in strength to an ultimate or 'residual' strength. In normally consolidated clays, the reduction in strength from peak to residual may be slight, while in overconsolidated clays the reduction is marked giving a brittle behaviour.

In a homogeneous soil which is not loaded, direct shear tests on specimens sampled along a circular arc through the soil will have different shear stress–deformation relationships due to the variation of strength with depth and with orientation. Distorting the soil in the area of the circular arc by a local surface loading will modify the stress–deformation relationships. During loading the stresses vary in the zone of influence and hence the shear strength around the circular arc varies according to the effective stresses normal to the circular arc.

If the circular arc becomes a slip surface some local regions will fail. Here the shear stresses tangential to the slip surface have exceeded the local soil strength generated by the local normal effective stresses. The strength in these failed regions reduces according to the strain softening behaviour. In the pre-failure regions, on the other hand, the shear stresses, increased by the load shedding of the strain softening zones will not have exceeded the soil strength available.

A state of limiting equilibrium is attained when the reduction in strength of the elements of the slip surface in the failure zone just begins to exceed the increase in stress taken by those elements in the pre-failure zone.[37,38] Some of the failed elements may have reduced in strength to the residual value but this is not strictly necessary for limiting equilibrium in a strain softening material. Indeed, the large displacements necessary to achieve ultimate residual strength in heavily overconsolidated clays[39] would suggest that failure conditions in such a material, in the sense of a factor of safety of unity, might obtain without any element of the slip surface reaching the residual strength.[38]

By way of illustrating this phenomenon, consider an embankment on a soft saturated strain-softening clay. The embankment is built up in three construction stages or 'lifts', as shown in Fig. 8. Consider further three locations, A, B and C on a hypothetical slip circle. The shear stress–displacement relationships are given for these locations. When the first lift is built the underlying soil will be subjected to shear stresses which will vary with orientation and location in a similar way to that in a loaded elastic solid. The induced stresses will cause the soil to deform and the variation in shear strain at locations A, B, and C will mobilise, say, pre-peak soil strengths corresponding to position 1. Now, let the embankment be built up with a second lift of fill, again increasing shear stresses in the ground. Let peak strength be attained at location A which is now within a zone of failure or

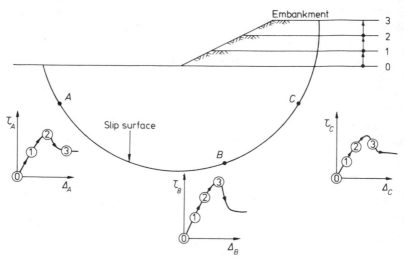

FIG. 8.   Diagrammatic representation of progressive failure due to soil strain softening of the saturated clay foundation of an embankment built in three lifts of fill.

plastic yield at the toe of the slip which has not yet fully formed. Locations B and C are still within a pre-failure zone at pre-peak stresses represented by position 2.

Now a third lift is constructed and further shear stresses applied to the soil with consequent shear strains developing. At location A this associated deformation exceeds the peak value and the load-carrying capacity of this location is diminished according to the strain-softening of the soil. This shedding of load puts additional shear stress on the pre-failed regions with the consequence that a further failure zone develops at location C where the peak stress is exceeded. In turn, the load shedding from location C causes location B to fail, also attaining peak shear stress at position 3. In this way a strain-softening soil fails in a progressive way.

Conventional stability analyses do not model progressive failure. To take the peak strength as acting on all elements simultaneously around the slip surface overestimates the factor of safety. On the other hand, to take the residual shear strength as acting simultaneously around the slip surface underestimates the factor of safety and is clearly inappropriate in any event as the residual shear strength holds on an established shear surface, i.e. after failure has occurred.

The rigid–plastic shear stress–displacement relationship implied by a traditional analysis clearly does not accommodate this strain softening behaviour (Fig. 9). Using the residual factor[40] goes some of

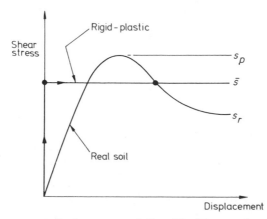

FIG. 9. Shear stress–displacement relationship for a real soil fitted with a rigid–plastic idealisation. The residual factor[40] is $R = (s_p - \bar{s})/(s_p - s_r)$.

the way towards compensating for this effect (Fig. 9) but the value of the residual factor needs to be estimated from field failures and the analytical model implies, inappropriately, a rigid–plastic idealisation of the shear stress–displacement relationship. An analytical model which simulates progressive failure has been suggested by Lo and Lee.[41]

Conventional limit analyses may be carried out following the procedures outlined by Janbu et al.,[42] using the following equations:

(a)   For slip circles and equal width slices

$$F = \frac{\Sigma \dfrac{s_u}{\cos \alpha}}{\Sigma p \sin \alpha} \tag{8}$$

for the total stress analysis,

and

$$F = \frac{\Sigma \dfrac{c' + (p - u) \tan \phi'}{m_\alpha}}{\Sigma p \sin \alpha} \tag{9}$$

for the effective stress analysis.

(b)   For non-circular failure surfaces and equal width slices

$$F = f_0 \frac{\Sigma s_u (1 + \tan^2 \alpha)}{\Sigma p \tan \alpha} \tag{10}$$

for the total stress analysis,

and

$$F = \frac{f_0 \Sigma \dfrac{c' + (p - u) \tan \phi'}{n_\alpha}}{\Sigma p \tan \alpha} \tag{11}$$

for the effective stress analysis,

where

$$m_\alpha = \cos \alpha (1 + \tan \alpha \cdot \tan \phi'/F) \tag{12}$$

$$n_\alpha = \cos^2 \alpha (1 + \tan \alpha \cdot \tan \phi'/F) \tag{13}$$

$f_0 =$ an influence factor varying from 1 to 1·13, depending on the geometry of the failure surface and c' and ϕ',

and u and p are the pore pressure and vertical total stress respectively at the mid-point of the base of each slice.

11.3   CASE RECORDS

11.3.1 The Brent Knoll Trial Embankment

The Brent Knoll trial embankment* was built as part of the testing programme for the geotechnical design of the M5 motorway. This motorway from Birmingham to Exeter forms part of the principal road network linking major development areas in Britain. The section of motorway from Clevedon 10 km south of Bristol, to Huntworth 2 km south of Bridgwater, is built on the soft alluvial clays of the Somerset Levels where the depth of alluvium is typically about 26 m. These sediments are so soft that the stability of all embankments was low and large settlements were expected.

The site of the trial embankment is 9 km SSE of Weston-super-Mare and 1·6 km from East Brent. The National Grid reference co-ordinates of the site are ST 36 53.

Construction of the trial bank started at the end of October 1967 and continued for 3 months. It was 189 m long and 70 m wide, with side slopes of about 1 to 3, and was built of compacted quarry waste with not more than 15% passing a 75-mm BS sieve, except for the bottom 0·46 m where the fines were limited to less than 10% in order to form a drainage layer. After compaction by at least eight passes of a vibrating roller, the fill had a bulk unit weight of about 23 kN/m³. The intention was to construct the bank to a height of 9·1 m with a final crest width of 15·2 m. This was the expected profile of the motorway interchange embankments. A plan of the embankment is shown in Fig. 10.

In January 1968, with 7·9 m of fill in place a 61-m length of the bank slipped with a 1·2-m slump at the crest, the failure extending to 20 m beyond the toe of the bank. Numerous tension cracks opened up in the unslipped fill generally parallel to the slump at the crest. The opposite side of the bank appeared to be on the verge of failure, the side slope becoming slightly S-shaped and a construction peg near the toe of the bank moving outwards approximately 0·5 m. A section through the slip is shown in Fig. 11.

To prevent further movement of the slip and additional damage to the instrumentation, a 1·2-m thick, 15·2-m wide berm was immediately placed along the toe of the bank in the area of the slip. Construction then ceased.

---

*Previously known as the East Brent trial embankment.

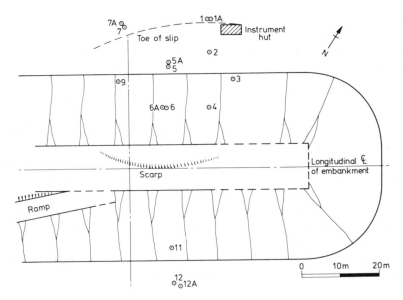

FIG. 10.   Plan of the Brent Knoll trial embankment.

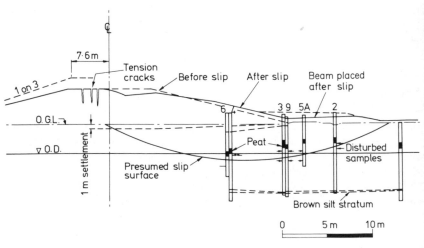

FIG. 11.   Cross-section through the Brent Knoll trial embankment after
failure.

Four borings (2, 3, 4 and 9) were made in the slipped area and continuous samples were taken. These were examined on site for signs of failure zones. Two 150-mm thick bands of disturbed clay were generally found in these holes and, although it was very difficult on site to differentiate between sample disturbance and failed material, a general pattern developed when these positions were plotted on a cross-section (Fig. 11). The evidence from borehole 9 is particularly conclusive as overlapping piston samples were taken in this hole.

Standard vane tests were carried out in three boreholes outside the bank (boreholes 1A, 7A and 12A). The results of two of these are shown in Fig. 12. It was found that the ground became disturbed very easily and that the vane tests had to be executed well below the bottom of the hole.

During July and August of 1972, the failed trial embankment was levelled to a height of approximately 2·5 m above original ground level. In November 1975 further field testing was carried out. A test

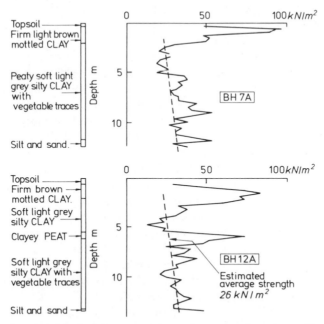

FIG. 12. Conventional shear vane measurements taken immediately after the slip of 1968, Brent Knoll trial embankment.

pit was excavated by digger to a depth of 2·5 m alongside the south-west corner of where the instrument hut had been sited. Two block samples were removed for subsequent laboratory testing. The block samples were cut with thin wire from the excavated clay retained in the digger bucket, trimmed on site to approximate cubes of side 400 mm, immediately wrapped in aluminium foil and sealed with melted wax.

From the floor of the test pit *in situ* shear vane tests were carried out using a conventional rectangular vane and a set of diamond vanes (Fig. 13). The concept of the diamond-shaped shear vane was introduced by Aas[43] who carried out shear tests using a diamond vane making an angle of 45° to the axis of rotation. Richardson *et al.*[35] and Menzies and Mailey[33] extended this concept separately by using diamond-shaped shear vanes of different angles to the axis of rotation, enabling the variation of shear strength with direction to be measured.

In common with the conventional vane, which was 140 mm long and 70 mm in diameter, the diamond vanes were symmetrical and cruciform in cross-section. The vanes were used in conjunction with a MHH B 800 torque indicator and were designed to give the same torque assuming strength isotropy. It was assumed that a diamond vane making an angle α with the vertical vane axis of rotation shears the soil on planes making an angle α with the vertical direction thus giving a measure of the undrained shear strength in the direction α.

The conventional and diamond shear vanes were progressively inserted in the base of the test pit and readings of strength with depth were carried out at different depths up to 2·0 m below the bottom of the pit. The excavation of the pit and the vane testing was carried out over a period of 8 h. A polar diagram showing the variation of strength with direction measured by the diamond vanes is given in Fig. 14. The curve fitted to the points has the eqn. (2).

It was felt that the low readings obtained by the 70° vane were anomalous (Fig. 14). To investigate the possibility that such an acutely pointed vane was not shearing the soil on circumscribing conical surfaces a 70° vane having eight blades was made (Fig. 15). Comparative tests gave results from the eight-bladed 70° vane which were indistinguishable from the four-bladed 70° vane. These tests were carried out down two boreholes within 1 m of the backfilled test pit and 2 weeks after the main testing had been completed. The results plot as shown in Fig. 16. The lower strengths obtained with the

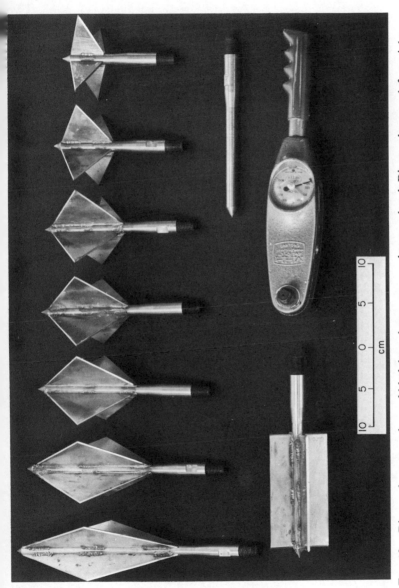

FIG. 13.   Diamond, rectangular and bladeless shear vanes and torque head. Diamond vanes, left to right, are 20, 30, 40, 45, 50, 60, and 70° with respect to the vertical axis of rotation.

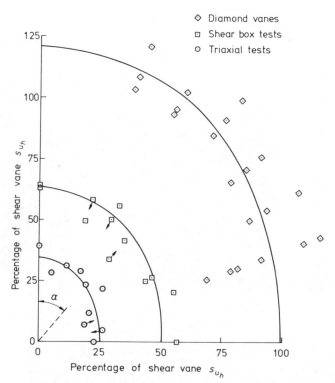

FIG. 14. Polar diagram of undrained shear strength measured by diamond vane, triaxial and direct shear tests as a percentage of the strength in the horizontal plane as measured by the diamond vane. The arrows show the direction of corrections for the effect of water content variations.

70° vane in the test pit may be due to the fact that these tests were the last of the series; the pit had been open for 8 h and pierced through the pit floor by several vane insertions and thus local swelling of the clay could possibly have occurred.

In the laboratory, triaxial test specimens 38 mm in diameter by 76 mm long and prismoidal direct shear test specimens 60 mm by 60 mm by 20 mm were trimmed from the blocks and tested in undrained shear. Polar diagrams showing the variation with direction of undrained shear strength are also given in Fig. 14. The polar diagram for the undrained triaxial test is not directly comparable with those for the shear vane and the direct shear tests. In the triaxial test results the direction given is of the test specimen axis and not of the

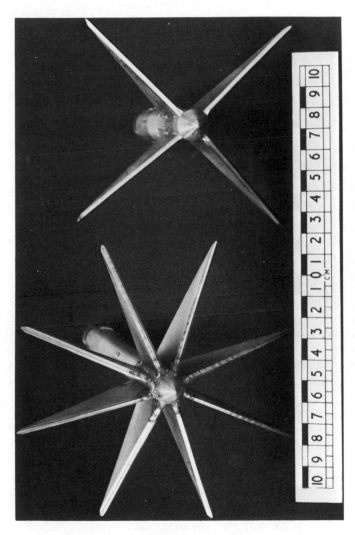

FIG. 15. Four-bladed and eight-bladed 70° diamond vanes.

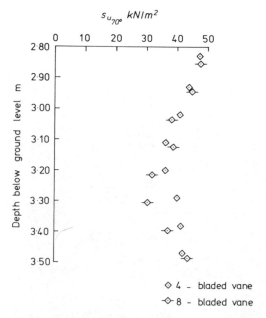

FIG. 16.　Variation of undrained shear strength with depth measured by the four-bladed and eight-bladed 70° diamond vanes.

failure plane. Distinct failure planes were not detected in the triaxial test specimens and so it was not possible to give the results with respect to the orientation of a sheared surface.

The measurements of strength anisotropy are summarised in Table 3. It may be seen that there are significant disparities in average strength magnitudes between different types of test. Similar disparities were observed by Madhloom[3] who carried out undrained tests on a soft clay of plasticity index 35% from King's Lynn, Norfolk, England. The direct shear test gave an undrained shear strength some 60% greater than that given by the triaxial compression test.[34]

For the soft clay of Brent Knoll some of this variation of strength with type of test may be attributed to the effects of type of soil distortion during the test, testing rate, time between sampling and testing, size of test specimen or zone, and degree of sample disturbance as summarised in Table 3. Equation (2) was fitted to the experimental points using the method of least squares to give the undrained shear strengths in the horizontal and vertical planes, as

TABLE 3

SUMMARY OF UNDRAINED SHEAR STRENGTH DATA FOR THE SOFT CLAY BENEATH THE BRENT KNOLL TRIAL EMBANKMENT

Type of test	Time to failure (min)	Time between sampling and testing (days)	Volume of soil tested as % of volume tested by vane	Degree of sample disturbance	Undrained shear strength (kN/m^2)		Degree of anisotropy $(R = s_{uv}/s_{uh})$	Anisotropy correction (μ_A)
					s_{uh}	s_{uv}		
Triaxial compression	10	27	15	Significant (inserting tubes into block sample —extruding specimens)	9·5	13·1	1·38	0·84
Direct shear	3	111	15	Moderate (hand trimming from block sample)	20·3	23·1	1·14	0·93
Diamond shear vane	0·25	0	100	Little (in situ test)	37·2	46·9	1·26	0·89
				Average			1·26	0·89

summarised in Table 3. The degree of anisotropy measured by the different types of test varies from about 1·1 to 1·4, the average being 1·26.

Using a conventional total stress stability limit analysis assuming strength isotropy, the field strength was deduced from the slip circle shown on Fig. 11 and found to be $19 \, kN/m^2$. The conventional shear vane measurements taken immediately after the slip in 1968 are given in Fig. 12. The conventional rectangular shear vane indicated an average undrained shear strength in the vicinity of the slip of about $26 \, kN/m^2$. Correcting this value of shear strength according to eqn. (7), using the data given in Table 4, gives a factor of safety of 1·02.

TABLE 4
STABILITY DATA RELATING TO THE BRENT KNOLL
TRIAL EMBANKMENT

Parameter	Value
Natural water content, $w(\%)$	60
Liquid limit, $w_L(\%)$	64
Plastic limit, $w_p(\%)$	28
Plasticity index $I_p(\%)$	36
Empirical correction factor, μ_B (Fig. 5)	0·84
Anisotropy correction factor, μ_A (Fig. 7)	0·89
No vane correction, F	1·37
With vane correction	
$\mu_B F$	1·15
$\mu_A \mu_B F$	1·02

This must be considered as fortuitously exact. The correction factor for anisotropy, μ_A, is itself only an approximation[28] and the empirical vane correction factor, μ_B, is based on stability data which were undoubtedly influenced by strength anisotropy to various unknown extents. Accordingly, it may be expected that applying both correction factors will over-correct for anisotropy and thus lead to conservatively low values of field strength. Nevertheless, making the correction for anisotropy together with the empirical vane correction leads to a good estimate of the field strength in this particular case.

11.3.2 The King's Lynn Trial Embankment
A trial embankment was constructed to failure in January 1970 on the low strength highly compressible organic deposits of the Fens at

King's Lynn in Norfolk. Information was required concerning the stability and settlement problems associated with the construction of motorway embankments and in particular the influence of vertical sand drains on the rate of settlement needed investigation. Only the strength and stability aspects are, however, discussed here. Fuller information is given by Wilkes.[21,44-46] King's Lynn is on the eastern edge of the fenland area bordering the Wash. The soil conditions at the site of the trial embankment consist of the post-glacial to recent alluvium and peat deposits of the fens overlying the Jurassic Kimmeridge clay.

The fen clays, apart from the peat and rootlets, occur mainly as uniform clays but occasionally contain partings and thin layers of fine sand and silt. Field vane and laboratory vane tests indicate the consistency of these clays to range from very soft to firm with an average sensitivity of 3·5. They exhibit a wide range of water contents and Atterberg limits, due to the effect of the peat and organic matter.

Peat occurs at approximately Newlyn Datum in thicknesses up to about 1·3 m, and, typically, is fully decomposed with liquid limits ranging from 260 to 435%.

A cross-section showing the soil types is given in Fig. 17 and the undrained shear strength as measured by the *in situ* vane is plotted on Fig. 18.

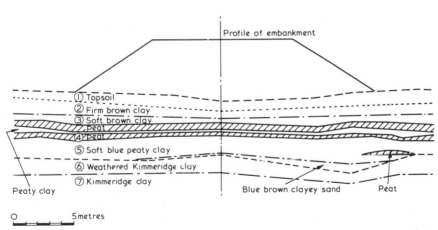

FIG. 17. Cross-section showing foundation soil type, King's Lynn trial embankment.

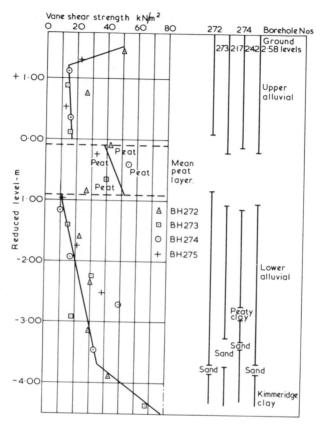

FIG. 18.   *In situ* vane undrained shear strengths, King's Lynn.

The results of undrained strength measurements carried out by Madhloom[3] to investigate the influence of anisotropy and type of test carried out are summarised in Table 5. It can be seen that the effect of anisotropy is not large. For the shear box tests, minimum strength was found for the vertical specimens; in triaxial compression, vertical specimens again showed the smallest strength, while in triaxial extension, the specimens inclined at 45° were the weakest.

It is clear from Table 5 however, that the type of test significantly affects the measured undrained shear strength. The shear box produces the greatest measured strength (approximately 30 kN/m²), the triaxial extension test gives the lowest results (about 10 kN/m²),

TABLE 5

VALUES OF STRENGTH RATIO RELATED TO TRIAXIAL COMPRESSION FOR DIFFERENT TYPES OF TEST, KING'S LYNN

Type of test	Orientation	Average strength (kN/m^2)	Strength ratio
Triaxial compression	Vertical	17·0	1·00
	Inclined at 45° and 56°	19·0	1·15
		20·1	1·17
	Horizontal	18·8	1·10
Extension	Vertical	10·8	0·63
	Inclined at 45°	9·1	0·53
	Horizontal	10·2	0·59
Direct shear box test	Vertical	24·8	1·48
	Inclined at 56°	31·8	1·86
	Horizontal	30·5	1·78
In situ vane test	Vertical	18·0	1·05

and these may be compared with the triaxial compression test (approximately $20 \, kN/m^2$). The average value from in situ vane tests was $18 \, kN/m^2$.

The results from King's Lynn have been compared in Table 1 with those published by Bjerrum[4] and significant differences in the behaviour of the King's Lynn clay compared with the other clays noted in the table are evident.

Firstly, for the King's Lynn clay, the highest value of s_u/p_0' was obtained from the shear box tests, while for the other clays, the shear box results lay within the range given by the triaxial compression and extension tests, apart from the Matagami clay when the shear box tests were even lower than the triaxial extension tests.

It is of interest to note that undrained tests carried out on remoulded normally consolidated kaolin at Cambridge, summarised by Chan,[47] show that the strength from the simple shear apparatus is greater than that from either triaxial compression or extension, the corresponding values of s_u/p_0' being 0·25, 0·21 and 0·18; the liquid limit was 71%, the plastic limit 41%, and the plasticity index 30%.

Secondly, for the clay from King's Lynn, the *in situ* vane test results were greater than the triaxial compression test values; for the other clays, however, the *in situ* vane test observations fell within the range noted from triaxial compression and extension tests.

Triaxial compression tests have also been carried out by Madhloom[3] to investigate the effect of rate of strain on the measured undrained shear strength. It was found that the decrease in shear strength per log cycle of time was approximately equal to 4%, suggesting that the strengths measured with a failure time of 15 min should be reduced by about 14% if applied to the analysis of a field failure occurring after about 40 days, as was the case for the trial embankment at King's Lynn.

A plan of the embankment is shown in Fig. 19. The embankment was sub-divided into two sections, one to act as a control and the other to be loaded until failure occurred, each section being monitored by separate lines of instruments. The second leg was used to test the suitability and effectiveness of sand drains. Two different diameters were tried, 300 mm and 150 mm at two different centres, 2·5 m and 3·75 m, thus giving four different sections each of which were instrumented identically, so that direct comparison could be made.

The shape of the slip surfaces developed during the failure of Section 2 can be deduced from the position of the cracks formed

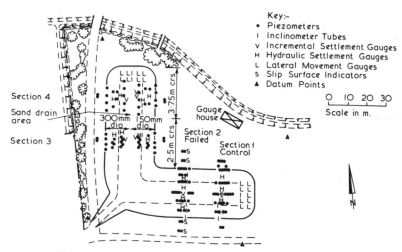

FIG. 19.   Plan of trial embankment site, King's Lynn.

around the concrete pipes, the position of the toe of the heaved zone at the base of the batter and from the behaviour of the various instruments. The slip surface indicators enable the extent and position of the slip zone to be deduced from the position of the jamming of the rod left at the base of the tube and also by the jamming of a second rod lowered from the surface. The position of the breaks in the inclinometer tubes was obtained by lowering the inclinometer and recording the depth to which it could be lowered. This position was checked by lowering the large settlement probe. As the latter was shorter than the inclinometer torpedo (150 mm as opposed to 300 mm) it was considered that a better assessment of the level of fracture point was obtained.

The various points were plotted on a cross-section of the embankment and an estimate of the shapes of the slip zones made. The slip zones obtained in this manner are shown in Fig. 20. As can be seen, the failure zone runs along the sand layer on top of the weathered Kimmeridge and is of a non-circular form.

Stability analyses have been carried out in terms of both total and effective stress for the following conditions and assumptions:

(a)   open vertical crack throughout the embankment in accordance with the observations of Wilkes;[45]

(b)   the total stress analyses are based on the *in situ* vane tests;

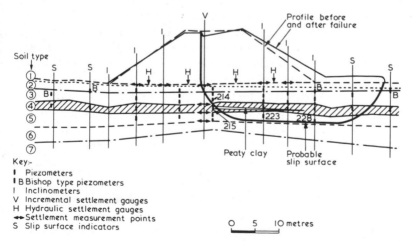

FIG. 20.   Cross-section showing deduced slip surface and instrumentation, King's Lynn trial embankment.

(c) for the foundation material, $c' = 4.8 \, \text{kN/m}^2$ (and zero) and $\phi' = 31.3°$;

(d) pore-water pressures as measured, or as predicted by the method proposed by Burland;[48]

(e) the bulk unit weights of the fill, alluvium and peat are 19·6, 16·8 and 11·8 kN/m^3 respectively.

The results of the stability calculations which have been performed by Haywood[49] and the writers are given in Table 6.

As far as the total stress method is concerned, it can be seen that the critical circle gives a factor of safety of 1·12 and if this is multiplied by a factor of 0·9, following Bjerrum,[4] then the value falls to 1·01. This must be considered partly fortuitous, because of the partial drainage which must have occurred during the construction period, and because the most critical circle was located above the peat layer, while the actual failure surface was found to descend almost to the Kimmeridge. A slip circle, or a non-circular surface, approximating to the observed failure surface, gave appreciably higher factors of safety, namely 1·41.

The effective stress analysis showed that the most critical slip circle and the most critical non-circular surface coincided closely with the observed failure surface. Based on measured pore-water pressures, factors of safety of 1·15 and 0·95 were obtained for $c' = 4.8 \, \text{kN/m}^2$ (as measured) and $c' = 0$, respectively. If predicted pore-water pressures are used, factors of safety less than unity are obtained.

It is clear that an excellent stability prediction is obtained by assuming a crack through the embankment, and then either using a total stress analysis based on *in situ* vane tests, corrected following Bjerrum,[4] or from an effective stress analysis based on measured pore-water pressures and taking $c' = 0$. This latter assumption is in agreement with the suggestion made by Parry.[9] Maximum measured pore-water pressures are plotted in Fig. 21. It can be seen that these are significantly less than those which can be predicted.

11.4 MAIN CONCLUSIONS

(a) For the King's Lynn trial embankment the measured pore-water pressures in the foundations at the end of the construction period were less than those which could have been predicted on a theoretical

TABLE 6

RESULTS OF TOTAL AND EFFECTIVE STRESS STABILITY CALCULATIONS, KING'S LYNN TRIAL EMBANKMENT

Assumed failure surface	Calculated factor of safety				
	Total stress	Measured pore-water pressure		Predicted pore-water pressure	
		$c' = 4.8 \, kN/m^2$	$c' = 0$	$c' = 4.8 \, kN/m^2$	$c' = 0$
Most critical circle	1·12	1·15	0·95	0·94	0·76
Most critical non-circular	1·31	1·54	1·22	1·27	0·99
Circle approximating to the observed failure surface	1·41	1·15	0·95	0·94	0·76
Non-circular approximating to the observed failure surface	1·41	1·54	1·22	1·27	0·99

Note: 1. Open vertical crack assumed through embankment.
2. Total stress analysis based on *in situ* vane tests.
3. Predicted pore-water pressures following Burland.[48]
4. $\phi' = 31.3°$ throughout foundation.

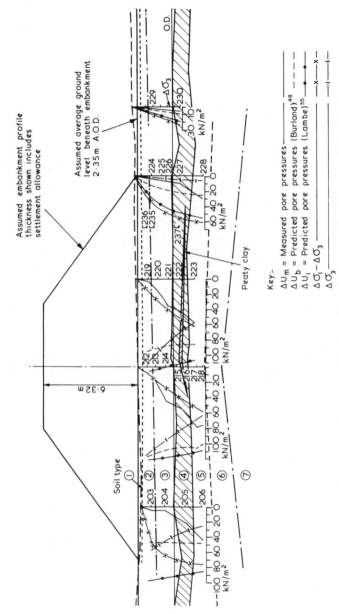

Fig. 21. Observed and computed pore-water pressures, King's Lynn trial embankment.

basis assuming undrained conditions. This is mainly due to some drainage which occurred during the construction period, for reasons discussed by Rowe,[50] but also possibly because of the initial degree of saturation being less than 100%, particularly near the surface.

(b) Satisfactory stability predictions can be made, provided a vertical crack is considered to exist in an embankment, and provided that due allowance is made for the presence of any pre-existing failure surfaces in the foundation.

(c) In spite of conclusion (a), a total stress analysis gives a good indication of the stability conditions, if the undrained strength is based on the *in situ* vane for soft clays, and on triaxial compression tests on undisturbed samples of sufficient size for stiff fissured clays, and if a correction factor depending on the plasticity index, as proposed by Bjerrum,[4] is applied.

For the Brent Knoll trial embankment, making this empirical correction together with an analytical anisotropy correction[28] leads to a good estimate of field stability. Further field studies are necessary to determine whether or not this approach is generally valid. Until sufficient data are acquired to enable this assessment to be made, it is considered advisable that both shear vane correction factors be used.

(d) The effective stress approach based on measured pore-water pressures and taking $c' = 0$ also gives a good indication of stability. If predicted pore-water pressures are used, then the calculated factors of safety are too low.

(e) Observations of pore-water pressures as embankment construction proceeds cannot be relied upon to show the onset of instability. Increases in the rate of lateral movement in the ground, as observed by inclinometers, seem to provide a more satisfactory indication of unstable conditions.[51]

ACKNOWLEDGEMENTS

The authors gratefully acknowledge the work of past research and final-year undergraduate project students, namely Madhloom,[3] Wilkes,[45] Roy,[20] Haywood,[49] Andrews,[52] Downham[53] and Nowak.[54]

The authors are also grateful to Mr J. M. McKenna and Mr M. A. Huxley for many valuable discussions, and to Mr B. Pimley and Mr L. J. Feldtman for field and laboratory assistance.

Data from the Brent Knoll trial embankment were kindly made

available by the Chief Engineer for the Somerset sub-unit of the South-Western Road Construction Unit.

The authors thank Julia Bentley and Linda Saunt who typed the script and Margaret Harris who drew the figures.

REFERENCES

1. ARTHUR, J. R. F. and MENZIES, B. K. (1972). Inherent anisotropy in a sand, *Géotechnique*, **22**, 115–128.
2. ARTHUR, J. R. F. and PHILLIPS, A. B. (1972). Discussion on: inherent anisotropy in a sand, *Géotechnique*, **22**, 537–538.
3. MADHLOOM, A. A. W. A. (1973). The undrained shear strength of a soft silty clay from King's Lynn, Norfolk, M.Phil Thesis, University of Surrey.
4. BJERRUM, L. (1972). Embankments on soft ground, *Proc. ASCE Spec. Conf., Performance of Earth and Earth Supported Structures*, Purdue University, 1972, **2**, 1–54.
5. BJERRUM, L., SIMONS, N. and TORBLAA, I. (1958). The effect of time on the shear strength of clay, *Proceedings of Brussels Conference, Earth Press Problems*, **1**, 148–158.
6. MARSLAND, A. (1972). The shear strength of stiff fissured clays, *Proceedings of Roscoe Memorial Symposium on Stress–Strain Behaviour of Soils*, Cambridge, 1971, 59–68.
7. DAVIS, E. H. and POULOS, H. G. (1967). Laboratory investigations of the effects of sampling, *Trans. IE Aust.*, **CE9**(1), 86–94.
8. SCHMERTMAN, J. H. (1955). The undisturbed consolidation behaviour of clay, *Trans. ASCE*, **120**, 1201–1233.
9. PARRY, R. H. G. (1972). *Stability Analysis for Low Embankments on Soft Clays. Stress–Strain Behaviour of Soils*, G. T. Foulis & Co Ltd, Henley-on-Thames, pp. 643–668.
10. PALMER, A. C. and RICE, J. R. (1973). The growth of slip surfaces in the progressive failure of overconsolidated clay, *Proc. R. Soc. Lond.* A. **332**, 527–548.
11. BISHOP, A. W. and BJERRUM, L. (1960). The relevance of the triaxial test to the solution of stability problems, *ASCE Research Conference on Shear Strength of Cohesive Soils*, Boulder, Colorado, (1960), 437–501.
12. PARRY, R. H. G. and McLEOD, J. H. (1967). Investigation of slip failure in flood levee at Launceston, Tasmania. *Proc. 5th Australia–NZ Conf. Soil Mech. Fdn. Engng.*, Auckland, (1967), 249–300.
13. EIDE, O. and HOLMBERG, S. (1972). Test fills to failure on the soft Bangkok clay, *Proc. ASCE Spec. Conf. Performance of Earth and Earth-Supported Structures*, Purdue University, 1972, **1**(1), 159–180.
14. GOLDER, H. Q. and PALMER, D. J. (1955). Investigation of a bank failure at Scrapsgate, Isle of Sheppey, Kent, *Géotechnique*, **5**, 55–73.
15. PILOT, G. (1972). Study of five embankment failures on soft soils, *Proc. ASCE Spec. Conf. Performance of Earth and Earth-Supported Structures*, Purdue University, 1972, **1**(1), 81–100.

16. EIDE, O. (1968). Geotechnical problems with soft Bangkok clay on the Nakhon Sawan highway project, *NGI Pub. No* 78.
17. PETERSON, R., IVERSON, N. L. and RIVARD, P. J. (1957). Studies of several dam failures on clay foundations. *Proc. 4th Int. Conf. Soil Mech. & Fdn. Engng.* London, 1957, **2**, 348–352.
18. DASCAL, O., TOURNIER, J. P. TAVENAS, F. and LA ROCHELLE, P. (1972). Failure of a test embankment on sensitive clay, *Proc. ASCE Spec. Conf. on Performance of Earth and Earth-Supported Structures*, Purdue University, 1972, **1**(1), 129–158.
19. SEROTA, S. (1966). Discussion, *Proc. ICE* **35**(9), 522.
20. ROY, M. (1975). Predicted and observed performance of motorway embankments on a soft alluvial clay in Somerset, M.Phil Thesis, University of Surrey.
21. WILKES, P. F. (1972). An induced failure at a trial embankment at King's Lynn, Norfolk, England, *Proc. ASCE Spec. Conf. Performance of Earth and Earth-Supported Structures*, Purdue University, 1972, **1**(1), 29–63.
22. LO, K. Y. and STERMAC, A. G. (1965). Failure of an embankment founded on varved clay, *Canadian Geotechnical Journal*, **2**(3), 243–253.
23. STAMATOPOULOS, A. C. and KOTZIAS, P. C. (1965). Construction and performance of an embankment in the sea on soft clay, *Proc. 6th Int. Conf. Soil Mech. & Fdn. Engng.*, Montreal, 1965, **2**, 566–571.
24. LA ROCHELLE, P., TRAK, B., TAVENAS, F. and ROY, M. (1974). Failure of a test embankment on a sensitive Champlain Clay deposit, *Canadian Geotechnical Journal*, **11**, 142–164.
25. FLAATE, K. and PREBER, T. (1974). Stability of road embankments on soft clay, *Canadian Geotechnical Journal*, **11**, 72–89.
26. LADD, C. C. (1972). Test embankment on sensitive clay, *Proc. ASCE Spec. Conf. on Performance of Earth and Earth-Supported Structures*, Purdue University, 1972, **1**(1), 101–128.
27. HAUPT, R. S. and OLSON, J. P. (1972). Case history—embankment failure on soft varied silt, *Proc. ASCE Spec. Conf. Performance of Earth and Earth-Supported Structures*, Purdue University, 1972, **1**, 29–64.
28. MENZIES, B. K. (1976). An approximate correction for the influence of strength anisotropy on conventional shear vane measurements used to predict field bearing capacity, *Géotechnique*, **26**(4), 631–634.
29. RAYMOND, G. P. (1967). The bearing capacity of large footings and embankments on clays, *Géotechnique*, **17**, 1–10.
30. TIMOSHENKO, S. (1934). *Theory of Elasticity*, McGraw-Hill, New York, p. 193.
31. CASAGRANDE, A. and CARRILLO, N. (1944). Shear failure of anisotropic materials, *Proc. Boston Soc. of Civ. Engrs.*, **31**, 74–87.
32. LO, K. Y. (1965). Stability of slopes in anisotropic soils, *J. Soil Mech. Fdn. Engng. Div.*, *ASCE*, **SM4**, 85–106.
33. MENZIES, B. K. and MAILEY, L. K. (1976). Some measurements of strength anisotropy in soft clays using diamond-shaped shear vanes, *Géotechnique*, **26**, 535–538.
34. SIMONS, N. E. and MENZIES, B. K. (1975). *A Short Course in Foundation Engineering*, Newnes–Butterworths, London, p. 159.

35. RICHARDSON, A. M., BRAND, E. W. and MEMON, A. (1975). *In situ* determination of anisotropy of a soft clay, *Proc. Conf. In Situ Measurement of Soil Properties*, ASCE, Raleigh, 1975, 336–349.
36. MENZIES, B. K. (1976). Strength, stability and similitude, *Ground Engineering*, **9**(5), 32–36.
37. BISHOP, A. W. (1972). Shear strength parameters for undisturbed and remoulded soil specimens, *Proc. Roscoe Memorial Symp. on Stress–Strain Behaviour of Soils*, Cambridge, 1971, 3–58.
38. BISHOP, A. W. (1971). The influence of progressive failure on the choice of method of stability analysis, *Géotechnique*, **21**, 168–172.
39. BISHOP, A. W., GREEN, G. E., GARGA, V. K., ANDRESEN, A and BROWN, J. D. (1971). A new ring shear apparatus, *Géotechnique*, **21**, 273–328.
40. SKEMPTON, A. W. (1964). Long-term stability of clay slopes, *Géotechnique*, **14**, 77–101.
41. LO, K. Y. and LEE, C. F. (1973). Stress analysis and slope stability in strain softening materials, *Géotechnique*, **23**, 1–11.
42. JANBU, N., BJERRUM, L. and KJAERNSLI, B. (1956). Veiledning ved løsning av fundamenteringsoppgaver, Norwegian Geotechnical Institute. *Publ. No 16*, 93pp.
43. AAS, G. (1967). Vane tests for investigation of anisotropy of undrained shear strength of clays, *Proc. Geotech. Conf.*, Oslo, 1967, **1**, 3–8.
44. WILKES, P. F. (1972). King's Lynn Trial Embankment, *J. Inst. Highway Engrs*, **XIX**(8), 9–16.
45. WILKES, P. F. (1974a). A geotechnical study of a trial embankment on alluvial deposits at King's Lynn, Ph.D Thesis, University of Surrey.
46. WILKES, P. F. (1974b). Instrumentation for the King's Lynn Southern Bypass. *Proc. BGS Symp. on Field Instrumentation in Geotechnical Engineering*, London, 1974, 448–481.
47. CHAN, K. C. (1975). Stresses and strains induced in soft clay by a strip footing, Ph.D. Thesis, Cambridge University.
48. BURLAND, J. B. (1972). A method of estimating the pore pressure and displacements beneath embankments on soft natural clay deposits, *Proc. Roscoe Memorial Symp. on Stress–Strain Behaviour of Soils*, Cambridge, 1971, 505–536.
49. HAYWOOD, C. L. (1974). The King's Lynn Trial Embankment, B.Sc. Project Dissertation, University of Surrey. Unpublished.
50. ROWE, P. W. (1972). The relevance of soil fabric to site investigation practice, *Géotechnique*, **22**, 193–300.
51. SIMONS, N. E. (1976). Field studies of the stability of embankments on clay foundations, *Laurits Bjerrum Memorial Volume*, Norwegian Geotechnical Institute, Oslo, pp. 183–209.
52. ANDREWS, K. R. (1976). The estimation of field stability from *in situ* measurements of strength using large diamond-shaped shear vanes, B.Sc. Project Dissertation, University of Surrey. Unpublished.
53. DOWNHAM, P. M. (1976). Some measurements of strength anisotropy in a soft clay, B.Sc. Project Dissertation, University of Surrey. Unpublished.

54. NOWAK, W. S. (1977). Embankments on clay foundations, B.Sc. Project Dissertation, University of Surrey. Unpublished.
55. LAMBE, T. W. (1962). Pore pressures in a foundation clay, *J. Soil Mech. & Fdn. Engng. Div., Proc. ASCE.,* **SM2**, 19–47.

INDEX